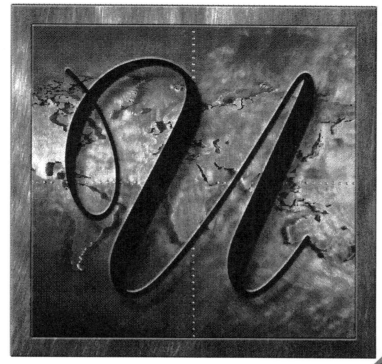

Stata User's Guide
Release 6

D1309204

Stata Press
College Station, Texas

Stata Press, 702 University Drive East, College Station, Texas 77840

The suggested citation for this software is

StataCorp. 1999. *Stata Statistical Software: Release 6.0*. College Station, TX: Stata Corporation.

Contents of User's Guide

ADVICE

Subject Table of Contents

Data manipulation and management

Utilities

Basic utilities

Error messages

Saved results

Internet

Data types and memory

Advanced utilities

Graphics

Statistics

Basic statistics

ANOVA and ANCOVA

Linear regression and related maximum-likelihood regressions

Logistic and probit regression

Survival analysis

Cross-sectional time series (panel data)

Auxiliary regression commands

Tables for epidemiologists

Analysis of survey data

Transforms and normality tests

Nonparametrics

Factor analysis

Do-it-yourself maximum likelihood estimation

Other statistics

Matrix commands

Basics

Programming

Other

Programming

Basics

Program control

Parsing and program arguments

Console output

Commonly used programming commands

Debugging

Advanced programming commands

File formats

Windows and Macintosh

Unix

Stata Basics

Chapters

1 Read this—it will help

Contents

The Stata Documentation Set contains over 3,000 pages of information. There are four parts to the Documentation Set:

> *Getting Started with Stata*
> *Stata User's Guide*
> 4-volume *Stata 6.0 Reference Manual* set
> *Stata Graphics Manual*

1.1 Getting Started with Stata

There are three *Getting Started* manuals:

> *Getting Started with Stata for Windows*
> *Getting Started with Stata for Macintosh*
> *Getting Started with Stata for Unix*

1. Locate your *Getting Started* manual.

2. Install Stata. The instructions are found in the *Getting Started* manual.

3. Learn how to invoke Stata and use it—read the *Getting Started* manual.

4. Now turn to the other manuals; see [U] **1.2 The User's Guide and the Reference Manual set**.

1.2   The User's Guide and the Reference Manual set

The *User's Guide* is divided into three sections: Basics, Elements, and Advice. At the beginning of each section is a list of the chapters found in that section. In addition to exploring the fundamentals of Stata—information that all users should know—this manual will guide you to other sources for Stata information.

Take a look at the *Stata Reference Manual*. This 4-volume set is organized like an encyclopedia—alphabetically. Every Stata command can be found in the *Reference Manual*. Look it up under the name of the command. If you do not find it, look in the index. There are a few commands that are so closely related that they are documented together, such as `signrank`, `signtest`, and `ranksum`, which are all documented in [R] **signrank**.

Not all of the entries in the *Reference Manual* are Stata commands; some contain technical information, like [R] **maximize**, which details Stata's iterative maximization process, or [R] **error messages**, which lists all the error messages and return codes. Others are a "quick reference", like [R] **estimation commands**. The quick reference entries summarize information that is discussed in more detail in either the *User's Guide* or other entries in the *Reference Manual*.

Like an encyclopedia, the *Reference Manual* is not designed to be read from cover(s) to cover(s). When you want to know what a command does, complete with all the details, qualifications, and pitfalls, or when a command produces an unexpected result, read its description. Each entry is written at the level of the command. The descriptions assume little knowledge of Stata's features when explaining simple commands, such as those for using and saving data. For more complicated commands, they assume you have a firm grasp of Stata's other features.

1.2.1   Cross-referencing

The *User's Guide*, *Reference Manual*, *Graphics Manual*, and *Getting Started* cross-reference each other. When you see

[U] **1.2.1 Cross-referencing**

that is an instruction to see section 1.2.1, which happens to be titled *Cross-referencing*, in this manual. When you see

[R] **egen**

that is an instruction to see **egen** in the *Reference Manual*, which, it so happens, comes after **edit** but before **eivreg**. When you see

[G] **axis labels**

that is an instruction to see **axis labels** in the *Graphics Manual*. When you see

[GSW] **A. Starting and stopping Stata for Windows 98/95/NT**
[GSM] **A. Starting and stopping Stata for Macintosh**
[GSU] **A. Starting and stopping Stata for Unix**

that is an instruction to see the appropriate section of the *Getting Started with Stata for Windows*; *Getting Started with Stata for Macintosh*; or *Getting Started with Stata for Unix* manuals.

1.2.2   The index

The *User's Guide* and the *Reference Manual* set are indexed together; that is, the same index is in all five volumes.

Look at the back of the *User's Guide* or any of the *Reference* manuals to find the index.

1.2.3   The subject table of contents

At the beginning of this manual and at the beginning of *Reference Manual* volume 1 is a subject table of contents.

If you look under "Functions and expressions", you will see

1.2.4   Typography

You will note that we mix the ordinary printing that you are reading now with a typewriter-style typeface `that looks like this`. When something is printed in the typewriter-style typeface, it means that something is a command or an option—it is something that Stata understands and something that you might actually type into your computer. Differences in typeface are important. If a sentence reads, "You could list the result ..." it is just an English sentence—you *could* list the result, but the sentence provides no clue as to how you might actually do that. On the other hand, if the sentence reads, "You could `list` the result ..." it is telling you much more—you could list the result and you could do that using the `list` command.

We will occasionally lapse into periods of inordinate cuteness and write, "We `described` the data and then `listed` it." You get the idea. `describe` and `list` are Stata commands. We purposely began the previous sentence with a lowercase letter. Because `describe` is a Stata command, it must be typed in lowercase letters. The ordinary rules of capitalization are temporarily suspended in favor of preciseness.

You will also notice that we mix in words printed in italic type, like "To perform the rank-sum test, type `ranksum` *varname* , `by`(*groupvar*)". Italicized words are not supposed to be typed; instead, you are to substitute another word for them.

We would also like users to note our rule for punctuation of quotes. We follow a rule that is often used in mathematics books and British literature. The punctuation mark at the end of the quote is only included in the quote if it is a part of the quote. For instance, the pleased Stata user said she thought that Stata was a "very powerful program". Another user said simply, "I love Stata."

In this manual, however, there is little dialogue, and we follow this rule to make precise what you are to type, as in, type '`cd c:`'. The period is outside the quotation mark since you should not type the period. If we had wanted you to type the period, we would have included two periods at the end of the sentence; one inside the quotation and one outside, as in, type '`use myfile.`'.

And then there is the word *data*. Dictionaries tell us the word is plural. "Where are those data?" In most cases, they are collected into something called a dataset and people who have spent (too much) time around computers often drop the *set* and then refer to the data in the singular. On occasion, we will do the same thing, and answer that the data is stored as `mydata.dta`.

We have tried not to violate the other rules of English. If you find such violations, they were unintentional and due to our own ignorance or carelessness. We would appreciate hearing about them.

We have heard from Dr. Nicholas J. Cox of the Department of Geography at the University of Durham in the U.K. and wish to express our appreciation. Nicholas' efforts have gone far beyond dropping us a note and—without Nicholas' assistance—there is no way with words that we can express our gratitude.

1.3   What's new

This section is intended for previous users of Stata. If you are new to Stata, you may as well skip it.

As always, Stata 6.0 is 100% compatible with the previous release of Stata, but this time we remind programmers that it is vitally important that you put `version 5.0` at the top of your old do-files and ado-files if they are to work. You were supposed to do that when you wrote them but, if you did not, go back and do it now. We have made a lot of changes (improvements) to Stata.

In addition, Stata's dataset format has changed. You will not care because Stata can still read old-format datasets, but if you need to send a dataset to someone still using Stata 5, remember to specify `save`'s `old` option. Among other things, specifying `old` will prevent the value labels from being saved because they can now be much longer.

Let us also note that the quality of the *Stata Technical Bulletin* (STB) submissions have been high and that we have been remiss in not adopting all that we should. We have, however, made it much easier for users to adopt the programs for themselves thanks to Stata's new `net` command. (Windows 98/95/NT users and Power Macintosh users can pull down **Help** and select **STB and User-written Programs** in addition to using `net`.)

1.3.1   Highlights of the new release

What is important varies from user to user, but here are a few of the changes we would like to call to your attention:

Stata is web-aware

Stata for Windows 98/95/NT, Stata for Power Macintosh, and Stata for Unix are now web-aware. (Stata for Windows 3.1 and Stata for 680x0 Macintosh are not.) See [U] **32 Using the Internet to keep up to date**.

You can `use` datasets over the web—try typing

```
. use http://www.stata.com/manual/oddeven.dta, clear
```

You can `update` your Stata over the web. Try typing

```
. update
```

or pull down **Help** and choose **Official Updates**.

You can obtain STB or other materials over the web. Try typing

```
. net from http://www.stata.com
```

or pull down **Help** and choose **STB and User-written Programs**.

You can create your own site to deliver additions to Stata—be they help files, ado-files, or data—and the new `checksum` command will confirm that the materials are delivered uncorrupted; see [R] **net** and [R] **checksum**.

For the latest information on what's available from us, type

. news

or pull down **Help** and choose **News**.

Scrolling Results window

Stata for Windows now has a scrolling results window, so you can look back at previous output, copy and paste output to other applications, etc.

New Do-file Editor

Stata for Windows 98/95/NT and Stata for Macintosh (both Power and 680x0) have a new do-file editor. Click on the **Do-file Editor** button or type doedit (one word) in the command window.

Long value labels

Stata's value labels can now be up to 80 characters long (as opposed to the previous maximum of 8), value labels may contain 65,536 mappings, and you may now label negative values. See [U] **1.3.5 New data-management features**.

Time-series features

Stata has added time-series analysis, estimation, and data management facilities. Time-series operators for differencing and lagging can now be used in expressions and variable lists for many commands. New time-series date formats are provided. New commands estimate ARIMA models and ARCH family models (ARCH, GARCH, EGARCH, ARCH-in-mean, ...). Other new commands graph and tabulate autocorrelations, partial autocorrelations, and cross-correlations. Commands for periodograms, unit-root tests, and white-noise tests have also been added. See [U] **1.3.2.4 New time-series features** for more information about these features and watch the *Stata Technical Bulletin* for additional features.

ANOVA: Repeated-measures, nested, and mixed designs

Up to four repeated-measure variables may now be specified with anova along with other categorical variables (providing repeated-measures ANOVA) and continuous variables (providing repeated-measures ANCOVA). Nested and mixed ANOVA and ANCOVA models are also now fully supported with an easier-to-use syntax. See [U] **1.3.2.1 New ANOVA features**.

New st survival analysis additions

There are four new parametric survival estimators that estimate lognormal, log-logistic, Gompertz, and generalized log-gamma models. Graphical and statistical tests of the proportional-hazards assumption can now be computed after Cox regression (stcox or cox). There are many more residuals available after Cox regression, and all these residuals are available after parametric survival estimators as well. The st system (stset, etc.) has been completely rewritten, and it allows for much more flexibility in the way that your data were collected and recorded, including allowing for multiple failure events in the same dataset. See [U] **1.3.2.2 New st survival analysis features** for a complete list of all the new st commands and features.

New xt panel estimators

There are 12 new xt estimators for use with panel data. For example, xtprobit now estimates random-effects probit models using Gauss–Hermite quadrature, in addition to estimating population-averaged models using GEE. In addition to random-effects probit, there is random-effects logit, tobit, interval regression, Poisson, negative binomial regression, and complementary log-log regression. There are also fixed-effects (conditional) Poisson and negative binomial estimators. See [U] **1.3.2.5 New xt panel estimators** for a complete list of the new panel estimators.

New svy survey data additions

There is a new `svytab` command that produces two-way contingency tables with tests of independence for survey data. There are also six new `svy` estimation commands. See [U] **1.3.2.3 New svy survey commands** for a complete list.

New ml command

The all new `ml` command for maximizing user-defined likelihood functions is easier to use, faster, and more robust. Even if you do not program your own MLEs, this change will affect you. Many of Stata's MLEs are programmed using `ml`, and so now they converge faster and more robustly. See [R] **ml**. Those interested in programming their own estimators will also want to see the new book *Maximum Likelihood Estimation with Stata* (Gould and Sribney 1999).

New matrix language

You can now write long, complicated matrix expressions; no longer are you restricted to one matrix operation per command. You can write expressions such as

```
matrix b = syminv(X´*X)*X´*y
```

This makes working with matrices in Stata much, much easier.

Ado-files now behave just like internal commands

Quotes now work with ado-file implemented commands. Previously, you could not type, for instance,

```
. logistic outcome x1 x2 if sex=="female"
```

because `logistic` was implemented as an ado-file and quoted strings confused ado-files. That is fixed.

Ado-files can now process datasets regardless of the number of variables in them. For instance, previously

```
. codebook
```

would not work if the dataset had more than 600 variables because the ado-file could not hold all the variable names in a single macro. That is fixed.

The result of these two changes is that ado-files now behave just like internal commands from the user's point of view.

New function returns the estimation sample

After running any estimation command, the new function `e(sample)` returns true (1) if the observation was used in estimation and false (0) otherwise. You can type, for instance,

```
summarize if e(sample)
```

to obtain summary statistics on the estimation sample. See [U] **23.4 Specifying the estimation subsample**.

New way of saving results

Run `summarize` and then type `return list`. You will see `r(N)`, `r(mean)`, `r(Var)`, and other `r(name)` items listed. `r(N)` contains the number of observations, `r(mean)` the mean, and `r(Var)` the variance. The new `r(name)` method of saving results replaces both `_result(#)` and `$S_#`. Hence, both ado-files and internal commands now save results in the same way.

Run `regress` and then type `estimates list`. You will see `e(N)` and other `e(`*name*`)` items. Some of the `e(`*name*`)` items are scalars, some are macros, some are matrices, and one is a function (`e(sample)`).

See [R] **saved results** for details on both `r()` and `e()`.

Stata is 5% faster

We have sped up the rate at which Stata can evaluate expressions along with making other speed improvements.

1.3.2   New statistical features

1.3.2.1   New ANOVA features

Repeated-measures ANOVA and ANCOVA: `anova` can now perform repeated-measures ANOVA and ANCOVA. Repeated-measure variables (up to four in one `anova`) are now fully supported. In addition to the regular ANOVA table, F tests based on the Box, Greenhouse–Geisser, and Huynh–Feldt corrections are also reported for terms involving a repeated-measures variable. See [R] **anova**.

Nested and mixed designs: `anova` now handles nested and mixed designs. It is now easy within `anova` to specify the appropriate error(s) for testing nested and mixed terms. This means that for most analyses you can now get the appropriate F tests for all terms in one ANOVA table with one command. The `test` command provides a simple way to obtain any other F tests of interest. This is possible because of new syntax that has been added to `anova` to make specification of various nonresidual error terms easier, and that same syntax is understood by `test` as well. See [R] **anova**.

1.3.2.2   New st survival analysis features

There are substantial additions and many changes to the `st` system of commands for survival analysis.

New st survival estimators

There are four new `st` survival estimators:

> **Lognormal parametric survival regression**
>
> **Log-logistic parametric survival regression**
>
> **Gompertz parametric survival regression**
>
> **Generalized log-gamma parametric survival regression**

In the `st` system, these four new estimators are obtained using the new `streg` command. In addition to lognormal, log-logistic, Gompertz, and generalized log-gamma models, `streg` also estimates the Weibull and exponential models. See [R] **st streg**. The old `stereg` and `stweib` commands are undocumented but continue to work.

For those not wanting to `stset` their data, the new stand-alone commands for estimating these parametric models are `lnormal`, `llogistic`, `gompertz`, and `gamma`, in addition to the previously existing `ereg` and `weibull` commands; all are documented under [R] **weibull**.

Other new st survival features

st has been rewritten: You can now have different types of failure events in the same dataset, and there is now a careful distinction made between analysis time and time as you measure it (analysis time = 0 corresponds to start of risk). These changes are all incorporated into `stset` and many new features are added as well. When you `stset` the data, you can specify when subjects became at risk (either by time or by event), when they came under observation (either by time or by event), and when they failed or were censored (either by time or by event). The new `streset` command allows varying a previous definition. See [R] **st stset**.

Because there have been so many STB inserts based on the old st system, if you type `version 5.0`, you will be running the old system.

Testing the proportional-hazards assumption: `stphtest`, for use after `stcox`, presents a test of the assumption; see [R] **st stcox**. `stphplot` and `stkmcox` (based on Garrett 1997) provide a graphical interpretation of the proportional-hazards assumption; see [R] **st stphplot**.

Ties in Cox regression: `stcox` and `cox` now provide three ways to handle ties in addition to the Breslow approximation: the exact partial likelihood method, the exact marginal likelihood method, and the Efron approximation. See [R] **st stcox** and [R] **cox**.

Cumulative baseline hazard: `stcox` and `cox` can now calculate the cumulative baseline hazard. See [R] **st stcox** and [R] **cox**.

Residuals after Cox regression: `stcox` and `cox` command can now calculate Schoenfeld residuals, scaled Schoenfeld residuals, Cox–Snell residuals, cumulative Cox–Snell residuals, cumulative martingale residuals, and deviance residuals, in addition to the martingale and efficient score residuals that were previously available. See [R] **st stcox** and [R] **cox**.

Residuals after parametric survival regression: After parametric survival model estimation with the `streg` command (exponential, Weibull, lognormal, log-logistic, Gompertz, or generalized log-gamma models), the same residuals are available as after `stcox` and `cox`; see [R] **st streg**.

Predicted survival and hazard functions: `stcurv`, after `streg`, will plot the predicted survival, hazard, and cumulative hazard functions; see [R] **st streg**.

Rates and SMRs: `strate` calculates and tabulates rates and SMRs by one or more categorical variables; see [R] **st strate**.

Stratified rate ratios: `stmc` command calculates and tests stratified rate ratios using Mantel–Cox methods. `stmh` command calculates and tests stratified rate ratios using Mantel–Haenszel methods. See [R] **st strate**.

Nested case-control datasets: `sttocc` command creates a nested case–control study dataset from a cohort-study dataset; see [R] **st sttocc**.

Splitting and joining time records: The features of `lexis` and `stlexis` (Clayton and Hills 1995a, 1997) have been incorporated into the new `stsplit` command. `stsplit` splits time records into two or more records at the time points specified. The new `stjoin` command (based on Weesie 1998) performs the inverse operation. See [R] **st stsplit**.

Snapshot data: `snapspan` makes it easier to convert snapshot data into time-span data; see [R] **snapspan**.

Changed st commands

stgen now has new functions for calculating earliest and latest times and times corresponding to an event; see [R] **st stgen**.

sts now provides the Nelson–Aalen estimator of the cumulative (integrated) hazard function in addition to what was previously provided such as the Kaplan–Meier estimate of the survivor function; see [R] **st sts**.

sts graph command now graphs the survival function from analysis time 0 rather than the time of the first failure and an option restores the previous behavior; see [R] **st sts graph**.

1.3.2.3   New svy survey commands

Two-way contingency tables: svytab produces two-way tabulations with tests of independence for complex survey data or other clustered data. The command can display estimated proportions with standard errors and confidence intervals. Tests of independence include the Rao-and-Scott second-order correction for the Pearson χ^2 and likelihood-ratio statistics. Wald tests, which historically have been used, can be also computed. See [R] **svytab**.

Censored and interval regression: the new svyintrg command is the parallel of intreg for survey data; see [R] **svy estimators**.

Instrumental variables regression: the new svyivreg command is the parallel of ivreg for survey data; see [R] **svy estimators**.

Multinomial logistic regression: the new svymlog command is the parallel of mlogit for survey data; see [R] **svy estimators**.

Ordered logistic regression: the new svyolog command is the parallel of ologit for survey data; see [R] **svy estimators**.

Ordered probit: the new svyoprob command is the parallel of oprobit for survey data; see [R] **svy estimators**.

Poisson regression: the new svypois command is the parallel of poisson for survey data; see [R] **svy estimators**.

1.3.2.4   New time-series features

Stata has added some new time-series estimators and developed several other features designed for time-series data including, importantly, changes to Stata's language to support time-series operators and additions of time-series date formats.

New time-series estimators

ARIMA: The new arima command estimates via maximum likelihood ARIMA models and models with ARMA disturbance structures; see [R] **arima**. Estimates and predictions are based on optimal filtering using the Kalman filter. Variance estimates for the parameters can be computed using either the standard method for MLEs (i.e., the inverse of the negative Hessian) or the robust Huber/White/sandwich variance estimator.

ARCH, GARCH, ARCH-in-mean: The new arch command estimates via conditional maximum likelihood a family of models with autoregressive conditional heteroscedastic disturbances: ARCH, GARCH, EGARCH, APARCH, NARCH, AARCH, GJR, and others; see [R] **arch**. Estimation is by conditional maximum likelihood, and coefficient variances can be estimated using either the standard method for MLEs (i.e., the inverse of the negative Hessian), or the outer product of gradients (OPG), or the robust Huber/White/sandwich variance estimator. In addition to conditional heteroscedasticity, `arch` can model multiplicative deterministic heteroscedasticity and ARMA structure in the disturbances.

Other new time-series features

Time-series varlists and time-series operators: Stata's new time-series features begin with the newly allowed time-series varlists and time-series operators in expressions—see [U] **14.4.3 Time-series varlists**, [U] **16.3.4 Time-series functions**, [U] **16.8 Time-series operators**, and [U] **27.3 Time-series dates**. You can use time-series operators in varlists; e.g., `L.gnp` means `gnp` lagged once and `L2.gnp` means `gnp` lagged twice. To use these new features, you must first `tsset` your data; see [R] **tsset**.

Many commands work with time-series data: Many new and existing commands now work with time-series data, meaning they allow you to specify a time-series varlist. These include, importantly, `regress`, `reg3`, `ivreg`, `prais`, and `sureg`, along with `generate` and `replace`, `graph`, `list`, `summarize`, and `correlate` (but you would probably prefer to use the new `xcorr` rather than `correlate`). You can tell when a command allows a time-series varlist because the note "varlist may contain time-series operators" appears at the end of the syntax diagram.

Graphs and tables of autocorrelations and partial autocorrelations: The new `corrgram`, `ac`, and `pac` commands (based on Becketti 1992) graph and displays tables of partial and autocorrelations; see [R] **corrgram**.

Graphs and tables of cross-correlations: The new `xcorr` command graphs and displays tables of cross-correlations; see [R] **xcorr**.

Periodograms: The new `pergram` command graphs periodograms; see [R] **pergram**.

Cumulative sample spectral density: The new `cumsp` command graphs the cumulative sample spectral density and optionally saves the values; see [R] **cumsp**.

Portmanteau test for white noise: The new `wntestq` command presents a portmanteau test for white noise, also known as the Box–Pierce test and the Box–Ljung test; see [R] **wntestq**.

Bartlett's periodogram test for white noise: The new `wntestb` command (based on Newton 1996) presents a Bartlett's periodogram test for white noise; see [R] **wntestb**.

Dickey–Fuller test for unit roots: The new `dfuller` command presents the augmented Dickey–Fuller test for unit roots; see [R] **dfuller**.

Phillips–Perron test for unit roots: The new `pperron` command presents the Phillips–Perron test for unit roots; see [R] **pperron**.

Prais–Winsten regression: `prais` now supports time-series operators, will produce heteroscedasticity robust variance estimates using White's method, provides the two-step method, and will calculate the autocorrelation coefficient in various ways. The existing `corc` and `hlu` commands are now undocumented and subsumed by new features of `prais`. See [R] **prais**.

Durbin–Watson statistic: The existing `regdw` command (regression with Durbin–Watson statistic) is now undocumented and replaced by the post-estimation command `dwstat` for use after `regress`; see [R] **regression diagnostics**.

New date formats: Stata has lots of new date formats for use with time-series data; see [U] **15.5.3 Time-series formats**.

A note about the Becketti time-series library: A side effect of the use of the new time-series operators (e.g., `L2.gnp`) is that periods are no longer allowed in variable names. Previously, a variable could be named `abc.def`, but now that would mean the result of applying the operator `abc` to the variable named `def`. This means the time-series commands in the Becketti library

(Becketti 1995) will no longer work. They will not work even if you set `version 5.0`. Stata's new time-series features mostly replace the Becketti functions, but in case you need to use an existing do-file to replicate old work, we have updated the Becketti library to work with Stata 6. To obtain the updated library,

type

. `net from http://www.stata.com`
. `net cd users`
. `net cd becketti`
. `net describe tslib`

or pull down **Help** and select **STB and User-written Programs**

click on *http://www.stata.com*
click on *users*
click on *becketti*
click on *tslib*

1.3.2.5 New xt panel estimators

Random-effects interval regression: new `xtintreg` command; see [R] **xtintreg**.

Hildreth–Houck random-coefficients regression: new `xtrchh` command; see [R] **xtrchh**.

Random-effects tobit: new `xttobit` command; see [R] **xttobit**.

Random-effects probit: see [R] **xtprobit**.

Random-effects logistic regression: new `xtlogit` command; see [R] **xtlogit**.

Random-effects complementary log-log regression: new `xtclog` command; see [R] **xtclog**.

Population-averaged complementary log-log regression: new `xtclog` command; see [R] **xtclog**.

Gaussian random-effects Poisson: see [R] **xtpois**.

Gamma random-effects Poisson: see [R] **xtpois**.

Fixed-effects (conditional) Poisson: see [R] **xtpois**.

Beta random-effects negative binomial regression: new `xtnbreg` command; see [R] **xtnbreg**.

Fixed-effects (conditional) negative binomial regression: new `xtnbreg` command; see [R] **xtnbreg**.

1.3.2.6 Other new estimators

3SLS: The new `reg3` command estimates systems of equations by 3SLS. It also estimates seemingly unrelated regression, multivariate regression, and 2SLS. It allows linear constraints within and across equations; allows iterated or two-step estimation; accepts new in-line equation syntax; and accepts time-series operated variable lists. See [R] **reg3**.

Instrumental-variable regression: The new `ivreg` command estimates instrumental-variable or 2SLS models that were previously estimated by an extension of `regress`'s syntax (which most users found confusing and which continues to work) and allows time-series operators. `predict` after `ivreg` now provides additional statistics. See [R] **ivreg**.

Interquantile regression: see [R] **qreg**.

Simultaneous quantile regression: see [R] **qreg**.

Bivariate probit: The new `biprobit` command estimates bivariate probit models and partial-observability bivariate probit models, with possibly different independent variables for each of the dependent variables; see [R] **biprobit**.

Probit with sample selection: The new `heckprob` command extends Heckman-style selection models to probit; see [R] **heckprob**.

Heteroscedastic probit: new `hetprob` command; see [R] **hetprob**.

Skewed logit: new `scobit` command; see [R] **scobit**.

Complementary log-log regression: new `cloglog` command (based on Hilbe 1996, 1998); see [R] **cloglog**.

Zero-inflated Poisson: new `zip` command; see [R] **zip**.

Zero-inflated negative binomial: new `zinb` command; see [R] **zip**.

1.3.2.7   Other new statistical commands

Adjusted predictions: The new `adjust` command (based on Garrett 1995, 1998) makes tables of predictions (or predicted probabilities) after estimation. The predictions can be adjusted to set levels of regressors, covariates, or terms. See [R] **adjust**.

Hausman test: The new `hausman` command provides a general implementation of the Hausman test. This includes the ability to test the independence of irrelevant alternatives (IIA) after multinomial logit or conditional logistic regression and tests exogeneity or over-identifying restrictions for 2SLS and 3SLS. See [R] **hausman**.

Tabulated odds and odds ratios: The new `tabodds` command (based on Clayton and Hills 1995b) calculates and tabulates odds and odds ratios for case–control or prevalence studies and performs a score test for linear trend in the log odds; see [R] **epitab**.

Mantel–Haenszel odds ratios: The new `mhodds` command (based on Clayton and Hills 1995b) calculates and reports Mantel–Haenszel odds ratios for case–control or prevalence studies; see [R] **epitab**.

Indirectly standardized rates: The new `istdize` command produces indirectly standardized rates using a standard population; see [R] **dstdize**.

Symmetry tests for tables: The new `symmetry` and `symmi` commands perform asymptotic symmetry and marginal homogeneity tests and exact symmetry tests on $n \times n$ tables where there is a one-to-one matching of cases and controls. They can also perform a test for linear trend in the log relative risks. See [R] **symmetry**.

Rootograms and histograms: The new `spikeplt` command (Cox and Brady 1997a, 1997b) graphs rootograms and histograms for both categorical and continuous variables; see [R] **spikeplt**.

Orthogonalization of variables: The new `orthog` command orthogonalizes a set of variables and creates a new set of orthogonal variables using a modified Gram–Schmidt procedure; see [R] **orthog**.

Checking quadrature results: The new `quadchk` command can be used to assess the stability of results from models estimated using Gauss–Hermite quadrature; see [R] **quadchk**.

1.3.3   Changes to existing commands

`alpha` incorporates the extensions of Weesie (1997a) adding item-test and item-rest correlations, average interitem covariance/correlation for test scale excluding item, and allowing pairwise computation of covariances as well as computations based on casewise deletion; see [R] **alpha**.

`brier` incorporates the extensions of Goldstein (1996): it now computes the mean probability of the forecast, the correlation between judgments and outcomes, the ROC area, and Spiegelhalter's test of the ROC area being greater than .5; see [R] **brier**.

`canon` now has new syntax; see [R] **canon**.

cc will now optionally perform the Breslow–Day test for homogeneity; see [R] **epitab**.

collapse allows pweights and iweights and provides the interquartile range; see [R] **collapse**.

corc is now undocumented and subsumed by new features of prais. See [R] **prais**.

decode has the new option maxlength(#), default maxlength(80), which specifies how many characters of the value label are to be kept in the newly created string; see [R] **encode**.

dydx and integ have been improved and now fit a cubic spline to the data and use that to produce the derivative and integral. Both commands also newly include a by() option so that calculations can be made within groups. See [R] **range**.

egen has several new functions; see [R] **egen**.

eq is now undocumented, but continues to work, and in its place multiple-equation estimators now accept a new in-line equation syntax or obtain the second equation as an option. heckman, for instance, does the latter; see [R] **heckman**. sureg, for instance, does the former; see [R] **sureg**. In all cases, the old syntax continues to work, but only if you first set version 5.0.

fit is undocumented because regress has been given all of its features; see [R] **regression diagnostics**.

for has been extensively updated. It is now more powerful, faster, and easier to use. The syntax is also different; for works the old way if you set version 5.0. See [R] **for**.

fracpoly, fracplot, and fracgen now provide mean adjustment to variables, provide component+residual plots, and provide other new features as well. See [R] **fracpoly**.

heckman has been extensively updated and is 3 to 6 times faster. It has a new and more flexible syntax; allows robust, cluster(*varname*), and pweights; and will optionally estimate Heckman's two-step model. The new **predict** used after heckman can produce Mills' ratio, probability of selected/observed, expected value of y given both selection and model equations, and the expected value of y conditional on selection. See [R] **heckman**.

hlu is now undocumented and subsumed by new features of prais. See [R] **prais**.

infile, insheet, and input now automatically widen the display format of variables when the automatic option is specified. This has to do with the new, longer value labels and basically allows these commands to work as you would expect them to work. See [R] **infile**, [R] **insheet**, and [R] **input**.

kap and kappa: The kap command can now handle two or more raters (as kappa always could); kap and kappa have been modified to deal with a large number of ratings; both commands now display tables with missing rows and columns more prettily, and a new absolute option has been added to kap for dealing with unobserved outcomes. See [R] **kappa**.

lfit, lroc, lsens, and lstat now work after logit as well as after logistic.

lincom allows longer expressions and has a new hazard rate hr option; see [R] **lincom**.

linktest has a new syntax that makes it easier and less error-prone to use; see [R] **linktest**.

logit and logistic have an asis option to suppress dropping variables and observations due to oneway causation. logit now allows iweights. See [R] **logit** and [R] **logistic**.

loneway incorporates the corrections of Gleason (1997) to the intraclass correlation coefficient for unbalanced and/or weighted data and provides asymptotic and exact confidence intervals for the correlation coefficient; see [R] **loneway**.

mcc and mcci will report the exact McNemar test in addition to the asymptotic χ^2 result; see [R] **epitab**.

`means` now reports confidence intervals for the arithmetic, geometric, and harmonic means (which additions are based on Carlin, Vidmar, and Ramalheira 1998); see [R] **means**.

`merge` has a new _merge(*newvarname*) option to specify the name for the _merge variable; see [R] **merge**.

`mlogit` allows `robust`, `cluster(`*varname*`)`, `pweights`, `iweights`, and `score()`. After `mlogit`, the matrix of coefficient estimates `e(b)` is now a row vector, just as the returned result would be after any other estimation command. Moreover, `get(_b)` after `mlogit` also returns a row vector unless you set `version` to 5.0 or before. See [R] **mlogit**.

`mvreg` now has new syntax; see [R] **mvreg**.

`nbreg` and `gnbreg` allow `robust`, `cluster(`*varname*`)`, `pweights`, `iweights`, and `score()`; see [R] **nbreg**.

`ologit` allows `robust`, `cluster(`*varname*`)`, `pweights`, `iweights`, and `score()`; see [R] **ologit**.

`oprobit` allows `robust`, `cluster(`*varname*`)`, `pweights`, `iweights`, and `score()`; see [R] **oprobit**.

`poisson` allows `robust`, `cluster(`*varname*`)`, `pweights`, `iweights`, and `score()`. The new command `poisgof` reports a goodness-of-fit test after `poisson`. See [R] **poisson**.

`prais` now supports time-series operators, will produce heteroscedasticity robust variance estimates using White's method, provides the two-step method, and will calculate the autocorrelation coefficient in various ways. The existing `corc` and `hlu` commands are now undocumented and subsumed by new features of `prais`. See [R] **prais**.

`predict` after estimation has been extensively reworked.

`predict` is now more tightly coupled to the estimation command. The default statistic calculated is now related to the dependent variable. For instance, `predict` after `weibull` gives predicted times, $\exp(E(\ln t|\mathbf{x}_j))$, and not $\mathbf{x}_j\mathbf{b}$.

`predict` now calculates more statistics after an estimation command. For instance, after linear-regression style estimators, `predict` can calculate the probability $a \leq y_j \leq b$, $E(y_j|a \leq y_j \leq b)$, and $E(y_j^*)$ where $y_j^* = \max(a, \min(y_j, b))$. `predict` can do this, for instance, after `tobit`, after `regress`, and after a number of other estimators.

What `predict` does after an estimation command is now documented with the estimation command so, if you wanted to know how `predict` works after `regress`, you would see [R] **regress** and if you wanted to know how `predict` works after `clogit`, you would see [R] **clogit**.

Given `predict`'s new capabilities, the old `fpredict`, `lpredict`, `nlpred`, `ologitp`, etc., commands are now undocumented (although they still work). `predict` replaces them all.

`probit` and `dprobit` have an `asis` option to suppress dropping variables and observations due to oneway causation. They now allow `iweights`. See [R] **probit**.

regress accepts the new time-series operators. For example, you can now estimate

> . regress irate gnp L.gnp L2.gnp

to model irate with predictors gnp, one-lagged gnp, and twice-lagged gnp. See [R] **regress** and [U] **14.4.3 Time-series varlists**.

regdw (regression with Durbin–Watson statistic) is now undocumented and replaced by the post-estimation command dwstat for use after regress; see [R] **regression diagnostics**.

reshape has an all new, easier-to-use syntax (Gould 1997, Weesie 1997b); see [R] **reshape**.

sampsi includes the additions of Seed (1997, 1998) which allow for repeated measurements.

serrbar has been improved; see [R] **serrbar**.

st has lots of changes; see [U] **1.3.2.2 New st survival analysis features** earlier in this chapter.

stack has a new group() option which provides an easier way to use stack. In addition, stack's into() option now understands variable ranges. See [R] **stack**.

sureg allows linear constraints and cross-equations constraints, accepts the new in-line equation syntax, optionally provides iterated maximum-likelihood estimates, and accepts time-series operators; see [R] **sureg**.

table and tabdisp now have a concise option which suppresses displaying rows with all missing values; see [R] **table** and [R] **tabdisp**.

testnl has a new syntax; see [R] **testnl**.

ttest, ttesti, sdtest, and sdtesti now display standard deviations and provide a level() option to specify the level for the confidence interval. In addition, string variables are now allowed with the by() option. See [R] **ttest** and [R] **sdtest**.

weibull converges more rapidly. See [R] **streg** and [R] **weibull**.

while now works interactively and in do-files. Previously, you could only use while in a program. See [R] **while**.

xtgee has more families and links. It allows inverse Gaussian and negative binomial families and complementary log-log, negative binomial, power, and odds-power links. xtgee now allows iweights. See [R] **xtgee**.

xtgls will now calculate the autocorrelation coefficient in various ways; see [R] **xtgls**.

xtreg, ml is now internal and fast. xtreg now allows iweights. See [R] **xtreg**.

1.3.4   New functions and formats

1.3.4.1   Date functions and formats

Weekly, monthly, quarterly, half-yearly, and yearly time variables: In addition to Stata's existing date format (0 = 1jan1960, 1 = 2jan1960, ...), new formats are provided for other time periods, such as 0 = first quarter of 1960, 1 = second quarter of 1960, etc. See [U] **27.3 Time-series dates**. Also see [U] **15.5.3 Time-series formats** for how the new display formats (called %t formats) work; for instance, 0 might be displayed as 1960q1 or 1960-1, etc.

Extensions to existing date format: Going along with the above, the %d format has picked up new features; see [U] **27.2.3 Displaying dates**.

String-to-date translation functions: There are new string-to-date translation functions `daily()`, `weekly()`, `monthly()`, `quarterly()`, `halfyearly()`, and `yearly()`. (`daily()` is a synonym for the previously existing `date()` function.) See [U] **27.3.5 Extracting components of time**.

New date-literal functions: New date-literal functions `d()`, `w()`, `m()`, `q()`, `h()`, and `y()` have been added. The `d()` function, for instance, lets you type things like `list if bdate>=d(15jul1982)`. See [U] **27.3.2 Specifying particular dates (date literals)**.

Y2K and the date function: The existing `date()` function now takes two or (new) three arguments. The third argument makes it easier to deal with two-digit years so that `15 Jan 02` could mean 15jan1902 or 15jan2002. The third argument specifies the maximum year that is to be assigned. If a third argument of 2050 is specified, then two-digit years will be interpreted as being in the range 1951–2050. See [U] **16.3.3 Date functions**.

Four-digit years by default: The default `%d` format is now `%dD1CY` and not `%dD1Y`. That means that, by default, the 7th of July, 2002, displays as 07jul2002 and not 07jul02. To obtain two-digit years, you must explicitly specify the previous `%dD1Y` format. See [U] **27.2.3 Displaying dates**.

1.3.4.2   Statistical and mathematical functions

New statistical and mathematical functions

`digamma()` returns the value of the digamma function $\Psi(x) = d\ln\Gamma(x)/dx$.

`trigamma()` returns the value of the trigamma function $d\Psi(x)/dx = d^2\ln\Gamma(x)/dx^2$.

`reldif(x,y)` returns $|x-y|/(|y|+1)$. Note that for very small values of y, `reldif()` is approximately the absolute difference $|x - y|$ and for very large values, `reldif()` is approximately the relative difference $|x - y|/|y|$. Programmers will find `reldif()` useful in measuring convergence.

`mreldif(X,Y)`, where $\mathbf{X}$ and $\mathbf{Y}$ are matrices, returns $\max_{i,j} |x_{ij} - y_{ij}|/(|y_{ij}| + 1)$. This allows comparing matrices.

`diag0cnt(X)` returns a count of the number of 0's on the diagonal of square matrix $\mathbf{X}$.

See [U] **16.3 Functions** and [U] **17.8 Matrix functions** for details about these functions.

Changed statistical and mathematical functions

`invnorm()` is now faster and more accurate.

`mod()` function now handles `mod(-a,b)`, $a > 0$, properly; it now returns a result that is greater than or equal to 0, not negative. `mod(a,-b)`, $b > 0$, now returns missing.

`normd()` now allows one (as previously) or two arguments. `normd(z)` returns the height of the $N(0, 1)$ density at z. `normd(z,s)` returns the height of a $N(0, s^2)$ density at z.

See [U] **16.3 Functions** for more information about these functions.

1.3.4.3   Other functions and formats

Other new functions and formats

Comma formats: Stata's display formats now support commas. The number 1,002 would still be displayed as 1002 by `%9.0g` but will be displayed as 1,002 by `%9.0gc`—note the c on the end of the format. See [U] **15.5 Formats: controlling how data is displayed**.

Left-justified string formats: The string "this" can be displayed left justified by the format `%-5s`; see [U] **15.5 Formats: controlling how data is displayed**.

Missing numeric or string expressions: The new missing(*numeric_or_string_exp*) function for use in expressions returns 1 if the argument evaluates to missing and 0 otherwise. Missing here is defined as a numeric argument being equal to . and a string argument being equal to "".

Other changed functions

cond(z_1,z_2,z_3) now allows z_2 and z_3 to both be string arguments or both be numeric arguments. z_2 and z_3 had to be numeric arguments previously. cond() returns z_2 if z_1 is true (not 0 and not missing) and z_3 otherwise.

string() now takes one (as previously) or two (this is new) arguments. The optional second argument specifies the format under which the first argument is to be translated to a string. See [U] **16.3.5 String functions**.

1.3.5   New data-management features

Longer value labels: Value labels may now be up to 80 characters in length (the previous maximum was 8). Value labels may contain up to 65,536 mappings (the previous maximum was 500). You may now label negative as well as positive values.

The new nofix option on label values and label define prevents the display format of a variable from being widened to accommodate the length of the longest value label. This has to do with the longer value labels. When you label a variable, the width of one of the value labels might be greater than the display format. In that case, the default action is to widen the display format. See [R] **label**.

Longer variable and dataset labels: Variable labels and the dataset label may now be up to 80 characters long, up from the previous maximum of 31.

Creating separate variables: The new separate command creates separate variables from a single variable for each category of another variable; see [R] **separate**.

contract opposite of expand: The new contract *varlist* command (Cox 1998) replaces the data in memory with a new dataset consisting of all combinations of *varlist* that exist in the data together with a new variable that contains the frequency of combination. Think of contract as the opposite of expand; see [R] **contract**.

New egen functions: The following new egen() functions—see [R] **egen**—are provided:

fill(*# # ...*) creates ascending, descending, or repeating patterns of numbers from the part of the pattern that is supplied.

rmin(*varlist*) returns the row minimum.

rmax(*varlist*) returns the row maximum.

rfirst(*varlist*) returns the first nonmissing value in a row.

rlast(*varlist*) returns the last nonmissing value in a row.

Concerning egen functions that previously existed:

mtr(*yearvar incvar*) now has tax rates through 1997.

pctile(*exp*) now calculates noninteger percentiles.

rank(*exp*) now allows a by() option.

rmiss(*varlist*) now works with strings as well as numeric variables.

1.3.6   New programming features

New matrix language

Full matrix parsing: Stata now has complete matrix parsing, meaning complicated expressions involving matrices are now understood. For example, you can now write

```
matrix b = syminv(X´*X)*X´*y
```

The matrix operations and functions in Stata remain for the most part unchanged, but now you can combine them in complicated expressions. This makes working with matrices in Stata much, much easier. See [U] **17 Matrix expressions**.

Other matrix changes:

   `matrix substitute` is gone and instead you perform matrix substitutions directly: `matrix A[`exp_1`,`exp_2`] = `exp_3.

You can no longer obtain the ith row by referring to `A[`i`,.]`; instead refer to `A[`i`,...]`.

The new `matrix rename` command allows renaming matrices; see [R] **matrix utility**.

`matrix score` now has an `eq()` option; see [R] **matrix score**.

`matrix accum`, `matrix vecaccum`, and `matrix glsaccum` now allow `iweights`.

New features for saved results

Estimation sample saved in `e(sample)`: After running any estimation command, the new function `e(sample)` returns true (1) if the observation was used in estimation and false (0) otherwise. You can type, for instance, `summarize if e(sample)` to obtain summary statistics on the estimation sample. See [U] **23.4 Specifying the estimation subsample**.

New way of saving results in `r(`*name*`)`: `_result(#)` and `$S_#` have been replaced by `r(`*name*`)`. Type `summarize` and then type `return list` and you will get the idea. Rather than the mean being stored in `_result(3)` and the variance in `_result(4)`, they are now stored in `r(mean)` and `r(Var)`. You can use `r(mean)` and `r(Var)` in subsequent expressions. See [R] **saved results**.

   Results are still saved in `_result(#)` and `$S_#` so old programs continue to work.

New way of saving estimation results in `e(`*name*`)`: Run any estimation command—`regress`, `logistic`, etc.—and then type `estimates list`. For instance, the number of observations will be found in `e(N)`. This too is described in [R] **saved results**.

New way of getting the coefficient vector and variance–covariance matrix: That results are saved in `e()` is carried forward even to the coefficient vector and the variance–covariance matrix of the estimators, which are now `e(b)` and `e(V)`. Instead of typing

```
. matrix b = get(_b)
. matrix V = get(VCE)
```

One simply uses

```
. matrix b = e(b)
. matrix V = e(V)
```

See [U] **17.5 Accessing matrices created by Stata commands** and [U] **21.9 Accessing results calculated by estimation commands**.

New way of posting estimation results: All issues of posting and redisplaying estimation results are now handled by `estimates`, not `matrix`. For instance, you use `estimates post` to post results, not `matrix post`. You use `estimates display` to display estimation results, not `matrix mlout`. The new `estimates repost` command makes it possible to change posted results. In addition, new options are available for `estimates display`. See [R] **estimates**.

Programs now have classes: Programs are now marked as being r, e, s, or n class, according to whether they save results in the new `r()`, `e()`, `s()`, or do not save results at all. If a program is r, e, or s class, you must specify the `rclass`, `eclass`, or `sclass` option on the `program define` statement. You may then use `return`, `estimates`, or `sreturn` commands in the program body. See [R] **return**.

Bigger macros and new style of quotes make ado-files just like internal commands

Bigger macros: Macros are now a maximum of 18,632 characters long. Note that $2,047 \times 9 = 18,423 < 18,632$, meaning that macros can hold the name of every variable in the dataset (2,047 is the maximum number of variables allowed in a dataset).

New double-quote characters: Stata has a new pair of double-quote characters, `` ` ``" and "´, to go along with its standard double-quote character, ". The new-style double quotes—called compound double quotes—can be used anywhere you could use the standard double quotes. Syntax diagrams continue to show the old-style double quotes, but it is implied that compound double quotes may be used. It is not anticipated that end-users of Stata will use the compound double quotes, but it is expected that programmers will use them. The advantage of the compound double quotes is that they nest. What does "A"B"C" mean? It could mean `` ` ``"A`` ` ``"B"´C"´ or it could mean `` ` ``"A"´B`` ` ``"C"´. See [U] **21.3.5 Double quotes**.

Ado-files now like internal commands: Taking the above two changes together, there is now no reason why an ado-file implemented command cannot be indistinguishable from an internally implemented command. The previous problems were (1) if there were too many variables, a single macro could not hold all of their names and (2) ado-files did not treat double-quoted strings correctly.

Longer limits elsewhere, too: Most of the other limits have changed, too. For instance, the maximum number of characters in a command is now 18,648 (the previous maximum was 6,144). Enter Stata and type `help limits`.

New commands for parsing, etc.

New syntax command replaces parse: The new `syntax` command replaces `parse` (which continues to work but is undocumented) for the parsing of standard Stata syntax. The new command is easier to use and more powerful. See [R] **syntax**.

New command for unloading arguments to a program: The new `args` command is the right way to receive positional arguments rather than referring to `` `1´ ``, `` `2´ ``, ... That is, rather than using

```
program define myprog
        local a `1´
        local b `2´
        local c `3´
        ...
```

you simply use

```
program define myprog
        args a b c
        ...
```

See [R] **syntax**.

New command for nonstandard parsing: The new `tokenize` command replaces `parse, parse()`. For example, you would use `tokenize` to parse on blanks or other special characters. `parse` continues to work, but is now undocumented. See [R] **tokenize**.

Original command line that invoked program is accessible: The macro `` `0´ `` now contains the original line, as typed by the user, with quotes, multiple blanks, and all, when a program is invoked. See [U] **21.4 Program arguments**.

Tokens can be unloaded one at a time: The new `gettoken` command allows fetching tokens one at a time from the input stream (`` `0´ ``) or from any macro. See [R] **gettoken**.

New tools for programming commands

New command for marking sample: The new `marksample` command is an easier-to-use and less error prone alternative to `mark` and `markout` when you use the new `syntax` command to parse input. In addition, both the existing `mark` and new `marksample` commands provide a new `zeroweight` option to include observations with zero weights. See [R] **mark**.

Programmers can make predict work with new estimation commands: Now `predict` is implemented as an ado-file, and, under the new scheme, it calls `` `e(predict)´ ``, the name of the prediction command saved by the estimation command. This command—another ado-file—is typically implemented in terms of `_predict`, the old built-in `predict` command. (When `version` is 5.0 or before, all of this is turned off and `_predict` becomes `predict`.) See [R] **_predict**.

Replaying estimation results: The new `replay()` function returns 1 if the first nonblank character in `` `0´ `` (what the user typed as the user typed it) is comma or if `` `0´ `` is "". This makes writing estimation commands easier. Early on you simply code 'if `replay()` {' and put the redisplay logic there. See [U] **16.3.6 Special functions**.

Command for handling new time-series varlists: The new `tsrevar` command assists in writing commands that use time-series varlists; see [R] **tsrevar**.

New command for unabbreviating varlists: The new `unab` command replaces `unabbrev` (`unabbrev` continues to work), but now, both are rarely used because of the new features of `syntax`. There is a `tsunab` command for time-series varlists, too. See [R] **unab** and [R] **syntax**.

New command for parsing numlists: The new `numlist` command helps parse numlists but, as with `unab`, it is rarely used because `syntax` can do that, too. See [R] **numlist** and [R] **syntax**.

Confirm numeric variable: The new `confirm numeric variable` command is just `confirm string variable` turned on its head; see [R] **confirm**.

Determining the version of caller: The new `_caller()` function returns the `version` number of the program or session that invoked the program currently being executed. This makes coding for backwards compatibility easier. For instance, we changed Stata's `st` commands significantly this release and yet wanted to still ensure that the old routines would work. Consider the nonexistent `st` command `stxyz`. We took the old `stxyz.ado` file and renamed it `stxyz_5.ado`. We coded our new `stxyz.ado` file as

```
program define stxyz
        version 6
        if _caller() <= 5 {
                stxyz_5 `0´
                exit
        }
        ...
end
```

New macro extended functions

`data label` returns the dataset label; see [U] **21.3.6 Extended macro functions**.

`label` extended macro function now has syntax:

local ... : label $\{$ *labelname* | (*varname*) $\}$ $\{$ maxlength | #$_{\text{val}}$ [#$_{\text{len}}$] $\}$

First, you can indirectly refer to the value label associated with a variable by enclosing the variable name in parentheses. Second, specifying `maxlength` returns the length of the longest label in the value label. Third, specifying the #$_{\text{len}}$ trims the result to being no more than #$_{\text{len}}$ characters long. See [U] **21.3.6 Extended macro functions**.

`piece` returns the piece of string, given a specified maximum length, the piece being at a word break; see [U] **21.3.6 Extended macro functions**.

`rowfullnames` and `colfullnames` return the "full" row and column names of a matrix. This has to do with the new time-series features. Row and column names now potentially consist of three parts: the equation name, the time-series operator, and the variable name, e.g., `eq1:L.gnp`. See [R] **matrix define** and the new [U] **17 Matrix expressions** chapter.

`subinstr` changes all occurrences of one substring to another; see [U] **21.3.6 Extended macro functions**.

`sysdir` returns the identities of various system directories; see [R] **sysdir**.

Other new programming features

Better way to get contents of characteristics: It is now possible to obtain the contents of characteristics by quoting them rather than first unloading them into a macro. For example, you can refer to `` `mpg[note0]´ `` and `` `_dta[tis]´ ``. See [U] **21.3.11 Referencing characteristics**.

New string formats for display: `display` now allows `%s` formats so you can write code such as

```
display %9s "`varname´"
```

to produce right-aligned strings. With the new `%-#s` format, you can produce left-aligned strings by coding

```
display %-9s "`varname´"
```

In addition, `display` (and only `display`) understands ~ to mean center. `display` can center strings:

```
display %~80s "My Title"
```

See [R] **display (for programmers)**.

Different characteristics for st datasets: Survival-time `st` datasets are now marked using different characteristics. Nevertheless, old `st` commands will continue to work as long as you set `version` 5.0. See [R] **st st_is** for a description of the new standard.

Underscore variables do not affect changed flag: Changes to variables beginning with an underscore no longer count as a change to the data in terms of the "dataset has changed since last saved" flag.

Utilities for writing certification scripts: Although not documented in the manuals, we have included the utility commands we use for writing certification scripts—do-files that prove Stata commands work as intended. If you write ado-files and want to write certification do-files to go along with them, start by seeing **help cscript**. `cscript` is a command to assist writing certification scripts, and in the help file we provide guidelines on how to do this along with links to the help files of other undocumented but useful commands such as `rcof`, `typeof`, and `old_ver`.

1.3.7 Other new features

Specifying lists of numbers: Stata now has a concept of a *numlist*. All throughout Stata, whenever you need to type a list of numbers, you can type that list using numlist syntax. numlist syntax includes 1/3 to mean 1, 2, 3; 2(2)6 to mean 2, 4, 6; 0 5 to 20 to mean 0, 5, 10, 15, 20; and so on. See [U] **14.1.8 numlist**.

Offsets: Many estimation commands now allow an offset(*varname*) option, specifying that the linear equation to be estimated is $\mathbf{x}_j\mathbf{b} + varname_j$.

New Help window: Stata for Windows 98/95/NT and Stata for Power Macintosh users: The new net features carry over to the Help window. Pull down **Help** and you will discover that the Help window has the look of a browser. The similarity is more than just in appearance. For instance, if a help file refers to a FAQ on our web site, if you click on the reference, your browser will be invoked to display the referenced FAQ. Back in Stata's Help window, you can also download and install official updates, or STB ado-files, etc., by pointing and clicking.

Searching for information on commands, STB articles, FAQs, etc.: lookup is now called search. It includes a new faq option for searching among the FAQs. See [R] **search**.

profile.do is automatically executed at start-up: Stata on all operating systems now looks for and, if found, executes profile.do when Stata is invoked. You can put commands in profile.do to tailor Stata to your tastes. See [GSW] **A.7 Executing commands every time Stata is started** (Windows 98/95/NT), [GSW] **B.7 Executing commands every time Stata is started** (Windows 3.1), [GSM] **A.6 Executing commands every time Stata is started** (Macintosh), or [GSU] **A.7 Executing commands every time Stata is started** (Unix).

New arrangement of Stata directories (folders): We have restructured the Stata directories (folders) and added a new command sysdir to make accessing them easier. See [R] **sysdir** for a description of the new scheme.

Copy files from inside Stata: The new copy command allows copying files from inside Stata. Moreover, copy, like all of Stata's file manipulation commands, can read files over the Internet. See [R] **copy**.

Create directories (folders) from inside Stata: The new mkdir command allows creating directories (folders) from inside Stata. See [R] **mkdir**.

1.4 References

Becketti, S. 1992. sts1: Autocorrelation and partial autocorrelation graphs. *Stata Technical Bulletin* 5: 27–28. Reprinted in *Stata Technical Bulletin Reprints*, vol. 1, pp. 221–223.

——. 1995. sts7.6: A library of time-series program for Stata. *Stata Technical Bulletin* 24: 30–35. Reprinted in *Stata Technical Bulletin Reprints*, vol. 4, pp. 233–241.

Carlin, J., S. Vidmar, and C. Ramalheira. 1998. sg75: Geometric means and confidence intervals. *Stata Technical Bulletin* 41: 23–25. Reprinted in *Stata Technical Bulletin Reprints*, vol. 7, pp. 197–199.

Clayton, D. and M. Hills. 1995a. ssa7: Analysis of follow-up studies. *Stata Technical Bulletin* 27: 19–26. Reprinted in *Stata Technical Bulletin Reprints*, vol. 5, pp. 219–227.

——. 1995b. ssa8: Analysis of case–control and prevalence studies. *Stata Technical Bulletin* 27: 26–31. Reprinted in *Stata Technical Bulletin Reprints*, vol. 5, pp. 227–233.

——. 1997. ssa10: Analysis of follow-up studies with Stata 5.0. *Stata Technical Bulletin* 40: 27–39. Reprinted in *Stata Technical Bulletin Reprints*, vol. 7, pp. 253–268.

Cox, N. J. 1998. dm59: Collapsing datasets to frequencies. *Stata Technical Bulletin* 44: 2–3.

Cox, N. J. and A. R. Brady. 1997a. gr25: Spike plots for histograms, rootograms, and time-series plots. *Stata Technical Bulletin* 36: 8–11. Reprinted in *Stata Technical Bulletin Reprints*, vol. 6, pp. 50–54.

——. 1997b. gr25.1: Spike plots for histograms, rootograms, and time series plots: update. *Stata Technical Bulletin* 40: 12. Reprinted in *Stata Technical Bulletin Reprints*, vol. 7, p. 58.

Garrett, J. M. 1995. sg33: Calculation of adjusted means and proportions. *Stata Technical Bulletin* 24: 22–25. Reprinted in *Stata Technical Bulletin Reprints*, vol. 4, pp. 161–165.

——. 1997. gr23: Graphical assessment of the Cox proportional hazards assumption. *Stata Technical Bulletin* 35: 9–14. Reprinted in *Stata Technical Bulletin Reprints*, vol. 6, pp. 38–44.

——. 1998. sg33.1: Enhancements for calculation of adjusted means and adjusted proportions. *Stata Technical Bulletin* 43: 16–24.

Gleason, J. R. 1997. sg65: Computing intraclass correlations and large ANOVAs. *Stata Technical Bulletin* 35: 25–31. Reprinted in *Stata Technical Bulletin Reprints*, vol. 6, pp. 167–176.

Goldstein, R. 1996. sg55: Extensions to the brier command. *Stata Technical Bulletin* 32: 21–22. Reprinted in *Stata Technical Bulletin Reprints*, vol. 6, pp. 133–134.

Gould, W. 1997. stata48: Updated reshape. *Stata Technical Bulletin* 39: 4–16. Reprinted in *Stata Technical Bulletin Reprints*, vol. 7, pp. 5–20.

Gould, W. and W. Sribney. 1999. *Maximum Likelihood Estimation with Stata*. College Station, TX: Stata Press.

Hilbe, J. 1996. sg53: Maximum-likelihood complementary log-log regression. *Stata Technical Bulletin* 32: 19–20. Reprinted in *Stata Technical Bulletin Reprints*, vol. 6, pp. 129–131.

——. 1998. sg53.2: Stata-like commands for complementary log-log regression. *Stata Technical Bulletin* 41: 23. Reprinted in *Stata Technical Bulletin Reprints*, vol. 7, pp. 166–167.

Newton, H. J. 1996. sts12: A periodogram-based test for white noise. *Stata Technical Bulletin* 34: 36–39. Reprinted in *Stata Technical Bulletin Reprints*, vol. 6, pp. 203–207.

Seed, P. 1997. sbe18: Sample size calculations for clinical trials with repeated measures data. *Stata Technical Bulletin* 40: 16–18. Reprinted in *Stata Technical Bulletin Reprints*, vol. 7, pp. 121–125.

——. 1998. sbe18.1: Update of sampsi2. *Stata Technical Bulletin* 45: 21.

Weesie, J. 1997a. sg66: Enhancements to the alpha command. *Stata Technical Bulletin* 35: 32–34. Reprinted in *Stata Technical Bulletin Reprints*, vol. 6, pp. 176–179.

——. 1997b. dm48: An enhancement of reshape. *Stata Technical Bulletin* 38: 2–4. Reprinted in *Stata Technical Bulletin Reprints*, vol. 7, pp. 40–43.

——. 1998. dm62: Joining episodes in multi-record survival time data. *Stata Technical Bulletin* 45: 5–6.

2 Resources for learning and using Stata

Contents

2.1   Overview

The *Getting Started*, *User's Guide*, *Graphics* manual, and *Reference* manuals are the primary tools for learning about Stata; however, there are many other sources of information as well. A few are

1. The Stata web site. Visit *http://www.stata.com* if you can. A large portion of the site contains user-support materials. Details are in [U] **2.2 The http://www.stata.com web site** below.

2. The Stata listserver. An active group of Stata users communicate over an Internet listserver. You can join and it is free. Details are in [U] **2.3 The Stata listserver** below.

3. The Stata software-distribution site and other user-provided software-distribution sites. Stata itself can download and install updates and additions. We provide official updates to Stata—type `update` or pull down **Help** and select **Official Updates**. We also provide user-written additions to Stata and links to other user-provided sites—type `net` or pull down **Help** and select **STB and user-written Programs**. Details are in [U] **2.5 Updating and adding features from the web** below.

4. The *Stata Technical Bulletin*. The journal contains articles as well as updated commands and additions to Stata. Even if you do not subscribe to the journal, the command updates are available via the Internet. Details are in [U] **2.4 The Stata Technical Bulletin** below.

5. NetCourses. We offer training via the Internet. Details are in [U] **2.6 NetCourses** below.

6. Books and support materials. Supplementary Stata materials are available. Details are in [U] **2.7 Books and other support materials** below.

7. Technical support. We provide technical support by telephone, fax, and email. Details are in [U] **2.8 Technical support** below.

In addition,

1. Stata has on-line tutorials; see [U] **9 Stata's on-line tutorials and sample datasets**.

2. Stata has an on-line help and search facility; see [U] **8 Stata's on-line help and search facilities**.

2.2   The http://www.stata.com web site

If you have access to the World Wide Web, point your browser to *http://www.stata.com* and then click on **User Support**. Over half of our web site is dedicated to providing support to users.

1. The web site provides FAQs (Frequently Asked Questions) on Windows, Macintosh, Unix, statistics, programming, graphics, and data management. These FAQs run the gamut from "I cannot save/open files" to "What does 'completely determined' mean in my logistic-regression output?" Everyone will find something of interest.

2. The web site provides additions to Stata in the form of new commands. Stata is programmable. Most of Stata is implemented in Stata so, even if you never write a Stata program, you make use of that feature; see [U] **20 Ado-files**. The web site is one source of new commands and some of them are spectacularly useful. Stata itself can download and install updates and additions, see [U] **2.5 Updating and adding features from the web** below. It turns out Stata for Windows 3.1 and Stata for 680x0 Macintosh cannot do that, however, so those users will want to use their browser to pull the materials for themselves.

3. Visiting the web site is one way you can subscribe to the Stata listserver. See [U] **2.3 The Stata listserver** below.

4. The web site provides detailed information about the NetCourses we offer, along with the current schedule; see [U] **2.6 NetCourses** below.

5. The web site provides an on-line bookstore for Stata-related books and useful links to other statistical sites.

In short, the web site provides up-to-date information on all support materials and, where possible, provides the materials themselves. Visit *http://www.stata.com* if you can.

2.3   The Stata listserver

The Stata listserver (Statalist) is an independently operated, real-time list of Stata users on the Internet. Anyone may join. Instructions for doing so can be found at *http://www.stata.com* or by emailing *stata@stata.com*.

Many knowledgeable users are active on the list, as are the StataCorp technical staff. We recommend new users subscribe, observe the exchanges, and, if it turns out not to be useful for you, unsubscribe.

2.4   The Stata Technical Bulletin

The *Stata Technical Bulletin* (STB) is a printed journal—there are six issues per year—that provides a forum for Stata users of all disciplines and levels of sophistication. The STB contains articles written by StataCorp staff members, Stata users, and others.

One purpose of the *Bulletin* is to provide a link between professional statisticians and professional researchers whose primary interest is not statistics, but who use statistics as a tool in pursuing their science.

Articles include new Stata commands (ado-files), programming tutorials, illustrations of data-analysis techniques, discussions on teaching statistics, debates on appropriate statistical techniques, reports on other programs, interesting datasets, announcements, questions, and suggestions.

One of the more attractive features about the STB is the new commands that it provides. Associated with each issue is a "disk" which contains programs written by Stata users, as well as official updates to Stata. The "disk" is available for free over the Internet even if you do not subscribe to the journal itself, or it is available for a charge as a real diskette.

We recommend that all users subscribe to the STB. Subscriptions are available either with or without disks.

Whether or not you subscribe, to obtain the software, either type

```
. net from http://www.stata.com
. net cd stb
```

or

1. Pull down **Help**
2. Select **STB and User-written Programs**
3. Click on *http://www.stata.com*
4. Click on *stb*

or point your browser to *http://www.stata.com* and click on *User support*.

The STB has been in continual publication since May 1991. After every six issues, the journals are reprinted into a book — *The Stata Technical Bulletin Reprints*. Those volumes are available for purchase and may also be available at your library.

2.5 Updating and adding features from the web

Stata itself is web-aware, or at least Stata for Windows 98/95/NT, Stata for Power Macintosh, and Stata for Unix are, which is to say, Stata for Windows 3.1 and Stata for 680x0 Macintosh are not. (3.1 and 680x0 users: point your browser to *http://www.stata.com*; you can get the updates described below from there.)

First, try this:

```
. use http://www.stata.com/manual/oddeven.dta, clear
```

That will load an uninteresting dataset into your computer from our web site. If you have a homepage, you can use this feature to share datasets with coworkers. Save a dataset on your homepage and researchers worldwide can use it. See [R] **net**.

We distribute updates to Stata every other month—in January, March, May, July, September, and November. Installing the updates is easy, type

```
. update
```

or pull down **Help** and select **Official Updates**. Do not be concerned; nothing will be installed unless and until you say so. Once you have installed the update, you can type

```
. help whatsnew
```

or pull down **Help** and select **What's New** to find out what has changed. We distribute official updates to fix bugs and add new features.

There are also "unofficial" updates—additions to Stata written by Stata users, which includes members of the StataCorp technical staff. Stata is programmable and even if you never write a Stata program, you may find these additions useful, and some of them spectacularly so. Anyway, you start by typing

```
. net from http://www.stata.com
```

or pulling down **Help** and selecting **STB and user-written Programs**.

Periodically you can type

```
. news
```

or pull down **Help** and select **News** to display a short message from our web site telling you what is newly available.

See [U] **32 Using the Internet to keep up to date**.

2.6 NetCourses

We offer courses on Stata at the introductory and advanced levels. Courses on software are typically expensive and time-consuming. They are expensive because, in addition to the direct costs of the course, participants must travel to the course site. We have found it successful to organize courses over the Internet—saving everyone time and money.

We offer courses over the Internet and call them Stata NetCourses™.

1. **What is a NetCourse?**

   A NetCourse is a course offered via email and varies in length from a few to twelve weeks. You must have an Internet email address to participate.

2. **How does it work?**

   Every Friday a "lecture" is emailed to the course participants. After reading the lecture over the weekend or perhaps on Monday, participants then email questions and comments back to the Course Leaders. Copies of these emailed questions are forwarded to all course participants by the NetCourse software. Course leaders respond to the questions and comments on Tuesday and Thursday. The other participants are encouraged to amplify or otherwise respond to the questions or comments as well. The next lecture is then emailed on Friday and the process repeats.

3. **Is this like a listserver?**

   It is similar to a listserver in that StataCorp establishes an email address such as *nc999@stata.com* and all email to the address is reflected to the participants. The discussion, however, is guided by the Course Leaders via the presentation of the weekly lecture. A lecture consists of approximately 20 printed pages.

4. **How much of my time does it take?**

   It depends on the course, but the introductory courses are designed to take roughly 3 hours per week.

5. **There are three of us here—can just one of us enroll and we redistribute the NetCourse materials ourselves?**

   We ask that you not. NetCourses are priced to cover the substantial time input of the Course Leaders. Moreover, enrollment is typically limited to prevent the discussion from becoming unmanageable. The value of a NetCourse, just like a real course, is the interaction of the participants, both with each other and with the Course Leaders.

6. **I've never taken a course by Internet before. I can see that it might work, but then again, it might not. How do I know I'll benefit?**

   All Stata NetCourses come with a 30-day satisfaction guarantee. The 30 days begin after the conclusion of the final lecture.

You can learn more about the current NetCourse offerings by visiting *http://www.stata.com*. Three courses we recommend are

> NC-101 An introduction to Stata
> NC-151 An introduction to Stata programming
> NC-152 Advanced Stata programming

2.7 Books and other support materials

There are books published on Stata, both by us and others. Visit the bookstore at *http://www.stata.com* for an up-to-date listing and the table of contents of each.

2.8 Technical support

Stata Corporation is committed to providing superior technical support for Stata software. In order to assist you as efficiently as possible, please follow the procedures listed below.

2.8.1 Register your software

You must register your software in order to be eligible for technical support, updates, special offers, and other benefits. By registering, you will receive the *Stata News* and you may access our support staff for free with any question that you might encounter. You may register your software either electronically or by mail.

Electronic registration: Point your web browser at *http://www.stata.com* and visit the post office to register your copy of Stata.

Mail-in registration: Fill in the registration card that came with Stata and mail to Stata Corporation at the address on the back of the cover of this manual.

2.8.2 Before contacting technical support

Our technical department will need some information from you in order to provide detailed assistance. Most important is your serial number, but they will also need the following information:

1. The system information on the computer that you are using—this is especially important if you are having hardware problems. Included in this information is the make and model of various hardware components such as the computer manufacturer, the video driver, operating system and its version number, relevant peripherals, and version number of any other software with which you experience a conflict. See [U] **6 Troubleshooting starting and stopping Stata**.

2. The version of Stata that you are running. Type 'about' at the Stata prompt and Stata will display the information we are seeking.

3. The types of variables in your dataset and the number of observations.

4. The command that is causing the error along with the exact error message and return code (error number).

2.8.3 How to contact technical support

Before you spend the time gathering the information our technical support department needs, make sure that the answer does not already exist in the help files. You can use the `help` and `search` commands to find all of the entries in Stata that address a given subject. Check the manual for a particular command. Many times, there are examples that address questions and concerns. Another good source of information is our web site. You should keep a bookmark of our frequently asked questions page and check it occasionally for new information.

2.8.4   Technical support by phone

Our technical support telephone number is located on the back cover of the manuals. Please have your serial number handy for the operator. If your question involves an error message from a command, please note the error message and number exactly as this will greatly help us in assisting you.

2.8.5   Technical support by email or fax

Send email requests to *tech-support@stata.com*. Send fax requests to the fax number listed on the back cover of this manual.

Remember to specify your serial number. In email, put it on the subject line if you can. Then remember to include the other information we requested above.

If possible—this works best—collect the relevant information in a log file and include the log file in your email or fax.

2.8.6   Comments and suggestions for our technical staff

By all means, send in your comments and suggestions by fax or email. Your input is what determines the changes that occur in Stata between releases, so if we don't hear from you, we may not include your most desired new estimation command! Fax and email are preferred as this provides us with a permanent copy of your request. In making a request for new commands, please include any references that you would like us to review should we develop those new commands.

3 A brief description of Stata

Stata is a statistical package for managing, analyzing, and graphing data.

Stata is available for a variety of platforms but, regardless of platform, Stata is command-driven. Here's an extract of a Stata session:

```
. summarize mpg weight
Variable |     Obs        Mean   Std. Dev.       Min        Max
---------+-----------------------------------------------------
     mpg |      74     21.2973    5.785503         12         41
  weight |      74    3019.459    777.1936       1760       4840
```

The user typed **summarize mpg weight** and Stata responded with a table of summary statistics. Other commands would produce different results:

```
. correlate mpg weight
(obs=74)
         |      mpg    weight
---------+------------------
     mpg |   1.0000
  weight |  -0.8072    1.0000

. gen w_sq = weight^2

. regress mpg weight w_sq
  Source |       SS       df       MS                  Number of obs =      74
---------+------------------------------             F(  2,    71) =   72.80
   Model |  1642.52197     2   821.260986             Prob > F      =  0.0000
Residual |  800.937487    71   11.2808097             R-squared     =  0.6722
---------+------------------------------             Adj R-squared =  0.6630
   Total |  2443.45946    73   33.4720474             Root MSE      =  3.3587

------------------------------------------------------------------------------
     mpg |      Coef.   Std. Err.       t     P>|t|     [95% Conf. Interval]
---------+--------------------------------------------------------------------
  weight |  -.0141581    .0038835     -3.646   0.001    -.0219016    -.0064145
    w_sq |   1.32e-06    6.26e-07      2.116   0.038     7.67e-08     2.57e-06
   _cons |   51.18308    5.767884      8.874   0.000     39.68225     62.68392
------------------------------------------------------------------------------
```

The user-interface model is type a little, get a little, so that the user is always in control.

Stata's model for a dataset is that of a table—the rows are the observations and the columns are the variables:

```
. list mpg weight in 1/10
          mpg    weight
  1.       22     2,930
  2.       17     3,350
  3.       22     2,640
  4.       20     3,250
  5.       15     4,080
  6.       18     3,670
  7.       26     2,230
  8.       20     3,280
  9.       16     3,880
 10.       19     3,400
```

Observations are numbered; variables are named.

Stata is very fast. Partly, that speed is due to clever programming and, partly, it is because Stata keeps the data in memory. Stata's data model is that of a word processor: a dataset may exist on disk, but that is just a copy. The data is loaded into memory, where it is worked on, analyzed, changes made, and then perhaps it is stored back on disk.

Having the data in memory makes Stata fast.

Working on a copy of the data in memory makes Stata safe for interactive use. The only way to harm the permanent copy of your data on disk is if you explicitly save over it.

Having the data in memory means that the dataset size is limited by the amount of memory. Stata stores the data in memory in a very compressed format—you will be surprised how much data can fit into a given region of memory. Nevertheless, if you work with large datasets, you may run into memory constraints. There are two solutions to this problem:

1. Stata comes out of the box allocating one megabyte to its data areas. You can increase this; see [U] **7 Setting the size of memory**.

2. You will want to learn how to compress your data as much as possible, see [R] **compress**.

4 Flavors of Stata

Contents

4.1   Platforms

Stata is available for a variety of different computers, including

Stata for Windows 98/95/NT
Stata for Windows 3.1

Stata for Power Macintosh
Stata for 680x0 Macintosh

Stata for DEC Alpha
Stata for HP-9000
Stata for Linux
Stata for RS/6000
Stata for SPARC

Which version of Stata you run does not matter—Stata is Stata. You instruct Stata in the same way and Stata produces the same results, right down to the random-number generator.

Even files can be shared. For instance, a dataset created using, say, Stata for Macintosh can be used on any other computer, and the same goes for graphs, programs, and any other file Stata uses or produces.

Moving a dataset or any other file across platforms requires no translation. If you do share datasets or graphs with other users across platforms, be sure you make exact binary copies.

4.2   Intercooled Stata and Small Stata

Stata for Windows and Stata for Macintosh are available in two "flavors": Small Stata and Intercooled Stata. Stata for Unix is available only in the Intercooled flavor. Both Small Stata and Intercooled Stata have the same features, but Intercooled Stata is able to work with larger datasets and it is faster. How much faster depends on the platform, but the advantage ranges from 50 to 600 percent.

Intercooled Stata is Stata. It is the version we recommend for all users doing serious data analysis and statistics.

Small Stata would perhaps be better named Stata for Small Computers.

4.2.1   Determining which version you own

Included with every copy of Stata is a paper license that contains important codes that you will input during installation. This license also determines which flavor of Stata you have—Intercooled or Small. Look at the license and see if it says Intercooled Stata or Small Stata.

If you purchased Small Stata and now suspect you want Intercooled, your Small version can be upgraded to Intercooled. In fact, we put both Small and Intercooled Stata on the same installation disks so you won't have to wait to receive another set of disks. All you need is an upgraded paper license with the appropriate codes.

By the way, even if you purchased Intercooled Stata, you may use Small Stata with your Intercooled license. This might be useful if you had a large computer at work and a smaller computer at home. Please remember, however, you have only one license (or however many licenses you purchased). You may, both legally and morally, use one, the other, or both, but you should not subject the pair to simultaneous use.

4.2.2   Determining which version is installed

If Stata is already installed, you can find out which Stata you are using by entering Stata as you normally do and typing about:

```
. about
Intercooled Stata 6.0 for Windows 98/95/NT
Born 01jan1999
Copyright (C) 1985-1999

10-user Windows (network) perpetual license:
        Serial number:  196040000
          Licensed to:  Marsha Martinez
                        StataCorp
```

You are running Intercooled Stata 6.0 for Windows 98/95/NT.

4.3   Size limits comparison of Intercooled Stata and Small Stata

Here are some of the different size limits for Intercooled Stata and Small Stata. See [R] **limits** for a longer list.

Maximum size limits for Intercooled Stata and Small Stata

	Small	Intercooled
Number of observations	fixed at approx. 1,000	limited only by memory
Number of variables	fixed at 99	2,047
Width of a dataset	200	8,192
Maximum matrix size (matsize)	fixed at 40	800
Number of characters in a macro	1,000	18,632
Number of characters in a command	1,100	18,648

That is, Intercooled Stata allows larger datasets and estimates models with more independent variables, has longer macros, and allows a longer command line (required because of the increased number of variables allowed).

4.4   Speed comparison of Intercooled Stata and Small Stata

Why the difference in speed between Intercooled and Small Stata? In part, it is due to different options we specified when we compiled Stata. In part, however, it is due to differences in the code.

For instance, in Stata's `test` command, there comes a place where it must compute the matrix calculation $\mathbf{RZR'}$ (where $\mathbf{Z} = (\mathbf{X'X})^{-1}$). Intercooled Stata makes the calculation in a straightforward way, which is to form $\mathbf{T} = \mathbf{RZ}$ and then calculate $\mathbf{TR'}$. This requires temporarily storing the matrix $\mathbf{T}$. Small Stata, on the other hand, goes into more complicated code to form the result directly—code that requires temporary storage of only one scalar! This code, in effect, recalculates intermediate results over and over again, and so it is slower.

Another difference is that Small Stata, since it is designed to work with smaller datasets, uses different memory-management routines. These memory-management routines use 2-byte rather than 4-byte offsets and therefore require only half the memory to track locations.

In any case, the differences are all technical and internal. From the user's point of view, Intercooled Stata and Small Stata work the same way.

5 Starting and stopping Stata

Contents

Below we assume that Stata is already installed. If you have not yet installed Stata, follow the installation instructions in the *Getting Started* manual.

5.1   Starting Stata

To start Stata:

Stata for Windows 98/95/NT:	Click on **Start**
	Click on **Programs**
	Select **Intercooled Stata, Small Stata,**
	Pseudo-Intercooled Stata, or **Small Stata**
	as appropriate; see [U] **4 Flavors of Stata**
Stata for Windows 3.1:	Double-click on the **Stata** icon
Stata for Macintosh:	Double-click on the **Stata.do** file icon
Stata for Unix:	Type `stata` at the Unix command prompt

Unix users will see something like this,

```
% stata

     ___  ____  ____  ____  ____ tm
    /__    /   ____/   /   ____/
   ___/   /   /___/   /   /___/    6.0    Copyright 1984-1999
      Statistics/Data Analysis            Stata Corporation
                                          702 University Drive East
                                          College Station, Texas 77840 USA
                                          800-STATA-PC        http://www.stata.com
                                          409-696-4600        stata@stata.com
                                          409-696-4601 (fax)

  Registration information appears

  Other information appears, such as Notes

  . _
```

and Stata for Windows 98/95/NT users will see something like,

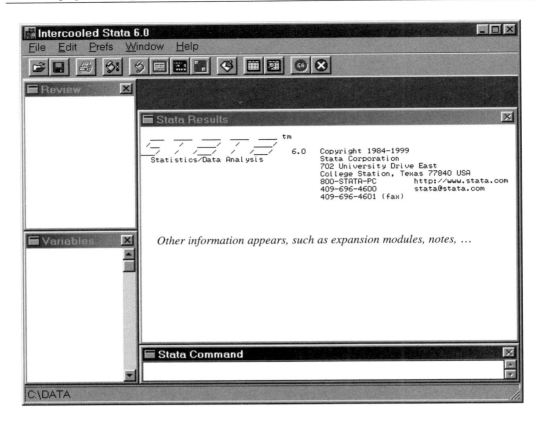

and Stata for Windows 3.1 and Stata for Macintosh users will see something similar.

If Stata does not come up, see [U] **6.1 If Stata does not start**.

Stata is now waiting for you to type something, at its dot prompt in the case of Stata for Unix and in the Command window in the case of the other Statas. From now on, when we say, for instance, type `verinst`, which we might even show as

```
. verinst
```

we mean type `verinst` and press *Enter* and do that in the Command window or, in the case of Stata for Unix, at Stata's dot prompt.

5.2   Verifying Stata is correctly installed

The first time you start Stata, you should verify that it is installed correctly. Type `verinst` and you should see something like

```
. verinst
You are running Intercooled Stata 6.0 for ...
Stata is correctly installed.
You can type exit to exit Stata.
```

If you do not see the message "Stata is correctly installed", see [U] **6.2 verinst problems**.

Remember the `verinst` command. If you ever change your computer setup and are worried that you somehow damaged Stata in the process, you can type `verinst` and obtain the reassuring "Stata is correctly installed" message.

5.3   Exiting Stata

To exit Stata,

Stata for Windows 98/95/NT:	Click on the close box or type `exit`
Stata for Windows 3.1:	Double-click on the close box or type `exit`
Stata for Macintosh:	Click on the close box or type `exit`
Stata for Unix:	type `exit`

Stata works with a copy of the data in memory. If (1) there is data in memory and if (2) it has changed and if (3) you click to close, a box will appear asking if it is okay to exit without saving the changes.

If you instead type `exit` and there is changed data in memory, Stata will refuse and instead say, "no; data in memory would be lost". In this case, either you must save the data on disk (see [R] **save**) or you can type `exit, clear` if you do not want to save the changes.

All of this is designed to prevent you from accidentally losing your data.

As you will discover, the `clear` option is allowed with all potentially destructive commands, of which `exit` is just one example. The command to bring data into memory, `use`, is another example. `use` is destructive since it loads data into memory and, in the process, eliminates any data already there.

If you type a destructive command and the data in memory has been safely stored on disk, Stata performs your request. If your data has changed in some way since it was last saved, Stata responds with the message "no; data in memory would be lost". If you want to go ahead anyway, you can retype the command and add the `clear` option. Once you become familiar with Stata's editing keys, you will discover it is not necessary to physically retype the line. You can press the *PrevLine* key (*PgUp*) to retrieve the last command you typed and append `, clear`. (On some Unix computers, you may have to press *Ctrl-R* instead of *PgUp*; see [U] **13 Keyboard use** for more details.)

Of course, you need not wait for Stata to complain—you can add the `clear` option the first time you issue the command—if you do not mind living dangerously.

5.4   Features worth learning about

5.4.1   Windows 98/95/NT

1. The Windows Properties Sheet for Stata controls how Stata comes up. The properties sheet determines which will be the working (start-in) directory when Stata comes up and how much memory will be allocated to Stata, among other things. You can change this; see

[GSW] **A.4 The Windows 98/95/NT Properties Sheet**
[GSW] **A.5 Starting Stata from other folders**
[GSW] **A.6 Specifying the amount of memory allocated**

You can also change the amount of memory Stata has during a session using the `set memory` command; see [U] **7 Setting the size of memory**.

2. Stata can execute commands every time it is invoked if you record the commands in the ASCII file `profile.do`, which you could put in Stata's working (start-in) directory; see [GSW] **A.7 Executing commands every time Stata is started**.

3. You can arrange to start Stata without going through the **Start** button. On your standard Windows screen, do you see how **My Computer** just sits on the desktop? You can put a Stata icon on the desktop, too. See [GSW] **A.8 Making shortcuts**.

4. You can run large jobs in Stata in batch mode. See [GSW] **A.9 Executing Stata in background (batch) mode**.

5. You can launch Stata by double-clicking on a Stata `.dta` dataset; see [GSW] **A.10 Launching by double-clicking on a .dta dataset**.

6. You can launch Stata and run a do-file by double-clicking on the do-file; see [GSW] **A.11 Launching by double-clicking on a do-file**.

7. Each time you launch Stata, you invoke a new instance of it, so if you want to run multiple Stata sessions simultaneously, you may.

5.4.2   Windows 3.1

1. The Windows Properties Sheet for Stata controls how Stata comes up. The properties sheet determines which will be the working directory when Stata comes up and how much memory will be allocated to Stata, among other things. You can change this; see

    [GSW] **B.4 The Windows 3.1 Properties Sheet**
    [GSW] **B.5 Starting Stata from other directories**
    [GSW] **B.6 Specifying the amount of memory allocated**

2. Stata can execute commands every time it is invoked if you record the commands in the ASCII file `profile.do`, which you would put in Stata's working directory; see [GSW] **B.7 Executing commands every time Stata is started**.

3. Each time you launch Stata, you invoke a new instance of it, so if you want to run multiple Stata sessions simultaneously, you may.

5.4.3   Macintosh

1. You can change the amount of memory allocated to Stata; see [GSM] **A.5 Specifying the amount of memory allocated**.

2. You can start Stata from other folders; simply copy `Stata.do`—a small file—to whatever folders you wish. See [GSM] **A.4 Starting Stata from other folders**.

3. You can start Stata from the Apple Menu if you add `Stata.do` (not the Stata application) as an alias to it; see [GSM] **A.7 Creating aliases**.

4. Stata can execute commands every time it is invoked if you record the commands in the ASCII file `profile.do`, which you could put in the Stata application folder; see [GSM] **A.6 Executing commands every time Stata is started**.

4. You can launch Stata by double-clicking on a Stata `.dta` dataset.

5. You can launch Stata by double-clicking on a do-file.

6. Do not have more than one Stata application installed on the same computer.

   Do not install the current version in one folder and an older version in another.

   Do not install more then one of Intercooled Stata, Pseudo-Intercooled Stata, or Small Stata (`Stata`, `Stata.noFPU`, `SmStata`) on the same computer.

   If you do, when you double-click on `Stata.do`, your Macintosh will randomly choose one of the Statas to run and you will never know which.

   If you installed more than one Stata on your computer, drag the extra Statas to the Trash, empty the Trash, and restart your Macintosh. All will be well.

5.4.4 Unix

1. You must add `/usr/local/stata` to your shell's PATH. before you can use Stata.

2. You can specify the amount of memory to be allocated to Stata when you invoke it or, once Stata is running, at any time using the `set memory` command. To learn about the start-up options, see [GSU] **A.4 Advanced starting of Stata for Unix**. To learn about `set memory`, see [U] **7 Setting the size of memory**.

3. Stata can execute commands every time it is invoked if you record the commands in the ASCII file `profile.do` and place the file someplace along your PATH. See [GSU] **A.7 Executing commands every time Stata is started**.

4. Stata has a background (batch) mode; see [GSU] **A.8 Executing Stata in background (batch) mode**.

5. For notes on using X Windows, see [GSU] **A.9 Using X Windows** and [GSU] **A.10 Using X Windows remotely**.

6. For a summary of environment variables that affect Stata, see [GSU] **A.12 Summary of environment variables**.

6 Troubleshooting starting and stopping Stata

Contents

6.1　If Stata does not start

You tried to start Stata and it refused; Stata or your operating system presented a message explaining that something is wrong. Here are the possibilities:

Bad command or file name or **Command not found**

You are using Stata for Unix and forgot to add Stata to your shell's PATH.

Cannot find license file

This message means just what it says; nothing is too seriously wrong, Stata simply could not find what it is looking for, probably because you did not complete the installation process or Stata is not installed where it should be.

Did you insert the codes printed on your paper license to unlock Stata? If not, go back and complete the installation; see the *Getting Started* manual.

Assuming you did unlock Stata, Stata is merely mislocated or the location has not been filled in.

Stata for Unix must be installed in `/usr/local/stata`; verify that you have done this.

Error opening or reading the file

Something is distinctly wrong and for purely technical reasons. Stata found the file it was looking for, but either the operating system refused to let Stata open it or there was an I/O error. On Windows and Macintosh computers, about the only way this could happen would be a hard-disk error. Under Unix, the `stata.lic` file could have incorrect permissions. Verify `stata.lic` is in `/usr/local/stata` and that everybody has been granted read permission. To change the permissions, as superuser, type `chmod a+r /usr/local/stata/stata.lic`.

This is a *n*-user license and *n* users are currently using Stata

You are trying to use Stata for Unix, but the number of people using it at your site is already at the maximum allowed by the license at your site; see *How Stata counts license positions* in [GSU] **1. Installation** for more information.

License not applicable

Stata has determined that you have a valid Stata license, but it is not applicable to the version of Stata which you are trying to run. You would get this message if, for example, you tried to run Stata for Macintosh using a Stata for Windows license.

The most common reason for this message is that you have a license for Small Stata, but you are trying to run Intercooled Stata. If this is the case, reinstall Stata, making sure to choose the appropriate version.

Other messages

The other messages indicate that Stata thinks you are attempting to do something you are not licensed to do. Most commonly, you are attempting to run Stata over a network when you do not have a network license, but there are a host of other alternatives. There are two possibilities: either you really are attempting to do something you are not licensed to do or Stata is wrong. In either case, you are going to have to call us. Your license can be upgraded or, if Stata is wrong, we can provide codes over the telephone to make Stata stop thinking you are violating the license.

6.2 verinst problems

Once Stata is running, you can type `verinst` to check if it is correctly installed. If the installation is correct, you will see something like

```
. verinst
You are running Intercooled Stata 6.0 for Windows.

Stata is correctly installed.
You can type exit to exit Stata.
```

If, however, there is a problem, `verinst` will report it. In most cases, `verinst` itself tells you what is wrong and how to fix it. There is one exception:

```
. verinst
unrecognized command
r(199);
```

This indicates that Stata could not even find the `verinst` command. The most likely cause is that, somehow, you have more than one `stata.lic` file on your computer. There can only be one and that one must be in the directory in which Stata is installed. Exit Stata (type `exit`), erase the copies, and try again.

6.3 Troubleshooting Windows

Crashes are called Application Faults in Windows even though the dreaded Application Fault is often not the application's fault. Most commonly, it is caused by configuration problems, bugs in device drivers, memory conflicts, and even hardware problems.

If you experience an Application Fault, first look at the Frequently Asked Questions (FAQ) for Windows in the user support section of the Stata web site *http://www.stata.com*. You may find the answer to the problem there. If not, we can help, but you must give us as much information as possible. See [GSW] **C. Troubleshooting starting and stopping Stata**.

6.4 Troubleshooting Macintosh

Once in a while, Macintosh users may run into the cryptic Type 1, Type 10, or Type 11 error messages: Application has quit due to error of type x—a crash by any other name. Often these errors are due to low memory but they can also be caused by bugs or incompatibilities in extensions, control panels, or other applications.

Type 10 is easy: you are trying to run Intercooled Stata for the 680x0 on a PowerMac. Drag the `Stata` folder to the trash can, empty the trash, and reinstall the PowerMac version of Stata.

Troubleshooting the others can be more difficult. See [GSM] **B. Troubleshooting starting and stopping Stata**.

7 Setting the size of memory

Contents

7.1 Memory size considerations

Stata works with a copy of the data that it loads into memory.

By default, small Stata allocates about 400K to Stata's data areas and you cannot change it.

By default, Intercooled Stata allocates 1 megabyte to Stata's data areas and you can change it.

You can even change the allocation to be larger than the physical amount of memory on your computer because Windows, Macintosh, and Unix systems provide virtual memory.

Virtual memory is slow but adequate in rare cases when you have a dataset that is too large to load into real memory. If you use large datasets frequently, we recommend that you add more memory to your computer.

One way to change the allocation is when you start Stata. Instructions for doing this are provided in

Windows 98/95/NT	[GSW]	**A.6 Specifying the amount of memory allocated**
Windows 3.1	[GSW]	**B.6 Specifying the amount of memory allocated**
Macintosh	[GSM]	**A.5 Specifying the amount of memory allocated**
Unix	[GSU]	**A.6 Specifying the amount of memory allocated**

In addition, if you use Stata for Windows 98/95/NT or Stata for Unix, you can change the total amount of memory allocated while Stata is running. That is the topic of this chapter.

Understand that it does not much matter which method you use. Being able to change the total on the fly is convenient, but even if you cannot do this, it just means you specify it ahead of time and if later you need more, you must exit Stata and reinvoke it with the larger total.

7.2 Setting the size on the fly: Windows 98/95/NT and Unix only

Assume you have changed nothing about how Stata starts so you get the default 1 megabyte of memory allocated to Stata's data areas. You are working with a large dataset and now wish to increase it to 32 megabytes. You can type

```
. set memory 32000
(32000k)
```

and, if your operating system can provide the memory to Stata, Stata will work with the new total. You type `set memory 32000` because you specify the total in K: 32,000K is roughly 32 megabytes. Later in the session, if you want to release that memory and work with only 2 megabytes, you could type

```
. set memory 2000
(2000k)
```

There is only one restriction on the `set memory` command: whenever you change the total, there cannot be any data already in memory. If you have a dataset in memory, you save it, clear memory, reset the total, and then use it again. We are getting ahead of ourselves, but you might type

```
. save mydata, replace
file mydata.dta saved
. drop _all
. set memory 32000
(32000k)
. use mydata
```

When you request the new allocation, your operating system might refuse to provide it:

```
. set memory 128000
op. sys. refuses to provide memory
r(909);
```

If that happens, you are going to have to take the matter up with your operating system. In the above example, Stata asked for 128 megabytes and the operating system said no.

7.3   The memory command

`memory` helps you figure out whether you have sufficient memory to do something.

```
. memory
    Total memory                 1,048,576 bytes    100.00%
    overhead (pointers)            114,136           10.88%
    data                           913,088           87.08%
                                 ------------
    data + overhead              1,027,224           97.96%
    programs, saved results, etc.    1,552            0.15%
                                 ------------
    Total                        1,028,776           98.11%
    Free                            19,800            1.89%
```

19,800 bytes free is not much. You might increase the amount of memory allocated to Stata's data areas by specifying `set memory 2000`. (Stata for Macintosh users would need to exit Stata and change Stata's preferred memory size. Stata for Windows 3.1 users would need to exit Stata and change the `k` option in the properties for **Stata**'s icon.)

```
. save nlswork
file nlswork.dta saved
. set memory 2000
(2000k)
. use nlswork
(National Longitudinal Survey.  Young Women 14-26 years of age in 1968)
. memory
    Total memory                 2,048,000 bytes    100.00%
    overhead (pointers)            114,136            5.57%
    data                           913,088           44.58%
                                 ------------
    data + overhead              1,027,224           50.16%
    programs, saved results, etc.    1,552            0.08%
                                 ------------
    Total                        1,028,776           50.23%
    Free                         1,019,224           49.77%
```

Over 1 megabyte free; that's better. See [R] **memory** for more information.

7.4 Virtual memory and speed considerations

When you use more memory than is physically available on your computer, Stata slows down. If you are only using a little more memory than on your computer, performance is probably not too bad. On the other hand, when you are using a lot more memory than is on your computer, performance will be noticeably affected. In these cases, we recommend that you

```
. set virtual on
```

Virtual memory systems exploit locality of reference which means that keeping objects closer together allows virtual memory systems to run faster. `set virtual` controls whether Stata should perform extra work to arrange its memory to keep objects close together. By default, `virtual` is set `off`. `set virtual` can only be specified if you are using Stata for Unix or Intercooled Stata for Windows or Macintosh.

In general, you want to leave `set virtual` set to the default of `off`. Stata will run faster.

When you `set virtual on`, you are asking Stata to arrange its memory so that objects are kept closer together. This requires Stata doing a substantial amount of work. We recommend setting virtual on only when the amount of memory in use drastically exceeds what is physically available. In these cases, setting virtual on will help, but keep in mind that performance will still be slow. If you are using virtual memory frequently, you should consider adding memory to your computer.

7.5 An issue when returning memory to Unix

There is a surprising issue of returning memory that Unix users need to understand. Let's say that you set memory to 128 megabytes, went along for a while, and then, being a good citizen, returned most of it:

```
. set memory 2000
(2000k)
```

Theoretically, 126 megabytes just got returned to the operating system for use by other processes. If you use Windows, that is exactly what happens and, with some Unixes, that is what happens, too.

Other Unixes, however, are strange about returned memory in a misguided effort to be efficient: they do not really take the memory back, instead they leave it allocated to you in case you ask for it back later. Still other Unixes sort of take the memory back: they put it in a queue for your use but, if you do not ask for it back in 5 or 10 minutes, then they return it to the real system pool!

The unfortunate situation is that we at Stata cannot force the operating system to take the memory back. Stata returns the memory to Unix and then Unix does whatever it wants with it.

So, let's review: You make your Stata smaller in an effort to be a good citizen. You return the memory so that other users can use it or perhaps so you can use it with some other software.

If you use Windows, the memory really is returned and all works exactly as you anticipated.

If you use Unix, it might go back immediately, it might go back in 5 or 10 minutes, or it might never go back. In the last case, the only way to really return the memory is to exit Stata. All Unixes agree on that: when a process ends, the memory really does go back into the pool.

To find out how your Unix works, you need to experiment. We would publish a table and just tell you, but we have found that within manufacturer the way their Unix works will vary by subrelease! The experiment is tedious but not difficult:

1. Bring up a Stata and make it really big; use a lot of memory, so much that you are virtually hogging the computer.

2. Go to another window or session and bring up another Stata. Verify that you cannot make it big—that you get the "system limit exceeded" message.

3. Go back to the first Stata, leaving the second running, and make it smaller.

4. Go to the second Stata and try again to make it big. If you succeed, then your Unix returns memory instantly.

5. If you still get the "system limit exceeded" message, wait 5 minutes and try again. If it now works, your system delays accepting returned memory for about 5 minutes.

6. If you still get the "system limit exceeded" message, wait another 5 minutes and try again. If it now works, your system delays accepting returned memory for about 10 minutes.

7. Go to the first Stata and exit from it.

8. Go to the second Stata and try to make it big again. If it now works, your system never really accepts returned memory. If it still does not work, start all over again. Some other process took memory and corrupted your experiment.

If you are one of the unfortunates who have a Unix that never accepts returned memory, you will just have to remember: you must exit and reenter Stata to really give memory back.

8 Stata's on-line help and search facilities

8.1   Introduction

Stata has help on-line and a lot of it. Windowed operating-system users have two ways to access this:

1. They can pull down **Help**.

2. They can type the `help` and `search` commands.

Stata for Unix users only have the second approach available to them.

Understand that both methods access the same underlying thing, but the first method of accessing it is better because it allows hypertext links. That is, if you are looking at one thing and it refers to something else, if that something else is in green, you can click on it to go directly to the help on it. Stata for Unix users instead have to type out `help` followed by the other thing.

8.2   help: Stata's on-line manual pages

The `help` command provides access to Stata's on-line manual pages. These pages are a shortened version of what is in the printed manuals.

It is easier to describe the line-by-line access on paper, so begin by entering Stata and typing `help help`. That will show you the on-line manual page for the `help` command itself:

```
. help help

-------------------------------------------------------------------------
help for help                                           (manual:  [R] help)
-------------------------------------------------------------------------

What to do when you see --more--
--------------------------------

The characters --more-- now appear at the bottom of the screen.  Stata pauses
and displays this message whenever the output from a command is about to scroll
off the screen.
```

```
        Action                          Result
----------------------  ------------------------------------------
Press Enter or Return   One more line of text is displayed
Press b                 The previous screen of text is displayed
Press any other key     The next screen of text is displayed
    (such as space bar)
PCs:
    Press Ctrl-Break    Stata stops processing the command ASAP
Macintosh:
    Press Command-.        "      "      "       "      "      "
Unix:
    Press Ctrl-C          "      "      "       "      "      "
--more--
```

If you now press the space bar, you will see the next page and if you instead press the *Break* key, Stata will stop showing you help on `help` and issue another dot prompt. We are going to press *Break*:

```
--Break--
r(1);
. _
```

Rule 1: If you know the name of the command and want to learn more about it, type `help` followed by the command name.

Try it. Two of Stata's commands are named `use` and `regress`. Type `help use`. Type `help regress`.

8.3   search: Stata's on-line index

Rule 1 works fine when you know the name of the command, but what if you do not?

In that case, you use `search`. `search`'s syntax is

search *anything you want*

or instead you can pull down **Help**, select **Search**, and then type *anything you want* in the Keywords input field. Either way, `search` is very understanding. Here is what happens when you `search` `logistic regression`:

```
. search logistic regression
[U]     Chapter 23 . . . . . . . . . Estimation and post-estimation commands
        (help est)
[U]     Chapter 29 . . . . . . . . . Overview of model estimation in Stata
        (help est)
[R]     clogit . . . . . . . Conditional logistic (fixed effects) regression
        (help clogit)
[R]     glogit . . . . . . . . . . . . . Logit and probit on grouped data
        (help glogit)
[R]     logistic . . . . . . . . . . . . . . . . . . Logistic regression
        (help logistic)
[R]     logit . . . . . . . . . . . . Maximum-likelihood logit estimation
        (help logit)
  (output omitted )
STB-36  sbe14 . OR´s & ci´s for logistic reg. models with effect modification
        (help effmod if installed) . . . . . . . . . . . . . . . J. Garrett
        3/97    STB Reprints Vol 6, pages 104--114
        calculate odds ratios and confidence intervals for logistic
        regression models with significant interaction terms
```

```
STB-35   sg63 .  Logistic regression: standardized coef. & partial correlations
         (help lstand if installed) . . . . . . . . . . . . . . . . . J. Hilbe
         1/97    STB Reprints Vol 6, pages 162--163
         command for use after logistic; displays predictor, its coefficient,
         odds ratio, standardized coefficient, partial correlation, and p-value
```
(*output omitted*)
```
FAQ      . . . . . . . Interpreting the cut points in ordered probit and logit
         . . . . . . . . . . . . . . . . . . . . . . . . . . . . . . . W. Gould
         4/97    In ordered probit and logit, what are the cut points?
                 http://www.stata.com/support/faqs/stat/oprobit.html
FAQ      . . . . . . Interpreting "outcome does not vary" when running logistic
         . . . . . . . . . . . . . . . . . . . . . . . . . . . . . . . . P. Lin
         11/96   Why do I get the message, "outcome does not vary" when
                 I perform a logistic or logit regression?
                 http://www.stata.com/support/faqs/stat/outcomes.html
```
(*output omitted*)

search responds by providing a list of references—references to the on-line help, references to the printed documentation, references to articles that have appeared in the *Stata Technical Bulletin* (STB), and references to FAQs at the *www.stata.com* web site. Moreover, if you install the official updates—see [U] **32 Using the Internet to keep up to date**—the references to the STB and FAQs will even be up to date.

Anyway, you are supposed to look over the list and find what looks relevant to you. If you are really interested in logistic regression, logistic and logit seem particularly appropriate, so you might next type help logistic and help logit.

8.4   Accessing help and search from the help pulldown

Stata for Windows and Stata for Macintosh users will get exactly the same output if they pull down **Help**, select **Search**, and type logistic regression in the Keyword(s) box.

There is an advantage because some of the resulting output will be highlighted in green to indicate clickables. *help logit* and *help logistic*, for instance, will be displayed in green. Click on either one and you will go to the on-line help for these commands.

There will be other places you can click, too. When search mentions an STB article, the insert number such as *sbe14* or *sg63* will be in green. Click on one and Stata will go to *www.stata.com* and show you a detailed description of the offered addition, leaving you just one click away from installing the addition if it really does seem to be of interest. (This feature is not available under Windows 3.1 or on the 680x0 Macintosh.)

When search mentions a FAQ, the URL will be in green. Click on that and Stata will launch your browser and you will be looking right at the answer to the frequently asked question. (This feature is not available under Windows 3.1 or on the 680x0 Macintosh.)

You can pull down **Help** at any time, not just when Stata is idle. You can leave the help up while you use Stata, which is especially convenient when viewing syntax diagrams and options.

8.5   More on search

However you access search—command or pulldown—it does the same thing. You tell search what you want information on and search searches Stata's on-line help files for relevant entries.

search can be used broadly or narrowly. For instance, if you want to perform the Kolmogorov–Smirnov test for equality of distributions, you could type

```
. search Kolmogorov-Smirnov test of equality of distributions
   [R]       ksmirnov . . . . . . Kolmogorov-Smirnov equality of distributions test
             (help ksmirnov)
```

In fact, we did not have to be nearly so complete—typing search Kolmogorov-Smirnov would have been adequate. Had we specified our request more broadly—looking up equality of distributions, say—we would have obtained a longer list that included ksmirnov.

Here are guidelines on using search.

1. Capitalization does not matter. Look up Kolmogorov-Smirnov or kolmogorov-smirnov.

2. Punctuation does not matter. Look up kolmogorov smirnov.

3. Order of words does not matter. Look up smirnov kolmogorov.

4. You may abbreviate, but how much depends. Break at syllables. Look up kol smir. search tends to tolerate a lot of abbreviation; it is better to abbreviate than to misspell.

5. The prepositions for, into, of, on, to, and with are ignored. Use them—look up equality of distributions—or omit them—look up equality distributions—it makes no difference.

6. search is tolerant of plurals, especially when they can be formed by adding an *s*. Even so, it is better to look up the singular. Look up normal distribution, not normal distributions.

7. Specify the search criterion in English, not computer jargon.

8. Use American spellings. Look up color, not colour.

9. Use nouns. Do not use -ing words. Look up median tests, not testing medians.

10. Use few words. Every word specified further restricts the search. Look up distribution and you get one list; look up normal distribution and the list is a sublist of that.

11. Sometimes words have more than one context. The following words can be used to restrict the context:

    a. data, meaning in the context of data management. Order could refer to the order of data or to order statistics. Look up order data to restrict order to the data management sense.

    b. statistics (abbreviation stat), meaning in the context of statistics. Look up order statistics to restrict order to the statistical sense.

    c. graph or graphs, meaning in the context of statistical graphics. Look up median graphs to restrict the list to commands for graphing medians.

    d. utility (abbreviation util), meaning in the context of utility commands. The search command itself is not data management, not statistics, not graphics; it is a utility.

    e. programs or programming (abbreviation prog), to mean in the context of programming. Look up programming scalar to obtain a sublist of scalars in programming.

search has other features as well; see [U] **8.8 search: All the details**.

8.6   More on help

help, unlike search, is not as understanding of mistakes. You may not abbreviate or misspell the command.

```
. help regres
help for regres not found
try  help contents  or  search regres
```

The command is regress and, whereas Stata will allow you to abbreviate the command when you use it (see [U] **14.2 Abbreviation rules**), help will not let you abbreviate it.

When help cannot find the command you are looking for, try the search facility. In this case, typing search regres will find the command (because regres is an abbreviation of the word regression) but, in general, that will not be the case. help is really suggesting that you go back and use search intelligently and that, of course, is always good advice.

For instance, Stata has a command with the inelegant name ksmirnov. You forget and think the command is called ksmir:

```
. help ksmir
help for ksmir not found
try  help contents  or  search ksmir
```

This is a case where help gives bad advice because typing search ksmir will do you no good. You should type search followed by what you are really looking for: search kolmogorov smirnov.

8.7   help contents: Stata's on-line table of contents

Typing help contents, or pulling down **Help** and selecting **Contents**, provides another way of locating entries in the documentation and on-line help. Either way, you will be presented with a long table of contents, organized topically.

```
. help contents
(output omitted )
```

8.8   search: All the details

The search command actually provides a few features not available from the **Help** pulldown. The full syntax of the search command is

search *word* [*word* ...] [, author entry exact historical or manual stb faq]

where underlining indicates minimum allowable abbreviation and [braces] indicate optional.

author specifies that the search is to be performed on the basis of author's name rather than keywords.

entry specifies that the search is to be performed on the basis of entry ids rather than keywords.

exact prevents matching on abbreviations.

historical adds to the search entries that are of historical interest only. By default, such entries are not listed. Past entries are classified historical if they discuss a feature that later became an official part of Stata. Updates to historical entries will always be found, even if historical is not specified.

or specifies that an entry should be listed if any of the words typed after search are associated with the entry. The default is to list the entry only if all the words specified are associated with the entry.

manual limits the search to entries in the *Reference Manual*.

stb limits the search to entries in the *Stata Technical Bulletin*.

faq limits the search to entries found in the FAQs at *www.stata.com*.

8.8.1 How search works

search has a database—a file—containing the titles, etc., of every entry in the *Graphics* manual, *Reference* manuals, the *Getting Started* manuals, articles in the *Stata Technical Bulletin*, and FAQs at *www.stata.com*. In this file is a list of words associated with each entry, called keywords.

When you type search *xyz*, search reads this file and compares the list of keywords with *xyz*. If it finds *xyz* in the list or a keyword that allows an abbreviation of *xyz*, it displays the entry.

When you type search *xyz abc*, search does the same thing, but displays an entry only if it contains both keywords. The order does not matter, so you can search linear regression or search regression linear or even search regression, linear because search ignores special characters such as commas.

Obviously, how many entries search finds depends on how the search database was constructed. We have included a plethora of keywords under the theory that, for a given request, it is better to list too much rather than risk listing nothing at all. Still, you are in the position of guessing the keywords. Do you look up normality test, normality tests, or tests of normality? Answer: normality test would be best, but all would work. In general, use the singular and strike the unnecessary words. We provide guidelines for specifying keywords in [U] **8.5 More on search** above.

8.8.2 Author searches

search ordinarily compares the words following search with the keywords for the entry. If you specify the author option, however, it compares the words with the author's name. In the search database, we have filled in author names for STB articles and FAQs.

For instance, in [R] **kdensity** you will discover that Isaías H. Salgado-Ugarte wrote the first version of Stata's kdensity command and published it in the STB. Assume you read his original and find the discussion useful. You might now wonder what else he has written in the STB. To find out, you type

```
. search Salgado-Ugarte, author
  (output omitted)
```

Names like Salgado-Ugarte are confusing to people in the U.S. search does not require you specify the entire name; what you type is compared with each "word" of the name and, if any part matches, the entry is listed. The dash, like the comma, is a special character and you can omit it. Thus, you can obtain the same list by looking up Salgado, Ugarte, or Salgado Ugarte without the dash.

Actually, to find all entries written by Salgado-Ugarte, you need to type

```
. search Salgado-Ugarte, author historical
  (output omitted)
```

Prior inserts in the STB that provide a feature that later were superseded by a built-in feature of Stata are marked as historical in the search database and, by default, not listed. The historical option ensures that all entries are listed.

8.8.3   Entry id searches

If you specify the `entry` option, `search` compares what you have typed with the entry id. The entry id is not the title—it is the reference listed to the left of the title that tells you where to look. For instance, in

```
[R]     regress . . . . . . . . . . . . . . . . . . . . Linear regression
        (help regress)
```

"[R] regress" is the entry id. In

```
GS       . . . . . . . . . . . . . . . . . . . . Getting Started manual
```

"GS" is the entry id. In

```
STB-28  dm36 . . . . . . . . . . . . . . . . Comparing two Stata data sets
        (help compdta if installed) . . . . . . . . . . . John R. Gleason
        11/95    STB Reprints Vol 5, pages 39--43
        compares the varlist from the dataset in memory with like-named
        variables in the Stata-format dataset on disk; alternative to cf
        command
```

"STB-28 dm36" is the entry id.

`search` with the `entry` option searches these entry ids.

Thus, one could generate a table of contents for the *Reference* manuals by typing

```
. search [R], entry
(output omitted )
```

One could generate a table of contents for the 16th issue of the STB by typing

```
. search STB-16, entry historical
(output omitted )
```

The `historical` option in this case is possibly important. STB-16 was published in November 1993 and perhaps some of its inserts have been marked historical.

One could obtain a complete list of all inserts associated with *dm36* by typing

```
. search dm36, entry historical
(output omitted )
```

Again, we include the `historical` option in case any of the relevant inserts have been marked historical.

8.8.4   FAQ searches

To search across the FAQs, specify the `faq` option:

```
. search logistic regression, faq
(output omitted )
```

FAQ is web-speak for Frequently Asked Questions.

8.8.5   Return codes

In addition to indexing the entries in the *User's Guide*, the *Reference* manual, the *Graphics* manual, and the *Getting Started* manuals, `search` also can be used to look up return codes.

To see information on return code 131, type

```
. search rc 131
[R]      error messages  . . . . . . . . . . . . . . . . . . . .  Return code 131
         not possible with test;
         You requested a test of a hypothesis that is nonlinear in the
         variables.  test tests only linear hypotheses.  Use testnl.
```

If you wanted to get a list of all Stata return codes, type

```
. search rc
(output omitted )
```

9 Stata's on-line tutorials and sample datasets

Contents

9.1   The on-line tutorials

Stata has 11 on-line tutorials which will help you learn about Stata's features. If you have not yet read the *Getting Started* manual, you should look at it before running the on-line tutorials.

To run the introductory tutorial (`intro.tut`), type

```
. tutorial intro
```

Here is a complete listing of Stata's on-line tutorials:

`intro.tut`	An introduction to Stata
`contents.tut`	Listing of on-line tutorials
`graphics.tut`	How to make graphs
`tables.tut`	How to make tables
`regress.tut`	Estimating regression models
`anova.tut`	Estimating one-, two-, and N-way ANOVA and ANCOVA
`logit.tut`	Estimating maximum-likelihood logit and probit models
`survival.tut`	Estimating maximum-likelihood survival models
`factor.tut`	Estimating factor and principal component models
`ourdata.tut`	Description of the example datasets we provide
`yourdata.tut`	How to input your own data into Stata

To run the other tutorials, type `tutorial` *filename*. For example, if you want to run `graphics.tut`, type

```
. tutorial graphics
```

9.2   Sample datasets

Various examples in this manual use what is referred to as the automobile data `auto.dta`. We previously created a dataset on the prices, mileages, weights, and other characteristics of 74 automobiles and saved it in a file called `auto.dta`. (This data originally came from the April 1979 issue of *Consumer Reports* and from the United States Government EPA statistics on fuel consumption; it was compiled and published by Chambers et al. 1983.)

In our examples, you will often see us type

```
. use auto
```

We include the `auto.dta` file with Stata. If you want to `use` it, you will have to add the Stata directory path. You probably need to type

Stata for Windows:	`use c:\stata\auto`
Stata for Unix:	`use /usr/local/stata/auto`
Stata for Macintosh:	`use ~:Stata:auto`

It all depends on where you installed Stata. The easy way to find out is to type

```
. sysdir
      STATA:  D:\STATA\
    UPDATES:  D:\STATA\ado\updates\
       BASE:  D:\STATA\ado\base\
       SITE:  D:\STATA\ado\site\
    STBPLUS:  D:\ado\stbplus\
   PERSONAL:  D:\ado\personal\
   OLDPLACE:  D:\ado\
```

Stata is installed in the STATA directory so, on the Windows computer we are using, we would type 'use d:\stata\auto'.

Windows and Macintosh users can also load `auto.dta` by pulling down **File** and choosing **Open**, but they still have to know where it is.

In the examples in the manuals, we copied `auto.dta` to the current directory so that we do not have to specify its path. You can do this, too:

```
. copy d:\stata\auto.dta auto.dta
```

Here is a list of the example datasets included with Stata:

`auto.dta`	1978 Automobile data
`cancer.dta`	Drug trial (survival analysis)
`census.dta`	1980 Census data by state
`hsng.dta`	1980 Census housing data by state
`kva.dta`	Failure-time experiment (survival analysis)
`machine.dta`	manufacturing machine experiment (nested ANOVA)
`sysage.dta`	Blood pressure data with age covariate
`systolic.dta`	Blood pressure data (two-way ANOVA)

9.3 References

Chambers, J. M., W. S. Cleveland, B. Kleiner, and P. A. Tukey. 1983. *Graphical Methods for Data Analysis*. Belmont, CA: Wadsworth International Group.

10   −more− conditions

Contents

10.1   Description

When you see --more-- at the bottom of the screen

Press ...	and Stata...
letter *l* or *Enter*	displays the next line
letter *q*	acts as if you pressed *Break*
space bar or any other key	displays the next screen

In addition, Stata for Windows and Stata for Macintosh users can press the *clear −more− condition* button, the button labeled **Go** with a circle around it under Windows 98/95/NT and Macintosh, and labeled **More** under Windows 3.1.

--more-- is Stata's way of telling you it has something more to show you, but showing you that something more will cause the information on the screen to scroll off.

10.2   set more off

If you type **set more off**, --more-- conditions will never arise—Stata's output will scroll by at full speed.

If you type **set more on**, --more-- conditions will be restored at the appropriate places.

Programmers: Do-file writers sometimes include **set more off** in their do-files because they do not care to interactively watch the output. They want Stata to proceed at full speed because they plan on making a log of the output which they will review later.

Do-filers need not bother to **set more on** at the conclusion of their do-file. Stata automatically restores the previous **set more** when the do-file (or program) concludes.

10.3   The more programming command

Ado-file programmers need take no special action to have --more-- conditions arise when the screen is full. Stata handles that automatically.

If, however, you wish to force a --more-- condition early, you can include the **more** command in your program. The syntax of **more** is

    more

more takes no arguments.

For additional information, see [R] **more**.

11 Error messages and return codes

Contents

11.1   Making mistakes

When an error occurs, Stata produces an error message and a *return code*. For instance:

```
. list myvar
no variables defined
r(111);
```

We ask Stata to list the variable named `myvar`. Stata responds with the message "no variables defined" and a line that reads "`r(111)`".

The "no variables defined" is called the error message.

The 111 is called the return code.

11.1.1   Mistakes are forgiven

Having said "no variables defined" and `r(111)`, all is forgiven; it is as if the error never occurred.

Typically, the message will be enough to guide you to a solution but, if it is not, the numeric return codes are documented in [R] **error messages**.

11.1.2   Mistakes stop user-written programs and do-files

Whenever an error occurs in a user-written program or do-file, the program or do-file immediately stops execution and the error message and return code are displayed.

For instance, consider the following do-file:

```
                                                                    top of myfile.do
    use auto
    decribe
    list
                                                                    end of myfile.do
```

Note the second line—you meant to type `describe` but typed `decribe`. Here is what happens when you execute this do-file by typing `do myfile`:

```
. do myfile

. use auto
(1978 Automobile Data)
```

```
. decribe
unrecognized command
r(199);

end of do-file
r(199);

. _
```

The first error message and return code were caused by the illegal `decribe`. This then caused the do-file itself to be aborted; the valid `list` command was never executed.

11.1.3   Advanced programming to tolerate errors

Errors are not only of the typographical kind; some are substantive. A command that is valid in one dataset might not be valid in another. Moreover, in advanced programming, errors are sometimes anticipated: use one dataset if it is there, but use another if you must.

Programmers can access the return code to determine whether an error occurred, which they can then ignore or, by examining the return code, code their programs to take the appropriate action. This is discussed in [R] **capture**.

11.2   The return message for obtaining command timings

In addition to error messages and return codes, there is something called a return message which you normally do not see. Normally, if you typed, say, `summarize tempjan`, you would see

```
. summarize tempjan
Variable |     Obs        Mean   Std. Dev.        Min        Max
---------+-----------------------------------------------------
 tempjan |    7632    35.74895    14.18162        2.2       72.6
```

If you were to type

```
. set rmsg on
```

sometime during your session, Stata will display return messages:

```
. summarize tempjan
Variable |     Obs        Mean   Std. Dev.        Min        Max
---------+-----------------------------------------------------
 tempjan |    7632    35.74895    14.18162        2.2       72.6
r; t=0.01 15:52:10
```

The line that reads `r; t=0.01 15:52:10` is called the return message.

The `r;` indicates that Stata successfully completed the command.

The `t=0.01` shows the amount of time, in seconds, it took Stata to perform the command (timed from the point you pressed *Return* to the time Stata typed the message). This command took a hundredth of a second. In addition, Stata shows the time of day using a 24-hour clock. This command completed at 3:52 p.m.

You will learn that Stata has the ability to run commands stored in files (called do-files) and the ability to log output. Some users find the detailed return message helpful with do-files. They construct a lengthy program and let it run overnight, logging the output. They come back the next morning, look at the output, and discover a mistake in some portion of the job. They can look at the return messages to determine how long it will take to rerun that portion of the program.

You may `set rmsg on` whenever you wish.

When you want Stata to stop displaying the detailed return message, type `set rmsg off`.

12   The Break key

12.1   Making Stata stop what it is doing

When you want to make Stata stop what it is doing and return to the Stata dot prompt, you press *Break*:

Stata for Windows 98/95/NT:	click the **Break** button (it is the button with the big red X) or hold down *Ctrl* and press the *Pause/Break* key
Stata for Windows 3.1:	click the **Break** button or hold down *Ctrl* and press the *Pause/Break* key
Stata for Macintosh:	Click the **Break** button or hold down *Command* and press period
Stata for Unix:	hold down *Ctrl* and press C

Elsewhere in this manual, we describe this action as simply pressing *Break*. Break tells Stata to cancel what it is doing and return control to you as soon as possible.

If you press *Break* in response to the input prompt or while you are typing a line, Stata ignores it, since you are already in control.

If you press *Break* while Stata is doing something—creating a new variable, sorting a dataset, making a graph, or the like—Stata stops what it is doing, undoes it, and issues an input prompt. The state of the system is the same as if you had never issued the command.

▷ Example

You are estimating a logit model, type the command, and, as Stata is working on the problem, realize that you omitted an important variable:

```
. logit for mpg weight
Iteration 0:   log likelihood =-1801.3284
Iteration 1:   log likelihood =-1197.7089
--Break--
r(1);

. _
```

When you pressed *Break*, Stata responded by typing `--Break--` and then typing `r(1);`. Pressing *Break* always results in a return code of 1—that is why return codes are called return codes and not error codes. The 1 does not indicate an error, but it does indicate that the command did not complete its task.

◁

12.2   Side-effects of pressing Break

In general, there are none. We said above that Stata undoes what it is doing so that the state of the system is the same as if you had never issued the command. There are two exceptions to this.

If you are reading data from disk using insheet, infile, or infix, whatever data has already been read will be left behind in memory; the theory being that perhaps you stopped the process so you could verify that you were reading the right data correctly before sitting through the whole process. If not, you can always drop _all.

```
. infile v1-v9 using workdata
(eof not at end of obs)
(4 observations read)
--Break--
r(1);
```

The other exception is sort. You have a large dataset in memory, decide to sort it, and then change your mind.

```
. sort price
--Break--
r(1);
```

If the data was previously sorted by, say, the variable prodid, it is no longer. When you press *Break* in the middle of a sort, Stata marks the data as unsorted.

12.3   Programming considerations

There are basically no programming considerations for handling Break because Stata handles it all automatically. If you write a program or do-file, execute it, and then press *Break*, Stata stops execution just as it would an internal command.

Advanced programmers may be concerned about cleaning up after themselves: perhaps, they have generated a temporary variable they intended later to drop or a temporary file they intended later to erase. If a Stata user presses *Break*, how can you ensure that these temporary variables and files will be erased?

If you obtain names for such temporary items from Stata's tempname, tempvar, and tempfile commands, Stata will automatically erase the temporary items; see [U] **21.7 Temporary objects**.

There are instances, however, when a program must commit to executing a group of commands without interruption, or the user's data would be left in an intermediate or undefined state. In these instances, Stata provides a

```
nobreak {
        ...
}
```

construct; see [R] **break**.

13 Keyboard use

Contents

13.1   Description

The keyboard should operate very much the way you would expect with a few additions:

1. There are some unexpected keys you can press to obtain previous commands you have typed. In addition, Stata for Windows and Stata for Macintosh users can click once on a command in the Review window to reload it or click twice to reload and execute; this feature is discussed in the *Getting Started with Stata for Windows* and *Getting Started with Stata for Macintosh* manuals.

2. There are a host of command-editing features for Stata for Unix users since their operating system does not offer such features. Mostly, these extra features are not included in Stata for Windows and Stata for Macintosh because they conflict with the actions you would normally expect.

3. Regardless of operating system, if there are *F*-keys on your keyboard, they have special meaning and you can change the definitions of the keys to match your preference.

13.2   F-keys

Note to Macintosh users: Not all Macintosh keyboards have *F*-keys.

Note to Windows users: *F10* means something special to Windows; you cannot program this key.

By default, Stata defines the *F*-keys to mean

F-key	Definition
F1	help
F2	#review;
F3	describe;
F7	save
F8	use

The semicolons at the end of some of the entries indicate the presence of an implied *Return*.

help shows a Stata manual page—you use it by typing help followed by the name of a Stata command; see [U] **8 Stata's on-line help and search facilities**. You can type out help or you can press *F1*, type the Stata command, and press *Return*.

#review is the command to show the last few commands you have issued. It is described in [U] **13.6 Editing previous lines**. Anyway, rather than typing out #review and pressing *Return*, you can simply press *F2*. You do not press *Return* following *F2* because the definition of F2 ends in a semicolon—Stata presses the *Return* key for you.

describe is the Stata command to report the contents of data loaded into memory. It is explained in [R] **describe**. Normally, you type describe and press *Return*. Alternatively, you can press *F3*.

save is the command to save the data in memory into a file and use is the command to load data. Both are described in [R] **save** and the syntax of each is the same: save or use followed by a filename. You can type out the commands or you can press *F7* or *F8* followed by the filename.

You can change the definitions of the *F*-keys. For instance, the command to list data is list; you can read about it in [R] **list**. The syntax is list to list all the data or list followed by the names of some variables to list just those variables (there are other possibilities).

If you wanted *F3* to mean list, you could type

. global F3 "list "

In the above, F3 refers to the letters *F* followed by *3*, not the *F3* key. Note the capitalization and spacing of the command.

You type global in lowercase, type out the letters F3, and then type "list ". The blank at the end of list is important. In the future, rather than typing out list mpg weight, you want to be able to press the *F3* key and then type only mpg weight. You put a blank in the definition of F3 so that you would not have to type a blank in front of the first variable name following pressing *F3*.

Now say you wanted *F5* to mean list all the data—list followed by *Return*. You could define

. global F5 "list;"

Now you would have two ways of listing all the data: (1) press *F3*, then press *Return* or (2) press *F5*. The semicolon at the end of the definition of *F5* will press the *Return* for you.

If you really want to change the definitions of *F3* and *F5*, you will probably want to change the definition every time you invoke Stata. One way would be to type out the two global commands every time you invoked Stata. Another way would be to type the two commands into an ASCII text file called profile.do. Stata executes the commands in profile.do every time it is launched if profile.do is placed in the appropriate directory:

Windows 98/95/NT:	put profile.do in the "start-in" directory; see [GSW] **A.7 Executing commands every time Stata is started**
Windows 3.1:	put profile.do in the "start-in" directory; see [GSW] **B.7 Executing commands every time Stata is started**
Macintosh:	put profile.do in the Stata application folder; see [GSM] **A.6 Executing commands every time Stata is started**
Unix:	put profile.do someplace along your shell's PATH; see [GSU] **A.7 Executing commands every time Stata is started**

You can use the *F*-keys anyway you desire: They contain a string of characters and pressing the *F*-key is equivalent to typing out those characters.

❏ Technical Note

(*Stata for Unix users.*) Sometimes Unix assigns a special meaning to the *F*-keys and, if it does, those meanings will take precedence over our meanings. Stata provides a second way to get to the *F*-keys. Hold down *Ctrl*, press *F*, release the keys, and then press one of the numbers 0 through 9. Stata interprets *Ctrl-F* plus 1 as equivalent to the *F1* key, *Ctrl-F* plus 2 as *F2*, and so on. *Ctrl-F* plus 0 means *F10*.

❏

13.3   Editing keys, Stata for Windows

Windows users have available to them the standard Windows editing keys. So, Stata for Windows should just edit what you type in the natural way—the Stata Command window is a standard edit window.

In addition, you can fetch commands from the Review window into the Command window. Click on a command in the Review window, and it is loaded into the Command window, where you can edit it. Alternatively, if you double-click on a line in the Review window, it is loaded and executed.

Another way to get lines from the Review window into the Command window is with the *PgUp* and *PgDn* keys. Tap *PgUp* and Stata loads the last command you typed into the Command window. Tap it again and Stata loads the line before that, and so on. *PgDn* goes the opposite direction.

Another editing key, which may be of interest to Stata for Windows users, is *Esc*. This key clears the Command window.

In summary:

Press	Result
PgUp	Steps back through commands and moves command from Review window to Command window
PgDn	Steps forward through commands and moves command from Command window to Review window
Esc	Clears Command window

13.4   Editing keys, Stata for Macintosh

Macintosh users have available to them the standard Macintosh editing keys. So, Stata for Macintosh should just edit what you type in the natural way—the Stata Command window is a standard edit window.

In addition, you can fetch commands from the Review window into the Command window. Click on a command in the Review window, and it is loaded into the Command window, where you can edit it. Alternatively, if you double-click on a line in the Review window, it is loaded and executed.

Another way to get lines from the Review window into the Command window is with the *PgUp* and *PgDn* keys. Tap *PgUp* and Stata loads the last command you typed into the Command window. Tap it again and Stata loads the line before that, and so on. *PgDn* goes the opposite direction.

Another editing key, which may be of interest to Stata for Macintosh users, is *Esc*. This key clears the Command window.

In summary:

Press	Result
PgUp	Steps back through commands and moves command from Review window to Command window
PgDn	Steps forward through commands and moves command from Command window to Review window
Esc	Clears Command window

13.5 Editing keys, Stata for Unix

Certain keys allow you to edit the line you are typing. Since Stata supports a variety of computers and keyboards, the location and the names of the editing keys are not the same for all Stata users.

Every keyboard has the standard alphabet keys (*QWERT* and so on) and every keyboard has a *Ctrl* key. Some keyboards go further, and have extra keys located to the right, above, or the left, labeled with names like *PgUp* and *PgDn*.

Throughout this manual we will refer to Stata's editing keys using names that appear on nobody's keyboard. For instance, PrevLine is one of the Stata editing keys—it retrieves a previous line. Hunt all you want, but you will not find it on your keyboard. So where is PrevLine? We have tried to put it where you would naturally expect it. On keyboards with a key labeled *PgUp*, *PgUp* is the PrevLine key; but on everybody's keyboard, no matter which version of Unix, brand of keyboard, or anything else, *Ctrl-R* also means PrevLine.

When we say press PrevLine, now you know what we mean: press *PgUp* or *Ctrl-R*. With that introduction, the editing keys are

Name for Editing Key	Editing Key	Function
Kill	*Esc* on PCs and *Ctrl-U*	Deletes the line and lets you start over.
Dbs	*Backspace* on PCs and *Backspace*, *Rubout*, or *Delete* on other computers	Backs up and deletes one character.
Lft	←, *4* on the numeric keypad for PCs, and *Ctrl-H*	Moves the cursor left one character without deleting any characters.
Rgt	→, *6* on the numeric keypad for PCs, and *Ctrl-L*	Moves the cursor forward one character.
Up	↑, *8* on the numeric keypad for PCs, and *Ctrl-O*	Moves the cursor up one physical line on a line that takes more than one physical line. Also see PrevLine
Dn	↓, *2* on the numeric keypad for PCs, and *Ctrl-N*	Moves the cursor down one physical line on a line that takes more than one physical line. Also see NextLine.
PrevLine	*PgUp* and *Ctrl-R*	Retrieves a previously typed line. You may press PrevLine multiple times to step back through previous commands.
NextLine	*PgDn* and *Ctrl-B*	The inverse of PrevLine
Seek	*Ctrl-Home* on PCs and *Ctrl-W*	Goes to the line number specified. Before pressing Seek, type the line number. For instance, typing *3* and then pressing Seek is the same as pressing PrevLine three times.
Ins	*Ins* and *Ctrl-E*	Toggles insert mode. In insert mode, characters typed are inserted into the line at the position of the cursor.
Del	*Del* and *Ctrl-D*	Deletes the character at the position of the cursor.
Home	*Home* and *Ctrl-K*	Moves the cursor to the start of the line.
End	*End* and *Ctrl-P*	Moves the cursor to the end of the line.
Hack	*Ctrl-End* on PCs, and *Ctrl-X*	Hacks off the line at the cursor.
Tab	→| on PCs, *Tab*, and *Ctrl-I*	Moves the cursor forward eight spaces.
Btab	|← on PCs, and *Ctrl-G*	The inverse of Tab.

▷ Example

It is difficult to demonstrate the use of editing keys on paper. You should try each of them. Nevertheless, here is an example:

. summarize price w̲aht

You typed summarize price waht and then tapped the *Lft* key (← key or *Ctrl-H*) three times to maneuver the cursor back to the a of waht. If you were to press *Return* right now, Stata would see the command summarize price waht, so where the cursor is does not matter when you press *Return*. If you wanted to execute the command summarize price, you could back up one more character and then press the Hack key. We will assume, however, that you meant to type weight.

If you were now to press the letter *e* on the keyboard, an e would appear on the screen to replace the a, and the cursor would move under the character h. We now have we̲ht. You press *Ins*, putting Stata into insert mode, and press *i* and *g*. The line now says summarize price weight̲, which is correct, so you press *Return*. Notice that we did not have to press *Ins* before every character we wanted to insert. The *Ins* key is a toggle: If we press it again, Stata turns off insert mode and what we type replaces what was there. When we press *Return*, Stata forgets all about insert mode, so we do not have to remember from one command to the next whether we are in insert mode.

◁

❏ Technical Note

Stata performs its editing magic based on the information about your terminal recorded in /etc/termcap(5) or, under System V, /usr/lib/terminfo(4). If some feature does not appear to work, it is probable that the entry for your terminal in the termcap file or terminfo directory is incorrect. Contact your system administrator.

If you are running on a Sun using Sunview or Open Look, Stata's editing facilities work only under shelltool(1)—they do not work under cmdtool(1). Moreover, if running from the Sun console, verify that the Unix environment variable $TERM contains sun and not sun-cmd. (Type echo $TERM to check the contents.) If you need to change the contents, type setenv TERM sun and unsetenv TERMCAP. Note carefully the capitalization.

❏

13.6   Editing previous lines, Stata for all operating systems

Stata for Windows and Stata for Macintosh users: What is said below is true but perhaps not important because the Review window shows the contents of the review buffer.

You may edit previously typed lines or at least any of the last 25 or so lines. Stata records every line you type in a wraparound buffer. A wraparound buffer is a buffer of finite length in which the most recent thing you type replaces the oldest thing stored in the buffer. Stata's buffer is 1,000 characters long for Small Stata and 6,144 characters long for the Intercooled version.

One way to retrieve lines is with the PrevLine and NextLine keys. Remember, PrevLine and NextLine are the names we attach to these keys—there are no such keys on your keyboard. You have to look back at the previous section to find out which keys correspond to PrevLine and NextLine on your computer. To save you the effort this time, PrevLine probably corresponds to *PgUp* and NextLine probably corresponds to *PgDn*.

Suppose you wanted to reissue the third line back. You could press PrevLine three times and then press *Return*. If you made a mistake and pressed PrevLine four times, you could press NextLine to go forward in the buffer. You do not have to count lines since, each time you press PrevLine or NextLine, the current line is displayed on your monitor. Simply tap the key until you find the line you want.

Another way to retrieve previous lines is with Seek. Typing a *3* and then pressing Seek tells Stata to fetch the third line back and show it to you.

A third method for reviewing previous lines, **#review**, is convenient when you want to see the lines in context.

▷ Example

Typing **#review** by itself causes Stata to list the last five commands you typed. (You need not type out **#review**—pressing *F2* has the same effect.) For instance:

```
. #review
5 list make mpg weight if abs(res)>6
4 list make mpg weight if abs(res)>5
3 tabulate foreign if abs(res)>5
2 regress mpg weight weight2
1 test weight2=0
. _
```

We can see from the listing that the last command typed by the user was **test weight2=0**.

◁

▷ Example

Perhaps the command you are looking for is not among the last five commands you typed. You can tell Stata to go back any number of lines. For instance, typing **#review 15** tells Stata to show you the last 15 lines you typed:

```
. #review 15
15 replace resmpg=mpg-pred
14 summarize resmpg, detail
13 drop predmpg
12 describe
11 sort foreign
10 by foreign: summarize mpg weight
9 * lines that start with a * are comments.
8 * they go into the review buffer too.
7 summarize resmpg, detail
6 list make mpg weight
5 list make mpg weight if abs(res)>6
4 list make mpg weight if abs(res)>5
3 tabulate foreign if abs(res)>5
2 regress mpg weight weight2
1 test weight2=0
. _
```

If you wanted to resubmit the tenth previous line, you could type **10** and press Seek or you could press PrevLine ten times. No matter which of the above methods you prefer for retrieving lines, you may edit previous lines using the editing keys.

◁

Elements of Stata

Chapters

14 Language syntax

Contents

14.1   Overview

With few exceptions, the basic Stata language syntax is

$$\left[\text{by } \textit{varlist}\text{:}\right] \quad \textit{command} \quad \left[\textit{varlist}\right] \quad \left[\text{=}\textit{exp}\right] \quad \left[\text{if } \textit{exp}\right] \quad \left[\text{in } \textit{range}\right] \quad \left[\textit{weight}\right] \quad \left[\text{, } \textit{options}\right]$$

where square brackets denote optional qualifiers. In this diagram, *varlist* denotes a list of variable names, *command* denotes a Stata command, *exp* denotes an algebraic expression, *range* denotes an observation range, *weight* denotes a weighting expression, and *options* denotes a list of options.

14.1.1   varlist

Most commands that take a subsequent *varlist* do not require that one be explicitly typed. If no *varlist* appears, these commands assume a *varlist* of _all, the Stata shorthand for indicating all the variables in the dataset. In commands that alter or destroy data, Stata requires that the *varlist* be specified explicitly. See [U] **14.4 varlists** for a complete description.

▷ Example

The `summarize` command lists the mean, standard deviation, and range of the variables specified. In [R] **summarize**, we see that the syntax diagram for summarize is

$$\left[\texttt{by } \textit{varlist:}\right] \text{ } \underline{\texttt{su}}\texttt{mmarize} \text{ } \left[\textit{varlist}\right] \text{ } \left[\textit{weight}\right] \text{ } \left[\texttt{if } \textit{exp}\right] \text{ } \left[\texttt{in } \textit{range}\right] \text{ } \left[\texttt{, } \underline{\texttt{d}}\texttt{etail } \underline{\texttt{f}}\texttt{ormat}\right]$$

Since everything but the word `summarize` is enclosed in square brackets, the simplest form of the command is 'summarize'. Typing `summarize` without arguments is equivalent to typing `summarize _all`; all the variables in the dataset are summarized. Underlining denotes the shortest allowed abbreviation: we could have typed just `su`; see [U] **14.2 Abbreviation rules**.

The `drop` command eliminates variables or observations from a dataset. The syntax diagram for the version that drops variables is

drop *varlist*

Typing `drop` by itself results in the error message "varlist or in range required". To drop all the variables in the dataset, you must type `drop _all`.

Even before looking at the syntax diagram, we could have predicted that the *varlist* would be required—`drop` is destructive, and hence we are required to spell out our intent. The syntax diagram informs us that the *varlist* is required since *varlist* is not enclosed in square brackets. Since `drop` is not underlined, it cannot be abbreviated.

◁

14.1.2 by varlist:

The `by` *varlist:* prefix causes Stata to repeat a command for each subset of the data for which the values of the variables in the *varlist* are equal. When prefixed with `by` *varlist:*, the result of the command will be the same as if you had formed separate datasets for each group of observations, saved them, and then given the command on each dataset separately. The data must already be sorted by *varlist*. See [U] **14.5 by varlist: construct** for more information.

▷ Example

Typing `summarize mrg dvc` produces a table of the mean, standard deviation, and range of `mrg` and `dvc` using all the observations in the data:

```
. summarize mrg dvc
Variable |    Obs       Mean    Std. Dev.        Min         Max
---------+-----------------------------------------------------
     mrg |     50     0.0187      0.0257      0.0104      0.1955
     dvc |     50     0.0080      0.0032      0.0040      0.0237
```

Typing `by region: summarize mrg dvc` produces a set of tables, one table for each region of the country:

```
. by region: summarize mrg dvc

-> region=N. Central
Variable |    Obs       Mean    Std. Dev.        Min         Max
---------+-----------------------------------------------------
     mrg |     12     0.0139      0.0016      0.0123      0.0181
     dvc |     12     0.0066      0.0016      0.0046      0.0103
```

```
-> region=N. East
Variable |    Obs        Mean    Std. Dev.        Min         Max
---------+-------------------------------------------------------
     mrg |      9      0.0121       0.0018      0.0104      0.0150
     dvc |      9      0.0058       0.0015      0.0040      0.0079

-> region=South
Variable |    Obs        Mean    Std. Dev.        Min         Max
---------+-------------------------------------------------------
     mrg |     16      0.0162       0.0037      0.0104      0.0247
     dvc |     16      0.0079       0.0018      0.0054      0.0112

-> region=West
Variable |    Obs        Mean    Std. Dev.        Min         Max
---------+-------------------------------------------------------
     mrg |     13      0.0307       0.0496      0.0120      0.1955
     dvc |     13      0.0108       0.0043      0.0064      0.0237
```

There is one restriction: The dataset must be sorted on the by variable(s):

```
. by region: summarize mrg dvc
not sorted
r(5);

. sort region

. by region: summarize mrg dvc
(output above appears)
```

by *varlist*: can be used with many Stata commands; you can tell which ones by looking at their syntax diagrams. For instance, we could obtain the correlations, by region, between mrg and dvc by typing by region: correlate mrg dvc.

◁

❑ Technical Note

The *varlist* in by *varlist*: may contain up to 2,048 variables; the maximum allowed in the dataset. For instance, if you had data on automobiles and wished to obtain means according to market category (market) broken down by manufacturer (origin), you could type by market origin: summarize. That *varlist* contains two variables: market and origin. If the data were not already sorted on market and origin, you would first type sort market origin.

❑

❑ Technical Note

The *varlist* in by *varlist*: may contain either string or numeric variables, or both. In the example above, region is a string variable, in particular, a str10. The example would have worked, however, if region were a numeric variable with values 1, 2, 3, and 4 or even 12.2, 16.78, 32.417, and 15.213.

❑

14.1.3   if exp

The if *exp* qualifier restricts the scope of a command to those observations for which the value of the expression is *true* (which is equivalent to the expression being nonzero; see [U] **16 Functions and expressions**).

▷ Example

Typing `summarize mrg dvc if region=="West"` produces a table for the western region of the country:

```
. summarize mrg dvc if region=="West"
Variable |    Obs        Mean   Std. Dev.        Min        Max
---------+--------------------------------------------------------
     mrg |     13      0.0307      0.0496     0.0120     0.1955
     dvc |     13      0.0108      0.0043     0.0064     0.0237
```

The double equal sign in `region=="West"` is not an error. Stata uses *double* equal signs to denote equality testing and a *single* equal sign to denote assignment; see [U] **16 Functions and expressions**.

A command may have at most one `if` qualifier. If you want the summary for the West restricted to observations with values of `mrg` in excess of 0.02, do *not* type `summarize mrg dvc if region=="West" if mrg>.02`. Type

```
. summarize mrg dvc if region=="West" & mrg>.02
Variable |    Obs        Mean   Std. Dev.        Min        Max
---------+--------------------------------------------------------
     mrg |      3      0.0793      0.1007     0.0211     0.1955
     dvc |      3      0.0155      0.0072     0.0104     0.0237
```

You may not use the word 'and' in place of the symbol '&' to join conditions. To select observations that meet one condition *or* another, you use the '|' symbol. For instance, `summarize mrg dvc if region=="West" | mrg>.02` summarizes all observations for which `region` is West *or* `mrg` is greater than 0.02.

◁

▷ Example

`if` may be combined with `by`. Typing `by region: summarize mrg dvc if mrg>.02` produces a set of tables, one for each region, reflecting summary statistics on `mrg` and `dvc` among observations for which `mrg` exceeds 0.02:

```
. by region: summarize mrg dvc if mrg>.02

-> region=N. Central
Variable |    Obs        Mean   Std. Dev.        Min        Max
---------+--------------------------------------------------------
     mrg |      0
     dvc |      0

-> region=N. East
Variable |    Obs        Mean   Std. Dev.        Min        Max
---------+--------------------------------------------------------
     mrg |      0
     dvc |      0

-> region=South
Variable |    Obs        Mean   Std. Dev.        Min        Max
---------+--------------------------------------------------------
     mrg |      2      0.0231      0.0023     0.0214     0.0247
     dvc |      2      0.0087      0.0035     0.0062     0.0112

-> region=West
Variable |    Obs        Mean   Std. Dev.        Min        Max
---------+--------------------------------------------------------
     mrg |      3      0.0793      0.1007     0.0211     0.1955
     dvc |      3      0.0155      0.0072     0.0104     0.0237
```

The results indicate that there are no states in the Northeast and North Central regions for which `mrg` exceeds 0.02, while there are two such states in the South and three states in the West.

◁

14.1.4   in range

The `in` *range* qualifier restricts the scope of the command to a specific observation range. A range specification takes the form $\#_1\left[/\#_2\right]$, where $\#_1$ and $\#_2$ are positive or negative. Negative integers are understood to mean "from the end of the data", with -1 referring to the last observation. The implied first observation must be less than or equal to the implied last observation.

The first and last observations in the dataset may be denoted by `f` and `l` (letter *el*) respectively. A range specifies absolute observation numbers within a dataset. As a result, the `in` qualifier may not be used when the command is preceded by the `by` *varlist*: prefix.

▷ Example

Typing `summarize mrg dvc in 5/25` produces a table based on the values of `mrg` and `dvc` in observations 5 through 25:

```
. summarize mrg dvc in 5/25
Variable |    Obs       Mean    Std. Dev.       Min        Max
---------+-----------------------------------------------------
     mrg |     21     0.0136      0.0033      0.0104     0.0247
     dvc |     21     0.0064      0.0016      0.0040     0.0098
```

This is, admittedly, a rather odd thing to want to do. It would not be odd, however, if we substituted `list` for `summarize`. If we wanted to see the states with the 10 lowest values of `mrg`, we could type `sort mrg` followed by `list mrg in 1/10`.

Typing `summarize mrg dvc in f/l` is equivalent to typing `summarize mrg dvc`—all observations are summarized.

◁

▷ Example

Typing `summarize mrg dvc in 5/25 if region=="South"` produces a table based on the values of the two variables in observations 5 through 25 for which the value of `region` is South:

```
. summarize mrg dvc in 5/25 if region=="South"
Variable |    Obs       Mean    Std. Dev.       Min        Max
---------+-----------------------------------------------------
     mrg |      4     0.0175      0.0056      0.0111     0.0247
     dvc |      4     0.0080      0.0018      0.0062     0.0098
```

The ordering of the `in` and `if` qualifiers is not significant. The command could also have been specified as `summarize mrg dvc if region=="South" in 5/25`.

◁

▷ Example

Negative `in` ranges can be usefully employed with `sort`. For instance, we have data on automobiles and wish to list the five with the highest mileage rating:

```
. sort mpg
. list make mpg in -5/l
                make        mpg
70.     Toyota Corolla       31
71.       Plym. Champ        34
72.            Subaru        35
73.         Datsun 210       35
74.         VW Diesel        41
```

◁

14.1.5   =exp

The *=exp* specifies the value to be assigned to a variable and is most often used with `generate` and `replace`. See [U] **16 Functions and expressions** for details on expressions and [R] **generate** for details on the `generate` and `replace` commands.

▷ Example

Expression	Meaning
`generate newvar=oldvar+2`	creates a new variable named `newvar` equal to `oldvar+2`
`replace oldvar=oldvar+2`	changes the contents of the existing variable oldvar
`egen newvar=rank(oldvar)`	creates `newvar` containing the ranks of oldvar (see [R] **egen**)

◁

14.1.6   weight

weight indicates the weight to be attached to each observation. The syntax of *weight* is

$$[weightword=exp]$$

where you actually type the square brackets and where *weightword* is one of

weightword	Meaning
<u>we</u>ight	default treatment of weights
<u>fw</u>eight *or* <u>frequ</u>ency	frequency weights
<u>pw</u>eight	sampling weights
<u>aw</u>eight *or* <u>cell</u>size	analytic weights
<u>iw</u>eight	importance weights

The underlining indicates the minimum acceptable abbreviation. Thus, `weight` may be abbreviated `w`, or `we`, etc.

▷ Example

Before explaining what the different types of weights mean, let's obtain the population-weighted mean of a variable called `medage` from data containing observations on 50 states of the U.S. The data also contains a variable named `pop` which is the total population of each state.

```
. summarize medage [weight=pop]
(analytic weights assumed)
Variable |    Obs      Weight      Mean    Std. Dev.      Min        Max
---------+----------------------------------------------------------------
  medage |     50   225907472     30.11       1.67       24.20      34.70
```

In addition to telling us that our data contains 50 observations, we are informed that the sum of the weight is 225,907,472, which was the number of people living in the U.S. as of the 1980 census. The weighted mean is 30.11. We were also informed that Stata assumed we wanted "analytic" weights.

◁

Stata understands four kinds of weights:

1. `fweight`s, or frequency weights, indicate duplicated observations. `fweight`s are always integers. If the `fweight` associated with an observation is 5, that means there are really 5 such observations, each identical.

2. `pweight`s, or sampling weights, denote the inverse of the probability that this observation is included in the sample due to the sampling design. A `pweight` of 100, for instance, indicates that this observation is representative of 100 subjects in the underlying population. The scale of these weights does not matter in terms of estimated parameters and standard errors, except when estimating totals and computing finite-population corrections with the `svy` commands.

3. `aweight`s, or analytic weights, are inversely proportional to the variance of an observation; i.e., the variance of the jth observation is assumed to be σ^2/w_j, where w_j are the weights. Typically, the observations represent averages and the weights are the number of elements that gave rise to the average. For most Stata commands, the recorded scale of `aweight`s is irrelevant; Stata internally rescales them to sum to N, the number of observations in your data, when it uses them.

4. `iweight`s, or importance weights, indicate the relative "importance" of the observation. They have no formal statistical definition; this is a catchall category. Any command that supports `iweight`s will define how they are treated. In most cases, they are intended for use by programmers who want to produce a certain computation.

See [U] **23.13 Weighted estimation** for a thorough discussion of weights and their meaning.

`weight` is each command's idea of what are the "natural" weights and is one of `fweight`, `pweight`, `aweight`, or `iweight`. When you specify the vague `weight`, the command informs you which kind it assumes. Not every command supports every kind of weight. A note below the syntax diagram for a command will tell you the types of weights the command supports.

❑ Technical Note

When you do not specify a weight, the result is equivalent to specifying `[fweight=1]`. The emphasis is on equivalent, since Stata may go to more work when you specify a weight.

❑

14.1.7   options

Many commands take command-specific options. These are described along with each command in the *Reference Manual*. Options are indicated by typing a comma at the end of the command, followed by a list of options.

▷ Example

Typing `summarize mrg` produces a table of the mean, standard deviation, minimum, and maximum of the variable `mrg`:

```
. summarize mrg
Variable |    Obs       Mean    Std. Dev.      Min       Max
---------+-----------------------------------------------------
     mrg |     50     0.0187     0.0257     0.0104     0.1955
```

The syntax diagram for `summarize` is

$$\left[\text{by } varlist:\right] \underline{\text{su}}\text{mmarize } [varlist] \; [weight] \; [\text{if } exp] \; [\text{in } range] \; [, \; \underline{\text{d}}\text{etail } \underline{\text{f}}\text{ormat }]$$

Thus, the options allowed by `summarize` are `detail` and `format`. The shortest allowed abbreviations for `detail` and `format` are `d` and `f`, respectively; see [U] **14.2 Abbreviation rules**.

Typing `summarize mrg, detail` produces a table that also includes selected percentiles, the four largest and four smallest values, the skewness, and the kurtosis.

```
. summarize mrg, detail

                             mrg
      -------------------------------------------------------------
           Percentiles      Smallest
      1%      0.0104         0.0104
      5%      0.0106         0.0104
     10%      0.0110         0.0106       Obs                50
     25%      0.0127         0.0107       Sum of Wgt.        50

     50%      0.0150                      Mean           0.0187
                             Largest      Std. Dev.      0.0257
     75%      0.0179         0.0212
     90%      0.0204         0.0214       Variance       0.0007
     95%      0.0214         0.0247       Skewness       6.6853
     99%      0.1955         0.1955       Kurtosis      46.4750
```

◁

❑ Technical Note

Once you have typed the *varlist* for the command, options may be placed anywhere in the command. You can type `summarize mrg dvc if region=="West", detail` or you can type `summarize mrg dvc, detail, if region=="West"`. Note the use of a second comma to indicate return to the command line as opposed to the option list. Leaving out the comma after the word `detail` would cause an error, since Stata would attempt to interpret the phrase `if region=="West"` as an option rather than as part of the command.

You may not type an option in the middle of a *varlist*. Typing `summarize mrg, detail, dvc` will result in an error.

Options need not be contiguously specified. You may type `summarize mrg dvc, detail, if region=="South", noformat`. Both `detail` and `noformat` are options.

❑

❑ Technical Note

Most options are toggles—they indicate that something either is or is not to be done. Sometimes it is difficult to remember which is the default. The following rule applies to all options: If *option* is an option, then no*option* is an option as well, and vice versa. Thus, if we could not remember whether `detail` or `nodetail` were the default for `summarize` but we knew that we did not want the detail, we could type `summarize, nodetail`. Typing the `nodetail` option is unnecessary, but Stata will not complain.

Some options take *arguments*. The Stata `kdensity` command has a `n(#)` option that indicates the number of points at which the density estimate is to be evaluated. When an option takes an argument, it is enclosed in parentheses.

Some options take more than one argument. In such cases arguments should be separated from one another by commas. For instance, you might see in a syntax diagram

> `saving(`*filename*`[, replace])`

In this case `replace` is the (optional) second argument. *Lists*, such as lists of variables (varlists) and lists of numbers (numlists) are considered to be one argument. If a syntax diagram reported

> `ylabel(`*numlist*`)`

the list of numbers would be one argument and so the elements would not be separated by commas. You would type, for instance, `ylabel(1 2 3 4)`. In fact, Stata will tolerate you typing commas in this case, so you could type `ylabel(1,2,3,4)`.

Some options take string arguments. The `graph` command's `title` option works this way—for instance, `title("Figure 1.")`. To play it safe, you should type the quotes surrounding the string, although it is not required. If you do not type the quotes, any sequence of two or more consecutive blanks will be interpreted as a single blank. Thus, `title(Figure   1.)` would be interpreted the same as `title(Figure 1.)`.

❑

14.1.8   numlist

A *numlist* is a list of numbers and allows certain shorthands to indicate ranges

numlist	meaning
1,2,3,4	1, 2, 3, 4
1 2 3 4	1, 2, 3, 4
2 4 to 8	2, 4, 6, 8
8 6 to 2	8, 6, 4, 2
2 4 : 8	2, 4, 6, 8
8 6 : 2	8, 6, 4, 2
2(2)8	2 4 6 8
8(-2)2	8, 6, 4, 2
10 12 14 18 to 26 28 to 32	10, 12, 14, 18, 22, 26, 28, 30, 32
10 12 14(4)26 28(2)32	10, 12, 14, 18, 22, 26, 28, 30, 32

`graph`'s `ylabel()` option has syntax `ylabel(`*numlist*`)`. Thus, you could type `ylabel(2 4 to 8)`, `ylabel(2(2)8)`, etc.

14.2   Abbreviation rules

Stata allows abbreviations. In this manual we usually avoid abbreviating commands, variable names, and options to ensure readability:

. summarize myvar, detail

Real Stata users, on the other hand, tend to abbreviate:

. sum myv, d

As a general rule, command, option, and variable names may be abbreviated to the shortest string of characters that uniquely identifies them.

This rule is violated if the command or option does something that cannot easily be undone: in that case, the command must be spelled out in its entirety.

In addition, a few common commands and options are allowed to have even shorter abbreviations than the general rule would allow.

The general rule is applied, without exception, to variable names.

14.2.1   Command abbreviation

The shortest allowed abbreviation for a command or option can be determined by looking at the command's syntax diagram. This minimal abbreviation is shown by underlining:

regress

rename

replace

rotate

run

Lack of underlining means no abbreviation is allowed. Thus, replace may not be abbreviated, the underlying reason being that replace changes the data.

regress can be abbreviated reg, regr, regre, regres, or can be spelled out in its entirety.

In the on-line help, highlighting or color is used in place of underlining in the syntax diagram.

As mentioned above, sometimes very short abbreviations are also allowed. Commands that begin with the letter *d* include decode, describe, dir, discard, display, do, and drop. This suggests that the shortest allowable abbreviation for describe is des. Since describe is such a commonly used command, you may abbreviate it with the single letter d. You may also abbreviate the list command with the single letter l.

The other exception to the general abbreviation rule concerns commands that alter or destroy data; such commands must be spelled out completely. Two of the commands that begin with the letter *d* above, discard and drop, are destructive in the sense that once you give one of these commands there is no way you can undo the result; therefore, both must be spelled out.

Another exception to the abbreviation rule occurs when you type help *command*. The *command* may not be abbreviated. For instance, although you may type su as an abbreviation for summarize, you must type help summarize to see the on-line help. Typing help su will result in the message "help for su not found".

The final exceptions to the general rule are commands implemented as ado-files. Such commands may not be abbreviated. Ado-file commands are external and their names correspond to the names of disk files.

14.2.2   Option abbreviation

Option abbreviation follows the same logic as command abbreviation: you determine the minimum acceptable abbreviation by examining the command's syntax diagram. The syntax diagram for summarize reads

$$\left[\text{by } \textit{varlist:}\right] \ \underline{\text{su}}\text{mmarize} \ \ldots, \underline{\text{d}}\text{etail} \ \underline{\text{f}}\text{ormat}$$

Option detail may be abbreviated d, de, det, ..., detail. Similarly, option format may be abbreviated f, fo, ..., format.

Options clear and replace occur with many commands. The clear option indicates that even though completion of this command will result in the loss of all data in memory, and even though the data in memory has changed since it was last saved on disk, you are aware of the situation and it is okay to continue. clear must be spelled out, as in use newdata, clear.

The replace option indicates that it's okay to save over an existing data set. If you type save mydata and the file mydata.dta already exists, you will receive the message "file mydata.dta already exists" and Stata will refuse to overwrite it. To allow Stata to overwrite the dataset, you type save mydata, replace. replace may not be abbreviated.

❑ Technical Note

replace is a stronger modifier than clear and one you should think about longer before specifying. With a mistaken clear, you can lose perhaps hours of work. With a mistaken replace, however, you can lose days of work.

❑

14.2.3   Variable-name abbreviation

Variable names may be abbreviated to the shortest string of characters that uniquely identifies them given the data currently loaded in memory.

If your dataset contained four variables, state, mrgrate, dvcrate, and dthrate, you could refer to the variable dvcrate as dvcrat, dvcra, dvcr, dvc, or dv. You might type list dv to list the data on dvcrate. You could not refer to the variable dvcrate as d, however, since that abbreviation does not distinguish dvcrate from dthrate. If you were to type list d, Stata would respond with the message "ambiguous abbreviation". (If you wanted to refer to *all* variables that started with the letter *d*, you could type list d*; see [U] **14.4 varlists**.)

14.3   Naming conventions

A name is a sequence of one to eight letters (A−Z and a−z), digits (0−9), and underscores (_).

Programmers: local macro names can have no more than 7 characters in the name; see [U] **21.3.1 Local macros**.

Stata reserves the following names:

_all	double	long	_rc
_b	float	_n	_se
byte	if	_N	_skip
_coef	in	_pi	using
_cons	int	_pred	with

You may not use these reserved names for your variables. Additionally, we advise that you avoid using the name **e** for your variables, since Stata will in some contexts have difficulty distinguishing between the variable **e** and the "e" in scientific notation; e.g., $e - 3$.

The first character of a name must be either a letter or an underscore. We recommend, however, that you do not begin your variable names with an underscore. All Stata built-in variables begin with an underscore, and we reserve the right to incorporate new _variables_ freely.

Stata respects case; that is, `myvar`, `Myvar`, and `MYVAR` are three distinct names.

All objects in Stata—not just variables—follow this naming convention.

14.4   varlists

A *varlist* is a list of variable names. The variable names in a *varlist* refer either exclusively to new (not yet created) variables or exclusively to existing variables.

14.4.1   Lists of existing variables

In lists of existing variable names, variable names may be repeated.

▷ Example

If you type `list state mrgrate dvcrate state` the variable `state` will be listed twice, once in the leftmost column and again in the rightmost column of the list.

◁

Existing variable names may be abbreviated as described in [U] **14.2 Abbreviation rules**. You may also append a '`*`' to a partial variable name to indicate all variables that start with that letter combination.

▷ Example

If the variables `poplt5`, `pop5to17`, and `pop18p` are in your dataset, you may type `pop*` as a shorthand way to refer to all three variables. For instance, `list state pop*` lists the variables `state`, `poplt5`, `pop5to17`, and `pop18p`.

◁

You may place a dash (–) between two variable names to specify all the variables stored between the two listed variables, inclusive. You can determine storage order using `describe`; it lists variables in the order in which they are stored.

▷ Example

If your data contains the variables `state`, `mrgrate`, `dvcrate`, and `dthrate`, in that order, typing `list state-dvcrate` is equivalent to typing `list state mrgrate dvcrate`. In both cases, three variables are listed.

◁

14.4.2 Lists of new variables

In lists of *new variables*, no variable names may be repeated or abbreviated.

You may specify a dash (-) between two variable names that have the same letter prefix and that end in numbers. This form of the dash notation indicates a range of variable names in ascending numerical order.

▷ Example

Typing input v1-v4 is equivalent to typing input v1 v2 v3 v4. Typing infile state v1-v3 ssn using rawdata is equivalent to typing infile state v1 v2 v3 ssn using rawdata.

◁

You may specify the storage type before the variable name to force a storage type other than the default. The numeric storage types are byte, int, long, float (the default), and double. The string storage types are str#, where # is replaced with an integer between 1 and 80, inclusive, representing the maximum length of the string. See [U] 15 Data.

For instance, the list var1 str8 var2 var3 specifies that var1 and var3 are to be given the default storage type, whereas var2 is to be stored as a str8—a string whose maximum length is eight characters.

The list var1 int var2 var3 specifies that var2 is to be stored as an int. You may use parentheses to bind a list of variable names. The list var1 int(var2 var3) specifies that both var2 and var3 are to be stored as ints. Similarly, the list var1 str20(var2 var3) specifies that both var2 and var3 are to be stored as str20s. The different storage types are listed in [U] 15.2.2 Numeric storage types and [U] 15.4.4 String storage types.

▷ Example

Typing infile str2 state str10 region v1-v5 using mydata reads the state and region strings from the file mydata.raw and stores them as str2 and str10, respectively, along with the variables v1 through v5, which are stored with the default storage type float (unless you have specified a different default with the set type command).

Typing infile str10(state region) v1-v5 using mydata would achieve almost the same result, except that the state and region values recorded in the data would both be assigned to str10 variables. (You could then use the compress command to shorten the strings. See [R] compress; it is well worth reading.)

◁

❏ Technical Note

You may append a colon and a *value label name* to numeric variables. (See [U] 15.6 Dataset, variable, and value labels for a description of value labels.) For instance, var1 var2:myfmt specifies that the variable var2 is to be associated with the value label stored under the name myfmt. This has the same effect as typing the list var1 var2 and then subsequently giving the command label values var2 myfmt.

The advantage of specifying the value label association with the colon notation is that value labels can then be assigned by the current command; see [R] input and [R] infile (free format).

❏

▷ Example

Typing `infile int(state:stfmt region:regfmt) v1-v5 using mydata, automatic` reads the state and region data from the file `mydata.raw` and stores them as `int`s, along with the variables `v1` through `v5`, which are stored with the default storage type.

In our previous example, both state and region were strings, so how can strings be stored in a numeric variable? See [U] **15.6 Dataset, variable, and value labels** for the complete answer. The colon notation specifies the name of the value label and the `automatic` option tells Stata to assign unique numeric codes to all character strings. The numeric code for state, which Stata will make up on the fly, will be stored in the `state` variable. The mapping from numeric codes to words will be stored in the value label named `stfmt`. Similarly, region will be assigned numeric codes which are stored in `region` and the mapping will be stored in `regfmt`.

If you were to `list` the data, the `state` and `region` variables would look like strings. `state`, for instance, would appear to contain things like `AL`, `CA`, and `WA`, but actually it contains only numbers like 1, 2, 3, and 4.

◁

14.4.3   Time-series varlists

Time-series varlists are a variation on varlists of existing variables. When a command allows a time-series varlist, you may include time-series operators. For instance, `L.gnp` refers to the lagged value of variable `gnp`. The time-series operators are

operator	meaning
`L.`	lag x_{t-1}
`L2.`	2-period lag x_{t-2}
. . .	
`F.`	lead x_{t+1}
`F2.`	2-period lead x_{t+2}
. . .	
`D.`	difference $x_t - x_{t-1}$
`D2.`	difference of difference $x_t - x_{t-1} - (x_{t-1} - x_{t-2}) = x_t - 2x_{t-1} + x_{t-2}$
. . .	
`S.`	"seasonal" difference $x_t - x_{t-1}$
`S2.`	lag-2 (seasonal) difference $x_t - x_{t-2}$
. . .	

Time-series operators may be repeated and combined. `L3.gnp` refers to the third lag of variable `gnp`. So does `LLL.gnp`, `LL2.gnp`, and `L2L.gnp`. `LF.gnp` is the same as `gnp`. `DS12.gnp` refers to the one-period difference of the 12-period difference. `LDS12.gnp` refers to the same concept, lagged once.

Note that `D1. = S1.` but `D2. ≠ S2.`, `D3. ≠ S3.`, and so on. `D2.` refers to the difference of the difference. `S2.` refers to the two-period difference. If you wanted the difference of the difference of the 12-period difference of `gnp`, you would write `D2S12.gnp`.

Operators may be typed in uppercase or lowercase. Most users would type `d2s12.gnp` instead of `D2S12.gnp`.

You may type operators however you wish; Stata internally converts operators to their canonical form. If you typed `1d2l1s12d.gnp`, Stata would present the operated variable as `L2D3S12.gnp`.

In addition to *operator#*, Stata understands *operator*(*numlist*) to mean a set of operated variables. For instance, typing L(1/3).gnp in a varlist is the same as typing 'L.gnp L2.gnp L3.gnp':

```
. list year gnp l(1/3).gnp

        year      gnp      L.gnp    L2.gnp    L3.gnp
  1.    1989    5452.8        .         .         .
  2.    1990    5764.9     5452.8       .         .
  3.    1991    5932.4     5764.9    5452.8       .
  4.    1992    6255.5     5932.4    5764.9    5452.8
 (output omitted )
```

In *operator#*, making # zero returns the variable itself. L0.gnp is gnp. Thus, the above listing could have been produced by typing 'list year l(0/3).gnp'.

The parentheses notation may be used with any operator. Typing D(1/3).gnp would return the first through third differences.

The parenthesis notation may be used in operator lists with multiple operators such as L(0/3)D2S12.gnp.

Operator lists may include up to one set of parentheses and the parentheses is to enclose a *numlist*; see [U] **14.1.8 numlist**. If individual elements are specified, they must be separated by commas. Type 'D2S(4,12).gnp', not 'D2S(4 12).gnp'.

Before you can use time-series operators in varlists, you must set the time variable using the **tsset** command:

```
. list l.gnp
time variable not set
r(111);

. tsset time
  (output omitted )

list l.gnp
  (output omitted )
```

See [R] **tsset**. The time variable must take on integer values. In addition, the data must be sorted on the time variable. **tsset** handles this but, later, you might encounter

```
. list l.mpg
not sorted
r(5);
```

In that case, type **sort time** to reestablish the order or type **tsset** which will do the same thing.

Note that the time-series operators respect the time variable. L2.gnp refers to gnp_{t-2} regardless of missing observations in the dataset. In the following dataset, the observation for 1992 is missing:

```
. list year gnp l2.gnp

        year      gnp     L2.gnp
  1.    1989   5,452.8       .
  2.    1990   5,764.9       .
  3.    1991   5,932.4    5,452.8
  4.    1993   6,560.0    5,932.4    ← note, filled in correctly
  5.    1994   6,922.4       .
  6.    1995   7,237.5    6,560.0
```

Operated variables may be used in expressions:

```
. gen gnplag2 = l2.gnp
(3 missing values generated)
```

Stata also understands panel (cross-sectional time-series) data as well as simple time-series data. If you have cross sections of time series, you indicate this at the time you `tsset` the data:

    . tsset country year

See [R] **tsset** and [U] **27.3 Time-series dates**.

14.5   by varlist: construct

by *varlist*: *command*

The `by` prefix causes *command* to be repeated for each unique set of values of the variables in the *varlist*. The data must already be `sort`ed by the *varlist*. *varlist* may contain numeric, string, or a mixture of numeric and string variables. (*varlist* may not contain time-series operators.)

`by` is an optional prefix to perform a Stata command separately for each group of observations where the values of the variables in the *varlist* are the same.

During each iteration, the values of the system variables _n and _N are set in relation to the first observation in the by-group; see [U] **16.7 Explicit subscripting**. The `in` *range* qualifier cannot be used in conjunction with `by` *varlist*: because ranges specify absolute rather than relative observation numbers.

❏ Technical Note

The inability to combine `in` and `by` is not really a constraint, since `if` provides all the functionality of `in` and quite a bit more besides. If you wanted to perform *command* for the first three observations in each of the by-groups, you could type

    . by *varlist*: *command* if _n<=3

❏

The results of *command* will be the same as if you had formed separate datasets for each group of observations, `save`d them, `use`d each separately, and issued *command*.

▷ Example

We provide some examples using `by` in [U] **14.1.2 by varlist:** above. We demonstrate the effect of `by` on _n, _N, and explicit subscripting in [U] **16.7 Explicit subscripting**.

`by` requires that the data first be sorted. For instance, if we had data on the average January and July temperatures in degrees Fahrenheit for 420 cities located in the Northeast and West and wanted to obtain the averages, `by region`, across those cities, we might type

    . by region: summarize tempjan tempjuly
    not sorted
    r(5);

Stata refused to honor our request since the data is not sorted by `region`. We must `sort` the data first; see [R] **sort**.

```
. sort region
. by region: summarize tempjan tempjuly
-> region=Northeast
Variable |    Obs        Mean   Std. Dev.       Min        Max
---------+-----------------------------------------------------------
 tempjan |    164       27.89        3.54       16.60      31.80
tempjuly |    164       73.35        2.36       66.50      76.80
-> region=West
Variable |    Obs        Mean   Std. Dev.       Min        Max
---------+-----------------------------------------------------------
 tempjan |    256       46.23       11.25       13.00      72.60
tempjuly |    256       72.11        6.48       58.10      93.60
```

◁

❏ Technical Note

Using the same data as in the example above, we estimate regressions, by region, of average January temperature on average July temperature. Both temperatures are specified in degrees Fahrenheit.

```
. by region:  regress tempjan tempjuly
-> region=      NE
  Source |       SS       df       MS                  Number of obs =      164
---------+------------------------------              F(  1,   162) =   479.82
   Model |  1529.74026      1  1529.74026             Prob > F      =   0.0000
Residual |  516.484453    162  3.18817564             R-squared     =   0.7476
---------+------------------------------              Adj R-squared =   0.7460
   Total |  2046.22471    163  12.5535258             Root MSE      =   1.7855

---------------------------------------------------------------------------
 tempjan |     Coef.   Std. Err.       t      P>|t|      [95% Conf. Interval]
---------+-----------------------------------------------------------------
tempjuly |   1.297424   .0592303     21.905   0.000      1.180461    1.414387
   _cons |  -67.28066   4.346781    -15.478   0.000     -75.86431     -58.697
---------------------------------------------------------------------------

-> region=     West
  Source |       SS       df       MS                  Number of obs =      256
---------+------------------------------              F(  1,   254) =     2.84
   Model |  357.161728      1  357.161728             Prob > F      =   0.0932
Residual |  31939.9031    254  125.74765             R-squared     =   0.0110
---------+------------------------------              Adj R-squared =   0.0072
   Total |  32297.0648    255  126.655156            Root MSE      =   11.214

---------------------------------------------------------------------------
 tempjan |     Coef.   Std. Err.       t      P>|t|      [95% Conf. Interval]
---------+-----------------------------------------------------------------
tempjuly |   .1825482   .1083166      1.685   0.093     -.0307648    .3958613
   _cons |   33.0621    7.84194       4.216   0.000      17.61859     48.5056
---------------------------------------------------------------------------
```

The regressions show that a one-degree increase in the average July temperature in the Northeast corresponds to a 1.3 degree increase in the average January temperature. In the West, however, it corresponds to a 0.18 degree increase, which is only marginally significant.

❏

14.6   File-naming conventions

Some commands require that you specify a *filename*. Filenames are specified in the way natural for your operating system:

Windows	Unix	Macintosh
mydata	mydata	mydata
mydata.dta	mydata.dta	mydata.dta
b:mydata.dta	~friend/mydata.dta	:mydisk:mydata.dta
"my data"	"my data"	"my data"
"my data.dta"	"my data.dta"	"my data.dta"
myproj\mydata	myproj/mydata	myproj:mydata
"my project\my data"	"my project/my data"	"my project:my data"
c:\analysis\data\mydata	~/analysis/data/mydata	~:analysis:data:mydata
"c:\my project\my data"	"~/my project/my data"	"~:my project:my data"
..\data\mydata	../data/mydata	..:data:mydata
"..\my project\my data"	"../my project/my data"	"..:my project:my data"

In most cases (the exceptions being copy, dir, ls, erase, rm, and type), Stata automatically provides a file extension if you do not supply one. For instance, if you type use mydata, Stata assumes you mean use mydata.dta, since .dta is the file extension Stata normally assumes for data.

Stata provides nine default file extensions that are used by various commands. They are

.ado	automatically loaded do-files
.dct	ASCII data dictionary
.do	do-file
.dta	Stata-format dataset
.gph	graph image
.log	log file
.out	file saved by outsheet
.raw	ASCII-format dataset
.sum	checksum files to verify network transfers

You do not have to name your data files with the .dta extension—if you type an explicit file extension, it will override the default. For instance, if your data was stored as myfile.dat, you could type use myfile.dat. If your data was stored as simply myfile with no file extension, type the period at the end of the filename to indicate that you are explicitly specifying the null extension. You type use myfile. to use this data.

All operating systems except Windows 3.1 allow blanks in filenames and so does Stata. However, if the filename includes a blank, you must enclose the filename in double quotes:

. save "my data"

would create the file my data.dta. Typing

. save my data

would be an error.

14.6.1 A special note for Macintosh users

Have you seen the notation `myfolder:myfile` before? It refers to the file `myfile` in the folder `myfolder`, which folder is contained in the current folder.

You do not have to use this notation if you do not like it. You could instead restrict yourself to using files only in the current directory. If that turns out to be too restricting, Stata for Macintosh provides enough pulldowns that you can probably get by. You may, however, find the notation convenient. In case you do, here is the rest of the definition.

`..:myfile` refers to `myfile` in the folder containing the current folder.

Thus, `..:nextdoor:myfile` refers to `myfile` in the folder `nextdoor` in the folder containing the current folder.

The notation `~:mydata:myfile` refers to `myfile` in the folder `mydata`, which is located in the folder containing the Stata folder.

Thus, `~:Stata:auto.dta` refers to the automobile data in the Stata folder.

14.6.2 A special note for Unix users

Stata understands `~` to mean your home directory. Stata understands this even if you do not use csh(1) as your shell.

15 Data

Contents

15.1 Data and datasets

Data is a rectangular table of numeric and string values in which each row is an observation on all the variables and each column contains the observations on a single variable. Variables are designated by *variable names*. Observations are numbered sequentially from 1 to _N. The following example of data contains the first five odd and first five even positive integers along with a string variable:

```
        odd   even    name
  1.      1      2    Bill
  2.      3      4    Mary
  3.      5      6     Pat
  4.      7      8   Roger
  5.      9     10    Sean
```

The observations are numbered 1 to 5 and the variables are named **odd**, **even**, and **name**. Observations are referred to by number, variables by name.

A *dataset* is *data* plus labelings, formats, notes, and characteristics.

All aspects of *data* and *datasets* are defined here.

15.2 Numbers

A *number* may contain a sign, an integer part, a decimal point, a fraction part, an e or E, and a signed integer exponent. Numbers may *not* contain commas; for example, the number 1,024 must be typed as 1024 (or 1024. or 1024.0). The following are examples of valid numbers:

```
5
-5
5.2
.5
5.2e+2
5.2e-2
```

❏ Technical Note

As a convenience for FORTRAN users, you may use d or D as well as e to indicate exponential notation. Thus, the number 520 may be written 5.2d+2, 5.2D+2, 5.2d+02, or 5.2D+02.

❏

15.2.1 Missing values

A number may also take on the special value *missing*, denoted by a single period (.). You may specify a missing value anywhere you may specify a number. Missing values differ from ordinary numbers in one respect: Any arithmetic operation on a missing value yields a missing value.

▷ Example

You have data on the income of husbands and wives recorded in the variables hincome and wincome, respectively. Typing the list command, you see that your data contains

```
. list
        hincome   wincome
  1.     35000     34000
  2.     28000     48000
  3.     47000         .
  4.     32000         0
```

The value of wincome in the third observation is *missing*, as distinct from the value of wincome in the last observation, which is known to be zero.

If you use the generate command to create a new variable, income, equal to the sum of hincome and wincome, one missing value would be produced:

```
. generate income=hincome+wincome
(1 missing value generated)
. list
        hincome   wincome   income
  1.     35000     34000    69000
  2.     28000     48000    76000
  3.     47000         .        .
  4.     32000         0    32000
```

generate produced a warning message that a missing value was created, and when we list the data, we see that 47,000 plus *missing* yields *missing*.

◁

❑ Technical Note

Stata stores numeric missing values as the largest number allowed by the particular storage type; see [U] **15.2.2 Numeric storage types**. There are two important implications. First, if you `sort` on a variable that has missing values, the missing values will be placed last.

```
. sort wincome
. list wincome
        wincome
  1.          0
  2.      34000
  3.      48000
  4.          .
```

The second implication concerns relational operators and missing values. Do not forget that a missing value will be larger than any numeric value.

```
. list if wincome > 40000
        hincome     wincome      income
  3.      28000       48000       76000
  4.      47000           .           .
```

Observation 4 is listed because '.' (missing) is greater than 40000. Relational operators are discussed in detail in [U] **16.2.3 Relational operators**.

❑

▷ Example

In producing statistical output, Stata ignores observations with missing values. Continuing with the example above, if we request summary statistics on `hincome` and `wincome` using the `summarize` command, we obtain

```
. summarize hincome wincome
Variable |    Obs        Mean    Std. Dev.        Min         Max
---------+-----------------------------------------------------------
 hincome |      4       35500     8185.353       28000       47000
 wincome |      3    27333.33     24684.68           0       48000
```

Note there are four observations on `hincome` but only three observations on `wincome`.

Some commands will discard the entire observation (known as *casewise deletion*) if one of the variables in the observation is missing. If we use the `correlate` command to obtain the correlation between `hincome` and `wincome`, for instance, we obtain

```
. correlate hincome wincome
(obs=3)
         |  hincome   wincome
---------+------------------
 hincome|   1.0000
 wincome|  -0.3614    1.0000
```

Note that the correlation coefficient is calculated over three observations.

◁

15.2.2   Numeric storage types

Numbers can be stored in one of five variable types: byte, int, long, float (the default), or double. bytes are, naturally enough, stored in 1 byte. ints are stored in 2 bytes, longs and floats in 4 bytes, and doubles in 8 bytes. The table below shows the minimum and maximum values for each storage type along with the encoding used internally by Stata to record *missing*.

Storage Type	Minimum	Maximum	Closest to 0 without being 0	missing
byte	-127	126	± 1	127
int	$-32{,}767$	32,766	± 1	32,767
long	$-2{,}147{,}483{,}647$	2,147,483,646	± 1	2,147,483,647
float	-10^{36}	10^{36}	$\pm 10^{-36}$	2^{128}
double	-10^{-308}	10^{308}	$\pm 10^{-323}$	2^{1023}

Do not confuse the term *integer*, which is a characteristic of a number, with int, which is a storage type. For instance, the number 5 is an integer no matter how it is stored; thus, if you read that an argument is required to be an integer, that does not mean that it must be stored as an int.

15.3   Dates

You can record dates any way you want, but there is one technique that Stata understands, called an elapsed date. An elapsed date is the number of days from January 1, 1960. In this format,

0	means	January 1, 1960
1		January 2, 1960
31		February 1, 1960
365		December 31, 1960
366		January 1, 1961
12,784		January 1, 1995
-1		December 31, 1959
-2		December 30, 1959
$-12{,}784$		December 31, 1924

Stata understands dates recorded like this from January 1, year 100 (elapsed date $-679{,}350$) to December 31, 9999 (elapsed date 2,936,549), although caution should be exercised in dealing with dates before Friday, October 15, 1582, when the Gregorian calendar went into effect.

Stata provides functions to convert dates into elapsed dates, formats to print elapsed dates in understandable forms, and other functions to manipulate elapsed dates.

In addition to elapsed dates, Stata provides five other date formats:

weekly	monthly	quarterly	half-yearly	yearly
-1 = 1959 week 52	-1 = Dec. 1959	-1 = 1959 quarter 4	-1 = 2nd half 1959	1959 = 1959
0 = 1960 week 1	0 = Jan. 1960	0 = 1960 quarter 1	0 = 1st half 1960	1960 = 1960
1 = 1960 week 2	1 = Feb. 1960	1 = 1960 quarter 2	1 = 2nd half 1960	1961 = 1961
. . .	. . .	. . .	. . .	. . .

For a full discussion of working with date variables, see [U] **27 Commands for dealing with dates**.

15.4   Strings

A *string* is a sequence of printable characters typically enclosed in double quotes. The quotes are not considered a part of the string: they merely delimit the beginning and end of the string. The following are examples of valid strings:

```
"Hello, world"
"String"
"string"
" string"
"string "
""
"x/y+3"
"1.2"
```

All of the strings above are distinct; that is, `"String"` is different from `"string"` which is different from `" string"` which is different from `"string "`. Also note that `"1.2"` is a string and not a number because it is enclosed in quotes.

All strings in Stata are of varying length, which means Stata internally records the length of the string and never loses track. There is never a circumstance in which a string cannot be delimited by quotes, but there are rare instances where strings do not have to be delimited by quotes, such as during data input. In those cases, nondelimited strings are stripped of their leading and trailing blanks. Delimited strings are always accepted as is.

The special string `""`, often called *null string*, is considered by Stata to be a *missing*. No special meaning is given to the string containing a single period, `"."`.

In addition to double quotes for enclosing strings, Stata also allows compound double quotes: `` `" `` and `"´`. You can type `"string"` or you can type `` `"string"´ ``, although users seldom type `` `"string"´ ``. Compound double quotes are of special interest to programmers because they nest and so provide a way for a quoted string to itself contain double quotes (either simple or compound). See [U] **21.3.5 Double quotes**.

15.4.1   Strings containing identifying data

String variables often contain identifying information, such as the patient's name or the name of the city or state. Such strings are typically listed but not used directly in statistical analysis, although the data might be sorted on the string or datasets might be merged on the basis of one or more string variables.

15.4.2   Strings containing categorical data

Occasionally, strings contain information that is to be used directly in analysis, such as the patient's sex, which might be coded `"male"` or `"female"`. Stata shows a decided preference for such information to be numerically encoded and stored in numeric variables. Stata's statistical routines treat string variables as if every observation records a numeric missing value.

All is not lost. Stata provides two commands for mapping string variables into numeric codes and back again: `encode` and `decode`; see [U] **26.2 Categorical string variables**.

15.4.3   Strings containing numeric data

If a string variable contains the character representation of a number—for instance, `myvar` contains "1", "1.2", and "-5.2"—you can convert it directly into a numeric variable using the `real()` function, as in `generate newvar=real(myvar)`.

Similarly, if you want to convert a numeric variable to its string representation, you can use the `string()` function: `generate str10 as_str=string(newvar)`.

See [U] **16.3.5 String functions**.

15.4.4   String storage types

Strings are stored in string variables with storage types `str1`, `str2`, ..., `str80`. The storage type merely sets the maximum length of the string, not its actual length; thus, "example" has length 7 whether it is stored as a `str7`, a `str10`, or even a `str80`. On the other hand, an attempt to assign the string "example" to a `str6` would result in "exampl".

The maximum length of a string in Stata is 80 characters. String literals may exceed 80 characters, but only the first 80 characters are significant.

15.5   Formats: controlling how data is displayed

Formats describe how a number or string is to be presented. For instance, how is the number 325.24 to be presented? As 325.2, or 325.24, or 325.240, or 3.2524e+2, or 3.25e+2, or how? The *display format* tells Stata exactly how you want this done. You do not have to specify display formats, since Stata always makes reasonable assumptions on how to display a variable, but you always have the option.

15.5.1   Numeric formats

A Stata numeric format is formed by

first type	%	to indicate the start of the format
then optionally type	–	if you want the result left-aligned
then type	a number w	stating the width of the result
then type	.	
then type	a number d	stating the number of digits to follow the decimal point
then type		
either	e	for scientific notation; e.g., 1.00e+03
or	f	for fixed format; e.g., 1000.0
or	g	for general format; Stata chooses based on the number being displayed
then optionally type	c	to indicate comma format (not allowed with e)

For example:

%9.0g	general format, 9 columns wide	
	sqrt(2) =	1.414214
	1,000 =	1000
	10,000,000 =	1.00e+07
%9.0gc	general format, 9 columns wide, with commas	
	sqrt(2) =	1.414214
	1,000 =	1,000
	10,000,000 =	1.00e+07
%9.2f	fixed format, 9 columns wide, 2 decimal places	
	sqrt(2) =	1.41
	1,000 =	1000.00
	10,000,000 =	10000000.00
%9.2fc	fixed format, 9 columns wide, 2 decimal places, with commas	
	sqrt(2) =	1.41
	1,000 =	1,000.00
	10,000,000 =	10,000,000.00
%9.2e	exponential format, 9 columns wide	
	sqrt(2) =	1.41e+00
	1,000 =	1.00e+03
	10,000,000 =	1.00e+07

Stata has three numeric format types: e, f, and g. The formats are denoted by a leading percent sign % followed by the string $w.d$, where w and d stand for two integers. The first integer, w, specifies the width of the format. The second integer, d, specifies the number of digits that are to follow the decimal point. Logic requires that d be less than w. Finally, a character denoting the format type (e, f, or g) and to that may optionally be appended a c indicating commas are to be included in the result. (c is not allowed with e.)

By default, every numeric variable is given a %w.0g format, where w is large enough to display the largest number of the variable's type. The g format is a complicated set of formatting rules that attempts to present values in as readable a fashion as possible without sacrificing precision. The g format changes the number of decimal places displayed whenever it improves the readability of the current value. The number after the decimal point specifies the *minimum* number of digits that are to follow the decimal point.

The default formats for each of the numeric variable types are

byte	%8.0g
int	%8.0g
long	%12.0g
float	%9.0g
double	%10.0g

You change the format of a variable using the format *varname* %*fmt* command.

Under the f format, values are always displayed with the same number of decimal places even if this results in a loss in the displayed precision. Thus, the f format is similar to the C f format. Stata's f format is also similar to the FORTRAN F format, but, unlike the FORTRAN F format, the width of the Stata f format is temporarily increased whenever a number is too large to be displayed in the specified format.

The e format is similar to the C e and the FORTRAN E format. Every value is displayed as a leading digit (with a minus sign, if necessary) followed by a decimal point, the specified number of digits, the letter e, a plus sign or a minus sign, and the power of ten (modified by the preceding sign) that multiplies the displayed value. When the e format is specified, the width must exceed the number of digits that follow the decimal point by at least seven. This space is needed to accommodate the leading sign and digit, the decimal point, the e, and the signed power of ten.

▷ Example

Below we concoct a five-observation dataset with three variables: e_fmt, f_fmt, and g_fmt. All three variables have the same values stored in them; only the display format varies. describe shows the display format to the right of the variable type:

```
. describe
Contains data
  obs:               5
  vars:              3
  size:             80 (99.9% of memory free)
-------------------------------------------------------------------------
   1. e_fmt       float   %9.2e
   2. f_fmt       float   %9.2f
   3. g_fmt       float   %9.0g
-------------------------------------------------------------------------
Sorted by:
```

The formats for each of these variables were set by typing

```
. format e_fmt %9.2e

. format f_fmt %9.2f
```

It was not necessary to set the format for the g_fmt variable, since Stata automatically assigned it the %9.0g format. Nevertheless, we could have typed format g_fmt %9.0g if we wished. Listing the data results in

```
. list
              e_fmt       f_fmt       g_fmt
   1.     2.80e+00        2.80     2.801785
   2.     3.96e+06  3962322.50      3962323
   3.     4.85e+00        4.85     4.852834
   4.    -5.60e-06       -0.00    -5.60e-06
   5.     6.26e+00        6.26     6.264982
```

◁

15.5.2   Date formats

Date formats are really a numeric format because Stata stores dates as the number of days from 01jan1960. See [U] **27 Commands for dealing with dates**.

%d is for displaying elapsed dates. The syntax of the %d format is

first type	%	to indicate the start of the format
then optionally type	–	if you want the result left-aligned
then type	d	
then optionally type	*other characters*	to indicate how the date is to be displayed

The %d format may be specified as simply %d or the %d may be followed by up to 11 characters that specify how the date is to be presented. Allowable characters are

c and C	display the century without/with a leading 0
y and Y	display the two-digit year without/with a leading 0
m and M	display Month, first letter capitalized, in three-letter abbreviation (m) or spelled out (M)
l and L	display month, first letter not capitalized, in three-letter abbreviation (l) or spelled out (L)
n and N	display month number 1–12 without/with a leading 0
d and D	display day-within-month number 1–31 without/with a leading 0
j and J	display day-within-year number 1–366 without/with leading 0s
h	display the half of year number 1 or 2
q	display quarter of year number 1, 2, 3, or 4
w and W	display week-of-year number 1–52 without/with a leading 0
_	display a blank
.	display a period
,	display a comma
:	display a colon
-	display a dash
/	display a slash
´	display a close single quote
!c	display character c (code !! to display an exclamation point)

Specifying %d by itself is equivalent to specifying %dD1CY. The first day of January 1999 is displayed 01jan1999 in this format. For examples of various date formats, see [U] **27 Commands for dealing with dates**.

15.5.3 Time-series formats

Time-series formats—also known as %t formats—are an extension of the date formats coded above. Stata's dates are coded 0 = 01jan1960 and 1 = 02jan1960. Stata also has dates where 0 represents the first week of 1960 (and 1 the second week), 0 represents the first month of 1960 (and 1 the second month), 0 represents the first quarter of 1960 (and 1 the second quarter), 0 represents the first half of 1960 (and 1 the second half), and 1960 represents the year 1960 (and 1961 the next year). %t formats are for displaying these quantities. The %t format is defined

first type	%	to indicate the start of the format
then optionally type	-	if you want the result left-aligned
then type	t	
then type		a character to indicate how the date is encoded:
type	d	if 0 = 01jan1960, same as %d format
or type	w	if 0 = 1960w1
or type	m	if 0 = 1960m1
or type	q	if 0 = 1960q1
or type	h	if 0 = 1960h1
or type	y	if 1960 = 1960 (it records the year itself)
then optionally type	*other characters*	to indicate how the date is to be displayed

where the optional characters are the same as for the %d format given in the table of the previous section, [U] **15.5.2 Date formats**.

In addition to the above is a %tg format—the g stands for generic. The %tg format is provided merely for completeness; it is equivalent to %9.0g. %tg is provided for users who want to put some sort of %t format on a time variable that is encoded differently than Stata understands; whether they do this makes no difference.

The minimal %t formats are %td, %tw, and so on. The default formats for each are

format	default	0 is displayed as	2,000 is displayed as
%td	%tdDlCY	01jan1960	23jun1965
%tw	%twCY!ww	1960w1	1998w25
%tm	%tmCY!mn	1960m1	2126m9
%tq	%tqCY!qq	1960q1	2460q1
%th	%thCY!hh	1960h1	2960h1
%ty	%tyCY	.	2000
%tg	%9.0g	0	2000

There are no mistakes in the table above. For %ty encoded data, the year range is 100–9999, so year 0 displays as missing. More typically, years will be in the range 1900–2100.

For more examples of the %t format, see [U] **27 Commands for dealing with dates**.

15.5.4 String formats

The syntax for a string format is

first type	%	to indicate the start of the format
then optionally type	-	if you want the result left-aligned
then type	a number	indicating the width of the result
then type	s	

For instance, %10s represents a string format of width 10.

For str$w$, the default format is %ws or %9s, whichever is wider. For example, a str10 variable receives a %10s format. Strings are displayed right-justified in the field unless the minus sign is coded; %-10s would display the string left-aligned.

15.6 Dataset, variable, and value labels

Labels are strings used to label things. Stata provides labels for datasets, variables, and values.

15.6.1 Dataset labels

Associated with every dataset is an 80-character *dataset label*. The dataset label is initially set to blanks. You can use the label data "*text*" command to define the dataset label.

▷ Example

You have just entered 1980 state data on marriage rates, divorce rates, and median ages. The describe command will describe the data in memory. The result of typing describe is

```
. describe
Contains data
  obs:           50
  vars:           4
  size:        1,200 (99.8% of memory free)
-------------------------------------------------------------------------------
    1. state     str8    %9s
    2. medage    float   %9.0g
    3. mrgrate   long    %12.0g
    4. dvcrate   long    %12.0g
-------------------------------------------------------------------------------
Sorted by:
     Note:  dataset has changed since last saved
```

describe shows that there are 50 observations on four variables. The four variables are named `state`, `medage`, `mrgrate`, and `dvcrate`. `state` is stored as a `str8`; `medage` is stored as a `float`; and `mrgrate` and `dvcrate` are both stored as `long`s. Each variable's display format (see [U] **15.5 Formats: controlling how data is displayed**) is shown. Finally, the data is not in any particular sort order, and the data has changed since it was last saved on disk.

You can label the data by typing `label data "1980 state data"`. You type this and then type `describe` again:

```
. label data "1980 state data"
. describe
Contains data
  obs:             50                           1980 state data
  vars:             4
  size:         1,200 (99.8% of memory free)
-----------------------------------------------------------------------------
   1. state      str8    %9s
   2. medage     float   %9.0g
   3. mrgrate    long    %12.0g
   4. dvcrate    long    %12.0g
-----------------------------------------------------------------------------
Sorted by:
     Note:  dataset has changed since last saved
```
◁

The dataset label is displayed by the `describe` and `use` commands.

15.6.2   Variable labels

In addition to the name, every variable has associated with it an 80-character *variable label*. The variable labels are initially set to blanks. You use the `label variable` *varname* "*text*" command to define a new variable label.

▷ Example

You have entered data on three variables: `state`, `medage`, `mrgrate`, and `dvcrate`. `describe` portrays the data you entered:

```
. describe
Contains data
  obs:             50                           1980 state data
  vars:             4
  size:         1,200 (99.8% of memory free)
-----------------------------------------------------------------------------
   1. state      str8    %9s
   2. medage     float   %9.0g
   3. mrgrate    long    %12.0g
   4. dvcrate    long    %12.0g
-----------------------------------------------------------------------------
Sorted by:
     Note:  dataset has changed since last saved
```

You can associate more informative labels with the variables by typing

```
. label variable medage "Median Age"
. label variable mrgrate "Marriages per 100,000"
. label variable dvcrate "Divorces per 100,000"
```

From then on, the result of describe will be

```
. describe
Contains data
  obs:            50                        1980 state data
  vars:            4
  size:        1,200 (99.8% of memory free)
-------------------------------------------------------------------------
    1. state      str8   %9s
    2. medage     float  %9.0g             Median Age
    3. mrgrate    long   %12.0g            Marriages per 100,000
    4. dvcrate    long   %12.0g            Divorces per 100,000
-------------------------------------------------------------------------
Sorted by:
    Note:  dataset has changed since last saved
```

◁

Whenever Stata produces output, it will use the variable labels rather than the variable names to label the results if there is room.

15.6.3   Value labels

Value labels define a correspondence or mapping between numeric data and the words used to describe what those numeric values represent. Mappings are named and defined by the label define *lblname # "string" # "string"*... command. The maximum length for the *lblname* is 8 characters. The maximum length of *string* is 80 characters. Named mappings are associated with variables by the label values *varname lblname* command.

▷ Example

The definition makes value labels sound more complicated than they are in practice. You create a dataset on individuals in which you record a person's sex, coding 0 for males and 1 for females. If your data also contained an employee number and salary, it might resemble the following:

```
. describe
Contains data
  obs:             7                        1998 Employee Data
  vars:            3
  size:          112 (99.9% of memory free)
-------------------------------------------------------------------------
    1. empno      float  %9.0g             Employee Number
    2. sex        float  %9.0g             Sex
    3. salary     float  %9.0g             Annual Salary
-------------------------------------------------------------------------
Sorted by:
    Note:  dataset has changed since last save

. list

           empno      sex     salary
    1.     57213        0      24000
    2.     47229        1      27000
    3.     57323        0      24000
    4.     57401        0      24500
    5.     57802        1      27000
    6.     57805        1      24000
    7.     57824        0      22500
```

You could create a mapping called `sexlbl` defining 0 as "Male" and 1 as "Female", and then associate that mapping with the variable `sex`, by typing

```
. label define sexlbl 0 "Male" 1 "Female"
. label values sex sexlbl
```

From then on, your data would appear as

```
. describe
Contains data
  obs:             7                          1998 Employee Data
  vars:            3
  size:          112 (99.9% of memory free)
-------------------------------------------------------------------------
  1. empno        float   %9.0g                 Employee Number
  2. sex          float   %9.0g      sexlbl     Sex
  3. salary       float   %9.0g                 Annual Salary
-------------------------------------------------------------------------
Sorted by:
      Note:  dataset has changed since last save
. list
            empno       sex      salary
  1.        57213      Male       24000
  2.        47229    Female       27000
  3.        57323      Male       24000
  4.        57401      Male       24500
  5.        57802    Female       27000
  6.        57805    Female       24000
  7.        57824      Male       22500
```

Notice not only that the value label is used to produce words when we `list` the data, but also that the association of the variable `sex` with the value label `sexlbl` is shown by the `describe` command.
◁

❑ Technical Note

Value labels and variables may share the same name. For instance, rather than calling the value label `sexlbl` in the example above, we could just as well have named it `sex`. We would then type `label values sex sex` to associate the value label named `sex` with the variable named `sex`.

❑

▷ Example

Stata's `encode` and `decode` commands provide a convenient way to go from string variables to numerically coded variables and back again. Let's pretend that in the example above, rather than coding 0 for males and 1 for females, you created a string variable recording either `"male"` or `"female"`. Your data looks like

```
. describe
Contains data
  obs:             7                          Employee Data
  vars:            3
  size:          126 (99.9% of memory free)
-------------------------------------------------------------------------
  1. empno        float   %9.0g                 Employee Number
  2. sex          str6    %9s                   Sex
  3. salary       float   %9.0g                 Annual Salary
-------------------------------------------------------------------------
Sorted by:
      Note:  dataset has changed since last save
```

```
. list
           empno       sex      salary
      1.    57213      male      24000
      2.    47229    female      27000
      3.    57323      male      24000
      4.    57401      male      24500
      5.    57802    female      27000
      6.    57805    female      24000
      7.    57824      male      22500
```

You now want to create a numerically encoded variable, we will call it **gender**, from the string variable. (You want to do this, say, because you typed `anova salary sex` to perform a one-way ANOVA of salary on sex and you were told that there were "no observations". You then remembered that all of Stata's statistical commands treat string variables as if they contain nothing but missing values. The statistical commands work only with numerically coded data.)

```
. encode sex, generate(gender)
. describe
Contains data
   obs:              7                    Employee Data
   vars:             4
   size:           140 (99.9% of memory free)
----------------------------------------------------------------------
   1. empno        float   %9.0g              Employee Number
   2. sex          str6    %9s                Sex
   3. salary       float   %9.0g              Annual Salary
   4. gender       int     %8.0g      gender  Sex
----------------------------------------------------------------------
Sorted by:
      Note:   dataset has changed since last save
```

encode adds a new **int** variable called **gender** to the data and defines a new value label called **gender**. The value label **gender** maps the numbers 1 to the string **male** and 2 to **female**, so if you were to **list** the data, you could not tell the difference between the **gender** and **sex** variables. But they are different. Stata's statistical commands know how to deal with **gender**. **sex** they do not understand. See [R] **encode**.

❑

❑ Technical Note

Perhaps rather than employee data, your data is on persons undergoing sex-change operations. As such, there would be two sex variables in your data, sex before the operation and sex after the operation. Assume the variables are named **presex** and **postsex**. You can associate the *same* value label to each variable by typing

```
. label define sexlbl 0 "Male" 1 "Female"
. label values presex sexlbl
. label values postsex sexlbl
```

❑

❑ Technical Note

Stata's input commands (**input** and **infile**) have the ability to go from the words in a value label back to the numeric codes. Remember that **encode** and **decode** can translate a string to a numeric mapping and vice versa, so you can map strings to numeric codes either at the time of input or later.

For example:

```
. label define sexlbl 0 "Male" 1 "Female"
. input empno sex:sexlbl salary, label
         empno      sex      salary
  1. 57213 Male 24000
  2. 47229 Female 27000
  3. 57323 0 24000
  4. 57401 Male 24500
  5. 57802 Female 27000
  6. 57805 Female 24000
  7. 57824 Male 22500
  8. end
```

The `label define` command defines the value label `sexlbl`. `input empno sex:sexlbl salary, label` tells Stata to input three variables from the keyboard (`empno`, `sex`, and `salary`), to attach the value label `sexlbl` to the `sex` variable, and to look up any words that are typed in the value label to try to convert them to numbers. To prove it works, we `list` the data that we recently entered:

```
. list
         empno      sex      salary
  1.     57213      Male     24000
  2.     47229    Female     27000
  3.     57323      Male     24000
  4.     57401      Male     24500
  5.     57802    Female     27000
  6.     57805    Female     24000
  7.     57824      Male     22500
```

Compare the information we typed for observation 3 with the result listed by Stata. We typed 57323 0 24000. Thus, the value of `sex` in the third observation is 0. When Stata listed the observation, it indicated the value is `Male`, because we told Stata in our `label define` command that zero is equivalent to `Male`.

Let's now add one more observation to our data:

```
. input, label
         empno      sex      salary
  8. 67223 FEmale 23000
´FEmale´ cannot be read as a number
  8. 67223 Female 23000
  9. end
```

At first we typed 67223 FEmale 23000, and Stata responded with "'FEmale' cannot be read as a number". Remember that Stata always respects case, so `FEmale` is not at all the same thing as `Female`. Stata prompted us to enter the line again, and we did so, this time correctly.

❏

❑ Technical Note

Coupled with the `automatic` option, Stata not only can go from words to numbers, but can create the mapping as well. Let's input this data again, but this time, rather than type it in at the keyboard, let's read it from a file. Assume you have an ASCII file called `employee.raw` stored on your disk. It contains

```
57213 Male 24000
47229 Female 27000
57323 Male 24000
```

```
57401 Male 24500
57802 Female 27000
57805 Female 24000
57824 Male 22500
```

The `infile` command can read this data *and* create the mapping automatically:

```
. label list sexlbl
sexlbl not found
r(111);
. infile empno sex:sexlbl salary using employee, automatic
(7 observations read)
```

Our first command, `label list sexlbl`, is only to prove that we had not previously defined the value label `sexlbl`. Stata `infiled` the data without complaint. We now have

```
. list
        empno      sex     salary
1.      57213     Male      24000
2.      47229   Female      27000
3.      57323     Male      24000
4.      57401     Male      24500
5.      57802   Female      27000
6.      57805   Female      24000
7.      57824     Male      22500
```

Of course, `sex` is just another numeric variable; it does not actually take on the values `Male` and `Female`—it takes on numeric codes which have been automatically mapped to `Male` and `Female`. We can find out what that mapping is by using the `label list` command:

```
. label list sexlbl
sexlbl:
          1 Male
          2 Female
```

We discover that Stata attached the codes 1 to `Male` and 2 to `Female`. Anytime we want to see what our data really looks like, ignoring the value labels, we can use the `nolabel` option:

```
. list, nolabel
        empno      sex     salary
1.      57213        1      24000
2.      47229        2      27000
3.      57323        1      24000
4.      57401        1      24500
5.      57802        2      27000
6.      57805        2      24000
7.      57824        1      22500
```

❑

15.7   Notes attached to data

A dataset may contain notes. These are nothing more than little bits of text you define and review with the `notes` command. Typing `note`, a colon, and the text defines a note:

```
. note:  Send copy to Bob once verified.
```

You can later display whatever notes you have previously defined by typing simply `notes`:

```
. notes
_dta:
  1.  Send copy to Bob once verified.
```

Notes are saved with the data, so once you save your dataset, you can replay this note in the future, too.

You can add more notes:

```
. note: Mary wants a copy, too.
. notes
_dta:
  1.  Send copy to Bob once verified.
  2.  Mary wants a copy, too.
```

The notes we have added so far are attached to the data generically, which is why Stata prefixes them with _dta when it lists them. You can attach notes to variables:

```
. note state: verify values for Nevada.
. note state: what about the two missing values?
. notes
_dta:
  1.  Send copy to Bob once verified.
  2.  Mary wants a copy, too.
state:
  1.  verify values for Nevada.
  2.  what about the two missing values?
```

When you **describe** your data, you can see whether notes are attached to the dataset or any of the variables:

```
. describe
Contains data
   obs:          50                          1980 state data
  vars:           4
  size:       1,200 (99.5% of memory free)   (_dta has notes)
---------------------------------------------------------------------
   1. state        str8    %9s          *
   2. medage       float   %9.0g        Median Age
   3. mrgrate      long    %12.0g       Marriages per 100,000
   4. dvcrate      long    %12.0g       Divorces per 100,000
                                      * indicated variables have notes
---------------------------------------------------------------------
Sorted by:
     Note:  dataset has changed since last save
```

See [R] **notes** for a complete description of this feature.

15.8 Characteristics

Characteristics are an arcane feature of Stata but of great use to Stata programmers. In fact, the notes command described above was implemented using characteristics.

Most users do not care about this detail.

The dataset itself and each variable within the dataset have associated with them a set of characteristics. Characteristics are named and referred to as *varname* [*charname*], where *varname* is the name of a variable or _dta. The characteristics contain text. Characteristics are stored with the data in the Stata-format .dta dataset, so they are recalled whenever the data is loaded.

How are characteristics used? The [R] **xt** commands need to know the name of the variable corresponding to time. These commands allow the variable name to be specified as an option but do not require it. When the user does not specify the variable, the commands somehow manage to remember it from last time even when the last time was a different Stata session. They do this with characteristics. When the user does not specify the variable name, the commands check the characteristic _dta[tis] for the name of the variable. If the time variable's name is stored there, they continue; if not, they issue an error because they need to know it. When the user specifies the option identifying the time variable, these commands store that name in the characteristic _dta[tis] so that they will know it next time. This use of characteristics is hidden from the user—no mention is made of how the commands remember the identity of the time variable.

Occasionally, commands identify their use of characteristics explicitly. The xi command (see [R] **xi**) states that it drops the first level of a categorical variable but that, if you wish to control which level is dropped, you can set the variable's omit characteristic. In the documentation, an example is provided where the user types

    . char agegrp[omit] 3

to set the default omission group to 3. As with the [R] **xt** commands, if the user saves the data after setting the characteristic, the preferred omission group will be remembered from one session to the next.

As a Stata user, you need only understand how to set and clear a characteristic for those few commands that explicitly reveal their use of characteristics. You set a variable *varname*'s characteristic *charname* to be *x* by typing

    . char *varname*[*charname*] *x*

You set the data's characteristic *charname* to be *x* by typing

    . char _dta[*charname*] *x*

You clear a characteristic by typing

    . char *varname*[*charname*]

where *varname* is either a variable name or _dta. You can clear a characteristic even if it has never been set.

The most important feature of characteristics is that Stata remembers them from one session to the next; they are saved with the data.

❏ **Technical Note**

Programmers will want to know more. A technical description is found in [R] **char**, but as an overview, you may reference *varname*'s *charname* characteristic by embedding its name in single quotes and typing ` *varname*[*charname*] ´; see [U] **21.3.11 Referencing characteristics**.

You can fetch the names of all characteristics associated with *varname* by typing

    . local *macname* : char *varname*[ ]

The maximum length of the contents of a characteristic is the same as for macros: 1,000 characters for Small Stata and 18,623 for Intercooled. The association of names with characteristics is by convention. If you, as a programmer, wish to create new characteristics for use in your ado-files, do so, but include at least one capital letter in the characteristic name. The current convention is that all lowercase names are reserved for "official" Stata.

❏

16 Functions and expressions

Contents

If you have not read [U] **14 Language syntax**, please do so before reading this entry.

16.1   Overview

Examples of expressions include

```
2+2
miles/gallons
myv+2/oth
(myv+2)/oth
ln(income)
age<25 & income>50000
age<25 | income>50000
age==25
name=="M Brown"
fname + " " + lname
substr(name,1,10)
_pi
val[_n-1]
L.gnp
```

Expressions like the ones above are allowed anywhere *exp* appears in a syntax diagram. One example is [R] **generate**:

> **generate** *newvar* = *exp* $\big[$if *exp*$\big]$ $\big[$in *range*$\big]$

The first *exp* specifies the contents of the new variable and the optional second expression restricts the subsample over which it is to be defined. Another is [R] **summarize**:

> **summarize** $\big[$*varlist*$\big]$ $\big[$if *exp*$\big]$ $\big[$in *range*$\big]$

The optional expression restricts the sample over which summary statistics are calculated.

Algebraic and string expressions are specified in a natural way using the standard rules of hierarchy. You may use parentheses freely to force a different order of evaluation.

▷ Example

myv+2/oth is interpreted as myv+(2/oth). If you wanted to change the order of the evaluation, you could type (myv+2)/oth.

◁

16.2   Operators

Stata has four different classes of operators: arithmetic, string, relational, and logical. Each of the types is discussed below.

16.2.1   Arithmetic operators

The *arithmetic operators* in Stata are + (addition), − (subtraction), * (multiplication), / (division), ^ (raise to a power), and the prefix − (negation). Any arithmetic operation on a missing value or an impossible arithmetic operation (such as division by zero) yields a missing value.

▷ Example

The expression $-(x+y^{\hat{}}(x-y))/(x*y)$ denotes the formula

$$-\frac{x + y^{x-y}}{x \cdot y}$$

and evaluates to *missing* if x or y is missing or zero.

◁

16.2.2   String operators

The + sign is also used to mean the string operator *concatenate* for the joining of two strings. Stata determines whether + means addition or concatenation by context.

▷ Example

The expression `"this"+"that"` results in the string `"thisthat"` whereas the expression 2+3 results in the number 5. Stata issues the error message "type mismatch" if the arguments on either side of the + sign are not of the same type. Thus, the expression `2+"this"` is an error, as is `2+"3"`.

The expressions on either side of the + can be arbitrarily complex:

$$\texttt{substr(string(20+2),1,1) + upper(substr("rf",1+1,1))}$$

The result of the above expression is the string `"2F"`. See [U] **16.3.5 String functions** below for a description of the `substr()`, `string()`, and `upper()` functions.

◁

16.2.3   Relational operators

The *relational operators* are > (greater than), < (less than), >= (greater than or equal), <= (less than or equal), == (equal), and ~= (not equal). Observe that the relational operator for equality is a pair of equal signs. This convention distinguishes relational equality from the *=exp* assignment phrase.

❏ Technical Note

As a convenience for persons familiar with C, you may use ! anywhere ~ would be appropriate. Thus, the not-equal operator may also be written !=. You will learn shortly that Stata includes a general *not* function '~'. You can also type '!', just as you would in C.

❏

Relational expressions are either *true* or *false*. Relational operators may be used on either numeric or string subexpressions; thus, the expression 3>2 is *true*, as is `"zebra">"cat"`. In the latter case, the relation merely indicates that `"zebra"` comes after the word `"cat"` in the dictionary. All uppercase letters precede all lowercase letters in Stata's book, so `"cat">"Zebra"` is also *true*.

Missing values may appear in relational expressions. If x were a numeric variable, the expression x==. is *true* if x is missing and *false* otherwise. A missing value is greater than any nonmissing value; see [U] **15.2.1 Missing values**.

▷ Example

> You have data on **age** and **income** and wish to list the subset of the data for persons aged 25 years or less. You could type

```
. list if age<=25
```

If you wanted to list the subset of data of persons aged exactly 25, you would type

```
. list if age==25
```

Note the doubled equal sign. It would be an error to type **list if age=25**.

◁

> Although it is convenient to think of relational expressions as evaluating to *true* or *false*, they actually evaluate to numbers. A result of *true* is defined as 1 and *false* as 0.

▷ Example

> The definition of *true* and *false* makes it easy to create indicator, or dummy, variables. For instance,

```
generate incgt10k=income>10000
```

creates a variable that takes on the value 0 when **income** is less than or equal to $10,000, and 1 when **income** is greater than $10,000. Since missing values are greater than all nonmissing values, the new variable **incgt10k** will also take on the value 1 when **income** is *missing*. It would be safer to type

```
generate incgt10k=income>10000 if income~=.
```

Now observations in which **income** is *missing* will also contain *missing* in **incgt10k**. Also see [U] **28 Commands for dealing with categorical variables** for more examples.

◁

❑ Technical Note

> Although you will rarely wish to do so, since arithmetic and relational operators both evaluate to numbers, there is no reason you cannot mix the two types of operators in a single expression. For instance, (2==2)+1 evaluates to 2, since 2==2 evaluates to 1, and 1 + 1 is 2.

> Relational operators are evaluated after all arithmetic operations. Thus, the expression (3>2)+1 is equal to 2, whereas 3>2+1 is equal to 0. Evaluating relational operators last guarantees the *logical* (as opposed to the *numeric*) interpretation. It should make sense that 3>2+1 is *false*.

❑

16.2.4 Logical operators

> The *logical operators* are **&** (and), **|** (or), and **~** (not). The logical operators interpret any nonzero value (including *missing*) as *true* and zero as *false*.

▷ Example

If you have data on **age** and **income** and wish to **list** data for persons making more than $50,000 along with persons under the age of 25 making more than $30,000, you could type

```
list if income>50000 | income>30000 & age<25
```

The **&** takes precedence over the **|**. If you were unsure, however, you could have typed

```
list if income>50000 | (income>30000 & age<25)
```

In either case, the statement will also **list** all observations for which **income** is *missing*, since *missing* is greater than 50,000.

◁

□ Technical Note

Like relational operators, logical operators return the value 1 for *true* and 0 for *false*. For example, the expression **5 & .** evaluates to 1. Logical operations, except for ~, are performed after all arithmetic and relational operations; the expression **3>2 & 5>4** is interpreted as **(3>2)&(5>4)** and evaluates to 1.

□

16.2.5   Order of evaluation, all operators

The order of evaluation (from first to last) of all operators is ~, ^, − (negation), /, *, − (subtraction), +, ~=, >, <, <=, >=, ==, &, |.

16.3   Functions

Functions may appear in expressions. Functions are indicated by the function name, an open parenthesis, an expression or expressions separated by commas, and a close parenthesis. For example, the square root of a variable named **x** is specified by **sqrt(x)**. The logarithm of the square root of **x** is **ln(sqrt(x))**. All numeric functions return *missing* when given missing values as arguments or when the result is undefined.

16.3.1   Mathematical functions

abs(x)
Domain: −8e+307 to 8e+307
Range: 0 to 8e+307
Description: absolute value $|x|$

acos(x)
Domain: −1 to 1
Range: 0 to π
Description: arc-cosine returning radians

asin(x)
Domain: −1 to 1
Range: −$\pi/2$ to $\pi/2$
Description: arc-sine returning radians

atan(x)
  Domain: −8e+307 to 8e+307
  Range: $-\pi/2$ to $\pi/2$
  Description: arc-tangent returning radians

comb(n,k)
  Domain n: integers 1 to 1e+305
  Domain k: integers 0 to n
  Range: 0 to 8e+307; missing
  Description: combinatorial function $n!/(k!(n-k)!)$

cos(x)
  Domain: −8e+307 to 8e+307
  Range: -1 to 1
  Description: cosine of radians

digamma(x)
  Domain: −1e+7 to 8e+307
  Range: −8e+307 to 8e+307; missing
  Description: $d\ln\Gamma(x)/dx$

exp(x)
  Domain: −8e+307 to 709
  Range: 0 to 8e+307
  Description: exponential e^x

ln(x)
  Domain: 1e–323 to 8e+307
  Range: −744 to 709
  Description: natural logarithm $\ln(x)$

lnfact(x)
  Domain: integers 0 to 1e+305
  Range: 0 to 8e+307
  Description: natural log of factorial; $\ln(x!)$

If you wish to calculate $x!$, use round(exp(lnfact(x)),1) to ensure the result is an integer. Logs of factorials are generally more useful than factorials themselves because of overflow problems.

lngamma(x)
  Domain: −2,147,483,648 to 1e+305 (excluding negative integers)
  Range: −8e+307 to 8e+307
  Description: $\ln[\Gamma(x)]$

lngamma(x) for $x < 0$ returns a number such that exp(lngamma(x)) is equal to the absolute value of the correct answer—that is, lngamma(x) always returns a real (as opposed to complex) result.

log(x)
  Domain: 1e–323 to 8e+307
  Range: −744 to 709
  Description: natural logarithm $\ln(x)$

log10(x)
  Domain: 1e–323 to 8e+307
  Range: −323 to 308
  Description: logarithm base 10

`mod(x,y)`
 Domain x: –8e+307 to 8e+307
 Domain y: 0 to 8e+307
 Range: 0 to 8e+307
 Description: modulus of x with respect to y

`sin(x)`
 Domain: –8e+307 to 8e+307
 Range: −1 to 1
 Description: sine of radians

`sqrt(x)`
 Domain: 0 to 8e+307
 Range: 0 to 1e+154
 Description: square root

`tan(x)`
 Domain: –8e+307 to 8e+307
 Range: −1e+17 to 1e+17; missing
 Description: tangent of radians

`trigamma(x)`
 Domain: −1e+7 to 8e+307
 Range: 0 to 8e+307; missing
 Description: $d^2 \ln\Gamma(x)/dx^2$

❏ Technical Note

The trigonometric functions are defined in terms of *radians*. There are 2π radians in a circle. If you prefer thinking in terms of *degrees*, since there are also 360 degrees in a circle, you may convert degrees into radians using the formula $r = d\pi/180$ where d represents degrees and r represents radians. Stata includes the built-in constant `_pi`, equal to π to machine precision. Thus, to calculate the sine of `theta`, where `theta` is measured in degrees, you could type

 `sin(theta*_pi/180)`

`atan()` similarly returns radians, not degrees. The arc-cotangent can be obtained as

 `_pi/2 - atan(x)`

❏

16.3.2 Statistical functions

`Binomial(n,k,π)`
 Domain n: 0 to 8e+307
 Domain k: 0 to n
 Domain π: 0 to 1
 Range: 0 to 1
 Description: probability of observing $\lfloor k \rfloor$ or more successes in $\lfloor n \rfloor$ trials
      when the probability of a success on a single trial is π

`binorm(h,k,ρ)`
 Domain h: –8e+307 to 8e+307
 Domain k: –8e+307 to 8e+307
 Domain ρ: −1 to 1

Range: 0 to 1

Description: joint cumulative distribution $\Phi(h, k, \rho)$ of bivariate normal with correlation ρ; cumulative over $(-\infty, h] \times (-\infty, k]$:

$$\Phi(h, k, p) = \frac{1}{2\pi\sqrt{1 - \rho^2}} \int_{-\infty}^{h} \int_{-\infty}^{k} \exp\left[-\frac{1}{2(1 - \rho^2)}\left(x_1^2 - 2\rho x_1 x_2 + x_2^2\right)\right] dx_1\, dx_2$$

chiprob(df, x)

Domain df: 1e–323 to 8e+307 (may be nonintegral)

Domain x: 0 to 8e+307

Range: 0 to 1

Description: upper-tail cumulative χ^2 with df degrees of freedom

fprob(df_1, df_2, F)

Domain df_1:   1e–323 to 8e+307 (may be nonintegral)

Domain df_2:   1e–323 to 8e+307 (may be nonintegral)

Domain F: 0 to 8e+307

Range: 0 to 1

Description: upper-tail cumulative F distribution with df_1 numerator and df_2 denominator degrees of freedom

gammap(a, x)

Domain a: 1e–323 to 8e+307

Domain x: 0 to 8e+307

Range: 0 to 1

Description: incomplete gamma function $P(a, x) = (1/\Gamma(a)) \int_0^x e^{-t} t^{a-1}\, dt$

Note that the cumulative Poisson (the probability of observing k or fewer events if the expected is x) can be evaluated as 1-gammap$(k+1, x)$. The reverse cumulative (the probability of observing k or more events) can be evaluated as gammap(k, x). See Press et al. (1986, 160–165) for a more complete description and suggested uses for this function.

ibeta(a, b, x)

Domain a: 1e–323 to 8e+307

Domain b: 1e–323 to 8e+307

Domain x: 0 to 1

Range: 0 to 1

Description: incomplete beta function $I_x(a, b) = (\Gamma(a + b)/\Gamma(a)\Gamma(b)) \int_0^x t^{a-1}(1 - t)^{b-1}\, dt$

Although Stata has a cumulative binomial function (see Binomial() above), the probability that an event occurs k or fewer times in n trials, when the probability of a single event is p can be cond$(k==n, 1, 1$-ibeta$(k+1, n-k, p))$. The reverse cumulative binomial (the probability that an event occurs k or more times) can be evaluated as cond$(k==0, 1,$ibeta$(k, n-k+1, p))$. See Press et al. (1986, 166–169) for a more complete description and suggested uses for this function.

invbinomial(n, k, p)

Domain n: 1 to 8e+307

Domain k: 1 to n

Domain p: 0 to 1 (exclusive)

Range: 0 to 1

Description: inverse binomial, but with a twist:

For $p \leq 0.5$, returns π, π = probability of success on a single trial, such that the probability of observing $\lfloor k \rfloor$ or more successes in $\lfloor n \rfloor$ trials is p; for $p > 0.5$, returns π such that the probability of observing $\lfloor k \rfloor$ or fewer successes in $\lfloor n \rfloor$ trials is $1 - p$.

`invchi(`*df*`,`*p*`)`
   Domain *df*:   1e–323 to 8e+307 (may be nonintegral)
   Domain *p*:   0 to 1
   Range: 0 to 8e+307
   Description: inverse of `chiprob()`;
               if `chiprob(`*df*`,`*x*`)` $= p$, then `invchi(`*df*`,`*p*`)` $= x$

`invfprob(`*df*$_1$`,`*df*$_2$`,`*p*`)`
   Domain *df*$_1$: 1e–323 to 8e+307 (may be nonintegral)
   Domain *df*$_2$: 1e–323 to 8e+307 (may be nonintegral)
   Domain *p*:   0 to 1
   Range: 0 to 8e+307
   Description: inverse upper-tail cumulative F distribution;
               if `fprob(`*df*$_1$`,`*df*$_2$`,`*F*`)` $= p$, then `invfprob(`*df*$_1$`,`*df*$_2$`,`*p*`)` $= F$

`invgammap(`*a*`,`*p*`)`
   Domain *a*:   1e–323 to 8e+307
   Domain *p*:   0 to 1
   Range: 0 to 8e+307
   Description: inverse incomplete gamma function;
               if `gammap(`*a*`,`*x*`)` $= p$, then `invgammap(`*a*`,`*p*`)` $= x$

`invnchi(`*df*`,`λ`,`*p*`)`
   Domain *df*:   integers 1 to 200
   Domain λ: 0 to 1,000
   Domain *p*:   0 to 1
   Range: 0 to 8e+307
   Description: inverse cumulative noncentral χ^2 distribution;
               if `nchi(`*df*`,`λ`,`*x*`)` $= p$, then `invnchi(`*df*`,`λ`,`*p*`)` $= x$;
               df must be an integer

`invnorm(`*p*`)`
   Domain *p*:   1e–323 to $1 - 2^{-53}$
   Range: -39 to 8.2095362
   Description: inverse cumulative normal;
               if `normprob(`*z*`)` $= p$, then `invnorm(`*p*`)` $= z$

`invt(`*df*`,`*p*`)`
   Domain *df*:   1e–323 to 1e+12 (may be nonintegral)
   Domain *p*:   0 to $1 - 2^{-52}$
   Range: 0 to 1e+10
   Description: inverse two-tailed cumulative t distribution;
               if `tprob(`*df*`,`*t*`)` $= p$, then `invt(`*df*`,`$1 - p$`)` $= t$

`nchi(`*df*`,`λ`,`*x*`)`
   Domain *df*:   integers 1 to 200
   Domain λ: 0 to 1,000
   Domain *x*:   0 to 8e+307
   Range: 0 to 1
   Description: cumulative noncentral χ^2 distribution defined by

$$\int_0^x \frac{e^{-t/2}\,e^{-\lambda/2}}{2^{\nu/2}} \sum_{j=0}^{\infty} \frac{t^{\nu/2+j-1}\,\lambda^j}{\Gamma(\nu/2+j)\,2^{2j}\,j!}\, dt$$

where ν denotes the degrees of freedom, λ is the noncentrality parameter, and x is the value of

χ^2. Note that $\mathtt{nchi}(\nu,0,x) = 1 - \mathtt{chi}(\nu,x)$, but $\mathtt{chi}()$ is the preferred function to use for the central χ^2 distribution. $\mathtt{nchi}()$ is computed using the algorithm of Haynam et al. (1970).

normd(z)
  Domain z: −8e+307 to 8e+307
  Range: 0 to .39894...
  Description: standard $N(0, 1)$ density

normd(z,σ)
  Domain z: −8e+307 to 8e+307
  Domain σ: 1e−308 to 8e+307
  Range: 0 to 8e+307
  Description: standard $N(0, \sigma^2)$ density;
            $\mathtt{normd}(z,1) = \mathtt{normd}(z)$ and $\mathtt{normd}(z,\sigma) = \mathtt{normd}(z)/\sigma$

normprob(z)
  Domain z: −8e+307 to 8e+307
  Range: 0 to 1
  Description: cumulative standard normal

npnchi(df,x,p)
  Domain df: integers 1 to 200
  Domain x: 0 to 8e+307
  Domain p: 1e−138 to $1 - 2^{-52}$
  Range: 0 to 1,000
  Description: noncentrality parameter λ for noncentral χ^2;
            if $\mathtt{nchi}(df,\lambda,x) = p$, then $\mathtt{npnchi}(df,x,p) = \lambda$

tprob(df,t)
  Domain df: 1e−323 to 8e+307
  Domain t: 0 to 8e+307
  Range: 0 to 1
  Description: two-tailed cumulative Student's t distribution; returns probability $|T| > |t|$

uniform()
  Range: 0 to nearly 1 (0 to $1 - 2^{-32}$)
  Description: uniform pseudo-random number function

$\mathtt{uniform}()$ returns uniformly distributed pseudo-random numbers on the interval $[0, 1)$. $\mathtt{uniform}()$ takes no arguments but the parentheses must be typed. $\mathtt{uniform}()$ can be seeded with the **set seed** command; see the technical note at the end of this subsection.

To generate pseudo-random numbers over the interval $[a, b)$, use $a\mathtt{+}(b\mathtt{-}a)\mathtt{*uniform}()$.

To generate pseudo-random integers over $[a, b]$, use $a\mathtt{+int}((b\mathtt{-}a\mathtt{+1})\mathtt{*uniform}())$.

To generate normally distributed random numbers with mean 0 and standard deviation 1, use $\mathtt{invnorm}(\mathtt{uniform}())$.

To generate normally distributed random numbers with mean μ and standard deviation σ, use $\mu\mathtt{+}\sigma\mathtt{*invnorm}(\mathtt{uniform}())$.

❑ Technical Note

The uniform pseudo-random number function $\mathtt{uniform}()$ is based on George Marsaglia's (1994) 32-bit pseudo-random number generator KISS (Keep It Simple Stupid). The KISS generator is composed of two 32-bit pseudo-random number generators and two 16-bit generators (combined to make one 32-bit generator). The four generators are defined by the recursions

$$x_n = 69069\, x_{n-1} + 1234567 \quad \mathrm{mod}\ 2^{32} \tag{1}$$
$$y_n = y_{n-1}(I + L^{13})(I + R^{17})(I + L^5) \tag{2}$$
$$z_n = 65184\big(z_{n-1}\ \mathrm{mod}\ 2^{16}\big) + \mathrm{int}\big(z_{n-1}/2^{16}\big) \tag{3}$$
$$w_n = 63663\big(w_{n-1}\ \mathrm{mod}\ 2^{16}\big) + \mathrm{int}\big(w_{n-1}/2^{16}\big) \tag{4}$$

In recursion (2), the 32-bit word y_n is viewed as a 1×32 binary vector; L is the 32×32 matrix that produces a left shift of one (L has 1s on the first left subdiagonal, 0s elsewhere); and R is L transpose, effecting a right shift by one. In recursions (3) and (4), $\mathrm{int}(x)$ is the integer part of x.

The KISS generator produces the 32-bit random number

$$R_n = x_n + y_n + z_n + 2^{16} w_n \quad \mathrm{mod}\ 2^{32}$$

`uniform()` takes the output from the KISS generator and divides it by 2^{32} to produce a real number on the interval $[0, 1)$.

The recursions (1)–(4) have, respectively, the periods

$$2^{32} \tag{1}$$
$$2^{32} - 1 \tag{2}$$
$$(65184 \cdot 2^{16} - 2)/2 \approx 2^{31} \tag{3}$$
$$(63663 \cdot 2^{16} - 2)/2 \approx 2^{31} \tag{4}$$

Thus, the overall period for the KISS generator is

$$2^{32} \cdot (2^{32} - 1) \cdot (65184 \cdot 2^{15} - 1) \cdot (63663 \cdot 2^{15} - 1) \approx 2^{126}$$

When Stata first comes up, it initializes the four recursions in KISS using the seeds

$$x_0 = 123456789 \tag{1}$$
$$y_0 = 521288629 \tag{2}$$
$$z_0 = 362436069 \tag{3}$$
$$w_0 = 2262615 \tag{4}$$

Successive calls to `uniform()` will then produce the sequence

$$\frac{R_1}{2^{32}},\ \frac{R_2}{2^{32}},\ \frac{R_3}{2^{32}},\ \dots$$

Hence, `uniform()` gives the same sequence of random numbers in every Stata session (measured from the start of the session) unless you reinitialize the seed. You can reinitialize the seed by issuing the command

        . set seed #

where # is any integer between 0 and $2^{31} - 1$ inclusive. When this command is issued the value of x_0 is set equal to #, and the other three recursions are restarted at the seeds y_0, z_0, and w_0 given above. The first 100 random numbers are discarded, and successive calls to `uniform()` give the sequence

$$\frac{R'_{101}}{2^{32}},\ \frac{R'_{102}}{2^{32}},\ \frac{R'_{103}}{2^{32}},\ \dots$$

However, if the command

        . set seed 123456789

is given, the first 100 random numbers are not discarded, and you get exactly the same sequence of random numbers that `uniform()` produces by default. See also [R] **generate**.

❑

❑ Technical Note

The formula used by `uniform()` to produce pseudo-random numbers changed between releases 3.1 and 4.0. If you `set version` to 3.1 or earlier (see [R] **version**), `set seed` and `uniform()` refer to the prior generator and so previously produced results are still reproducible. When `version` is set to 4.0 or higher, `set seed` and `uniform()` refer to the new function, but you can access the prior function using `set seed0` and `uniform0()`. The only reason to use the prior function, however, is reproduction of previous results; the new function is better.

❑

16.3.3   Date functions

Stata includes the following *date functions*. These functions are described, with examples, in [U] **27 Commands for dealing with dates**. What follows is a technical description. s is used to indicate a string subexpression—a string literal, a string variable, or another string expression—and e, m, d, and y are used to indicate numeric subexpressions—numbers, numeric variables, or other numeric expressions. The date functions interpret e, m, d, and y as $\text{int}(e)$, $\text{int}(m)$, $\text{int}(d)$, and $\text{int}(y)$, respectively.

An elapsed date is the number of days from 1jan1960; negative numbers indicate dates prior to 1jan1960. Allowable dates are between 1jan100 and 31dec9999, inclusive, but all functions are based on the Gregorian calendar and values do not correspond to historical dates prior to Friday, 15oct1582.

`date(`s_1`,`s_2`[,`y`])`
Domain s_1:   strings
Domain s_2:   strings
Domain y: integers 1000 to 9998 (but probably 2001 to 2099)
Range: dates 01jan0100 to 31dec9999 (integers $-679{,}350$ to $2{,}936{,}549$); missing
Description: returns the elapsed date corresponding to s_1 based on s_2 and y

s_1 contains the date, recorded as a string, in virtually any format. Months can be spelled out, abbreviated (to three characters), or indicated as numbers; years can include or exclude the century; blanks and punctuation are allowed.

s_2 is any permutation of m, d, and [##]y with their order defining the order that month, day, and year occur in s_1. ##, if specified, indicates the default century for 2-digit years in s_1. For instance, $s_2 = $ `"md19y"` would translate $s_1 = $ `"11/15/91"` as 15nov1991.

y provides an alternate way of handling two-digit years. y specifies the largest year that is to be returned when a two-digit year is encountered. For instance,

$$\text{date}(\text{"1/15/99"},\text{"mdy"},2050) = \text{15jan1999}$$
$$\text{date}(\text{"1/15/98"},\text{"mdy"},2050) = \text{15jan1998}$$

$$\vdots$$

$$\text{date}(\text{"1/15/51"},\text{"mdy"},2050) = \text{15jan1951}$$
$$\text{date}(\text{"1/15/50"},\text{"mdy"},2050) = \text{15jan2050}$$
$$\text{date}(\text{"1/15/49"},\text{"mdy"},2050) = \text{15jan2049}$$

$$\vdots$$

$$\text{date}(\text{"1/15/01"},\text{"mdy"},2050) = \text{15jan2001}$$
$$\text{date}(\text{"1/15/00"},\text{"mdy"},2050) = \text{15jan2000}$$

If neither ## nor y is specified, then date() returns missing value when it encounters a two-digit year.

d(*l*)
Domain l:	date literals 01jan0100 to 31dec9999
Range:	dates 01jan0100 to 31dec9999 (integers $-679{,}350$ to 2,936,549)
Description:	convenience function to make typing dates in expressions easier; e.g., typing d(2jan1960) is equivalent to typing 1

mdy(*m*,*d*,*y*)
Domain m:	integers 1 to 12
Domain d:	integers 1 to 31
Domain y:	integers 100 to 9999 (but probably 1800 to 2100)
Range:	dates 01jan0100 to 31dec9999 (integers $-679{,}350$ to 2,936,549); missing
Description:	returns the elapsed date corresponding to m, d, y

day(*e*)
Domain e:	dates 01jan0100 to 31dec9999 (integers $-679{,}350$ to 2,936,549)
Range:	integers 1 to 31; missing
Description:	returns the numeric day of the month corresponding to e

month(*e*)
Domain e:	dates 01jan0100 to 31dec9999 (integers $-679{,}350$ to 2,936,549)
Range:	integers 1 to 12; missing
Description:	returns the numeric month corresponding to e

year(*e*)
Domain e:	dates 01jan0100 to 31dec9999 (integers $-679{,}350$ to 2,936,549)
Range:	integers 100 to 9999 (but probably 1800 to 2100)
Description:	returns the numeric year corresponding to e

dow(*e*)
Domain e:	dates 01jan0100 to 31dec9999 (integers $-679{,}350$ to 2,936,549)
Range:	integers 0 to 6; missing
Description:	returns the numeric day of the week corresponding to e; 0 = Sunday, 1 = Monday, ..., 6 = Saturday

doy(*e*)
Domain e:	dates 01jan0100 to 31dec9999 (integers $-679{,}350$ to 2,936,549)
Range:	integers 1 to 366; missing
Description:	returns the numeric day of the year corresponding to e

week(*e*)
Domain e:	dates 01jan0100 to 31dec9999 (integers $-679{,}350$ to 2,936,549)
Range:	integers 1 to 52; missing
Description:	returns the numeric week of the year corresponding to e (the first week of a year is the first seven days of the year)

quarter(*e*)
Domain e:	dates 01jan0100 to 31dec9999 (integers $-679{,}350$ to 2,936,549)
Range:	integers 1 to 4; missing
Description:	returns the numeric quarter of the year corresponding to e

halfyear(*e*)
Domain e:	dates 01jan0100 to 31dec9999 (integers $-679{,}350$ to 2,936,549)
Range:	integers 1, 2; missing
Description:	returns the numeric half of the year corresponding to e

16.3.4 Time-series functions

In addition to elapsed dates for which $0 = 01\text{jan}1960$, Stata provides five other encodings for dates:

Description	Format	Coding
daily	%td	0 = 01jan1960 (same as elapsed date)
weekly	%tw	0 = 1960w1
monthly	%tm	0 = 1960m1
quarterly	%tq	0 = 1960q1
half-yearly	%th	0 = 1960h1
yearly	%ty	1960 = 1960 (records the year itself)

The following functions are for use with %t dates:

daily$(s_1,s_2[,y])$
   Domain s_1:   strings
   Domain s_2:   strings
   Domain y: integers 1000 to 9998 (but probably 2001 to 2099)
   Range: %td dates 01jan0100 to 31dec9999 (integers $-679,350$ to $2,936,549$); missing
   Description: returns the elapsed date corresponding to s_1;
                same as date() function; see [U] **16.3.3 Date functions** above

d(l)
   Domain l: date literals 01jan0100 to 31dec9999
   Range: %td dates 01jan0100 to 31dec9999 (integers $-679,350$ to $2,936,549$)
   Description: convenience function to make typing dates in expressions easier;
                e.g., typing d(2jan1960) is equivalent to typing 1

weekly$(s_1,s_2[,y])$
   Domain s_1:   strings
   Domain s_2:   strings "wy" and "yw"; y may be prefixed with ##
   Domain y: integers 1000 to 9998 (but probably 2001 to 2099)
   Range: %tw dates 0100w1 to 9999w52 (integers $-96,720$ to $418,079$); missing
   Description: returns the %tw date corresponding to s_1;
                y specifies top year; see date() in [U] **16.3.3 Date functions** above

w(l)
   Domain l: date literals 0100w1 to 9999w52
   Range: %tw dates 0100w1 to 9999w52 (integers $-96,720$ to $418,079$)
   Description: convenience function to make typing dates in expressions easier;
                e.g., typing w(1960w2) is equivalent to typing 1

monthly$(s_1,s_2[,y)])$
   Domain s_1:   strings
   Domain s_2:   strings "my" and "ym"; y may be prefixed with ##
   Domain y: integers 1000 to 9998 (but probably 2001 to 2099)
   Range: %tm dates 0100m1 to 9999m12 (integers $-22,320$ to $96,479$); missing
   Description: returns the %tm date corresponding to s_1;
                y specifies top year; see date() in [U] **16.3.3 Date functions** above

m(l)
   Domain l: date literals 0100m1 to 9999m12
   Range: %tm dates 0100m1 to 9999m12 (integers $-22,320$ to $96,479$)
   Description: convenience function to make typing dates in expressions easier;
                e.g., typing m(1960m2) is equivalent to typing 1

quarterly($s_1,s_2\big[,y\big]$)
Domain s_1:　　strings
Domain s_2:　　strings "qy" and "yq"; y may be prefixed with ##
Domain y:　　integers 1000 to 9998 (but probably 2001 to 2099)
Range:　　　　%tq dates 0100q1 to 9999q4 (integers $-7{,}440$ to 32,156); missing
Description:　returns the %tq date corresponding to s_1;
　　　　　　　y specifies top year; see date() in [U] **16.3.3 Date functions** above

q(l)
Domain l:　　date literals 0100q1 to 9999q4
Range:　　　　%tq dates 0100q1 to 9999q4 (integers $-7{,}440$ to 32,156)
Description:　convenience function to make typing dates in expressions easier;
　　　　　　　e.g., typing q(1960q2) is equivalent to typing 1

halfyearly($s_1,s_2\big[,y\big]$)
Domain s_1:　　strings
Domain s_2:　　strings "hy" and "yh"; y may be prefixed with ##
Domain y:　　integers 1000 to 9998 (but probably 2001 to 2099)
Range:　　　　%th dates 0100h1 to 9999h2 (integers $-3{,}720$ to 16,079); missing
Description:　returns the %th date corresponding to s_1;
　　　　　　　y specifies top year; see date() in [U] **16.3.3 Date functions** above

h(l)
Domain l:　　date literals 0100h1 to 9999h2
Range:　　　　%th dates 0100h1 to 9999h2 (integers $-3{,}720$ to 16,079)
Description:　convenience function to make typing dates in expressions easier;
　　　　　　　e.g., typing h(1960h2) is equivalent to typing 1

yearly($s_1,s_2\big[,y\big]$)
Domain s_1:　　strings
Domain s_2:　　string "y"; y may be prefixed with ##
Domain y:　　integers 1000 to 9998 (but probably 2001 to 2099)
Range:　　　　%ty dates 0100 to 9999 (integers 100 to 9999); missing
Description:　returns the %ty date corresponding to s_1;
　　　　　　　y specifies top year; see date() in [U] **16.3.3 Date functions** above

y(l)
Domain l:　　date literals 0100 to 9999
Range:　　　　%ty dates 0100 to 9999 (integers 100 to 9999)
Description:　convenience function to make typing dates in expressions easier;
　　　　　　　e.g., typing y(1961) is equivalent to typing 1961
　　　　　　　Note that y(61) would produce an error message; programmers find this useful.

mdy(m,d,y)
Domain m:　　integers 1 to 12
Domain d:　　integers 1 to 31
Domain y:　　integers 100 to 9999 (but probably 1800 to 2100)
Range:　　　　%td dates 01jan0100 to 31dec9999 (integers $-679{,}350$ to 2,936,549); missing
Description:　returns the elapsed date corresponding to m, d, y

yw(y,w)
Domain y:　　integers 100 to 9999 (but probably 1800 to 2100)
Domain w:　　integers 1 to 52
Range:　　　　%tw dates 0100w1 to 9999w52 (integers $-96{,}720$ to 418,079)
Description:　returns the %tw date corresponding to year y, week w

ym(y,m)
    Domain y: integers 100 to 9999 (but probably 1800 to 2100)
    Domain m: integers 1 to 12
    Range: %tm dates 0100m1 to 9999m12 (integers $-22{,}320$ to 96,479)
    Description: returns the %tm date corresponding to year y, month m

yq(y,q)
    Domain y: integers 100 to 9999 (but probably 1800 to 2100)
    Domain q: integers 1 to 4
    Range: %tq dates 0100q1 to 9999q4 (integers $-7{,}440$ to 32,156)
    Description: returns the %tq date corresponding to year y, quarter q

yh(y,h)
    Domain y: integers 100 to 9999 (but probably 1800 to 2100)
    Domain h: integers 1, 2
    Range: %th dates 0100h1 to 9999h2 (integers $-3{,}720$ to 16,079)
    Description: returns the %th date corresponding to year y, half h

dofd(e)
    Domain e: %td dates 01jan0100 to 31dec9999 (integers $-679{,}350$ to 2,936,549)
    Range: %td dates 01jan0100 to 31dec9999 (integers $-679{,}350$ to 2,936,549)
    Description: returns the %td date of e
                    (yes, this is the identity function)

dofw(e_w)
    Domain e_w: %tw dates 0100w1 to 9999w52 (integers $-96{,}720$ to 418,079)
    Range: %td dates 01jan0100 to 24dec9999 (integers $-679{,}350$ to 2,936,542)
    Description: returns the %td date of the start of e_w

wofd(e_d)
    Domain e_d: %td dates 01jan0100 to 31dec9999 (integers $-679{,}350$ to 2,936,549)
    Range: %tw dates 0100w1 to 9999w52 (integers $-96{,}720$ to 418,079)
    Description: returns the %tw date containing e_d

dofm(e_m)
    Domain e_m: %tm dates 0100m1 to 9999m12 (integers $-22{,}320$ to 96,479)
    Range: %td dates 01jan0100 to 01dec9999 (integers $-679{,}350$ to 2,936,519)
    Description: returns the %td date of the start of e_m

mofd(e_d)
    Domain e_d: %td dates 01jan0100 to 31dec9999 (integers $-679{,}350$ to 2,936,549)
    Range: %tm dates 0100m1 to 9999m12 (integers $-22{,}320$ to 96,479)
    Description: returns the %tm date containing e_d

dofq(e_q)
    Domain e_q: %tq dates 0100q1 to 9999q4 (integers $-7{,}440$ to 32,156)
    Range: %td dates 01jan0100 to 01oct9999 (integers $-679{,}350$ to 2,936,458)
    Description: returns the %td date of the start of e_q

qofd(e_d)
    Domain e_d: %td dates 01jan0100 to 31dec9999 (integers $-679{,}350$ to 2,936,549)
    Range: %tq dates 0100q1 to 9999q4 (integers $-7{,}440$ to 32,156)
    Description: returns the %tq date containing e_d

dofh(e_h)
   Domain e_h: %th dates 0100h1 to 9999h2 (integers −3,720 to 16,079)
   Range: %td dates 01jan0100 to 01jul9999 (integers −679,350 to 2,936,366)
   Description: returns the %td date of the start of e_h

hofd(e_d)
   Domain e_d: %td dates 01jan0100 to 31dec9999 (integers −679,350 to 2,936,549)
   Range: %th dates 0100h1 to 9999h2 (integers −3,720 to 16,079)
   Description: returns the %th date containing e_d

dofy(e_y)
   Domain e_y: %ty dates 0100 to 9999 (integers 100 to 9999)
   Range: %td dates 01jan0100 to 01jan9999 (integers −679,350 to 2,936,185)
   Description: returns the %td date of the start of e_y

yofd(e_d)
   Domain e_d: %td dates 01jan0100 to 31dec9999 (integers −679,350 to 2,936,549)
   Range: %ty dates 0100 to 9999 (integers 100 to 9999)
   Description: returns the %ty date (year) containing e_d

tin(d_1,d_2)
   Domain d_1:   date literals recorded in units of t previously tsset;
   Domain d_2:   date literals recorded in units of t previously tsset
   Range: 0 and 1, 1 ⇒ *true*
   Description: *true* if $d_1 \leq t \leq d_2$ where t is the time variable previously tsset

You must have previously tsset the data to use tin(); see [R] **tsset**. When you tsset the data, you specified a time variable t and the format on t states how it is recorded. You type d_1 and d_2 according to that format.

If t has a %td or %d format, you could type tin(5jan1992, 14apr2002).

If t has a %tw format, you could type tin(1985w1, 2002w15).

If t has a %tm format, you could type tin(1985m1, 2002m4).

If t has a %tq format, you could type tin(1985q1, 2002q2).

If t has a %th format, you could type tin(1985h1, 2002h1).

If t has a %ty format, you could type tin(1985, 2002).

Otherwise, t is just a set of integers and you could type tin(12, 38).

Note that the details of the %t format do not matter. If your t is formatted %tdn/d/y so that 5jan1992 displays as 1/5/92, you would still type the date in day–month–year order: tin(5jan1992, 14apr2002).

twithin(d_1,d_2)
   Domain d_1:   date literals recorded in units of t previously tsset
   Domain d_2:   date literals recorded in units of t previously tsset
   Range: 0 and 1, 1 ⇒ *true*
   Description: *true* if $d_1 < t < d_2$ where t is the time variable previously tsset
                see tin() function above; twithin() is similar except the range is exclusive

16.3.5   String functions

Stata includes the following *string functions*. In the display below, s is used to indicate a string subexpression—a string literal, a string variable, or another string expression—and n is used to indicate a numeric subexpression—a number, a numeric variable, or another numeric expression.

`index(`s_1`,`s_2`)`
  Domain s_1:   strings (to be searched)
  Domain s_2:   strings (to search for)
  Range: integers 0 to 80
  Description: position in s_1 at which s_2 is first found or else returns 0;
             `index("this","is")` $= 3$ and `index("this","it")` $= 0$

`length(`s`)`
  Domain s: strings
  Range: integers 0 to 80
  Description: length of string; `length("ab")` $= 2$

`lower(`s`)`
  Domain s: strings
  Range: strings with lowercased characters
  Description: lower-case string; `lower("THIS")` $=$ `"this"`

`ltrim(`s`)`
  Domain s: strings
  Range: strings without leading blanks
  Description: removes leading blanks; `ltrim(" this")` $=$ `"this"`

`real(`s`)`
  Domain s: strings
  Range: $-8e+307$ to $8e+307$; missing
  Description: converts s to numeric or returns missing value;
             `real("5.2")+1` $= 6.2$ and `real("hello")` $=$.

`rtrim(`s`)`
  Domain s: strings
  Range: strings without trailing blanks
  Description: removes trailing blanks; `rtrim("this ")` $=$ `"this"`

`string(`n`)`
  Domain n: $-8e+307$ to $8e+307$; missing
  Range: strings
  Description: converts n to a string;
             `string(4)+"F"` $=$ `"4F"`
             `string(1234567)="1234567"`
             `string(12345678)` $=$ `"1.23e+07"`
             `string(.)="."`

`string(`n`,`s`)`
  (see next page)

`string(`n`,`s`)`
  Domain n: $-8e+307$ to $8e+307$; missing
  Domain s: strings containing %*fmt*
  Range: strings
  Description: converts n to a string;
                 `string(4,"%9.2f") = "4.00"`
                 `string(123456789,"%11.0g") = "123456789"`
                 `string(123456789,"%13.0gc") = "123,456,789"`
                 `string(0,"%d") = "01jan1960"`
                 `string(225,"%tq") = "2016q2"`
                 `string(225,"not a format") = ""`

`substr(`s`,`n_1`,`n_2`)`
  Domain s: strings
  Domain n_1: integers 1 to 80 and -1 to -80
  Domain n_2: integers 0 to 80 and missing
  Range: strings
  Description: substring of s starting at column n_1 for a length of n_2;
                 if $n_1 < 0$, n_1 is interpreted as distance from the end of the string;
                 if $n_2 = .$ (missing), the remaining portion of the string is returned
                 `substr("abcdef",2,3) = "bcd"`
                 `substr("abcdef",-3,2) = "de"`
                 `substr("abcdef",2,.) = "bcdef"`
                 `substr("abcdef",-3,.) = "def"`
                 `substr("abcdef",2,0) = ""`
                 `substr("abcdef",15,2) = ""`

`trim(`s`)`
  Domain s: strings
  Range: strings without leading or trailing blanks
  Description: removes leading and trailing blanks; equivalent to `ltrim(rtrim(`s`))`
                 `trim(" this ") = "this"`

`upper(`s`)`
  Domain s: strings
  Range: strings with uppercased characters
  Description: upper-case string; `upper("this") = "THIS"`

16.3.6   Special functions

`autocode(`x`,`n`,`x_0`,`x_1`)`
  Domain x: $-8e+307$ to $8e+307$
  Domain n: integers 1 to $8e+307$
  Domain x_0: $-8e+307$ to $8e+307$
  Domain x_1: x_0 to $8e+307$
  Range: x_0 to x_1
  Description: places x into categories

`autocode()` partitions the interval from x_0 to x_1 into n equal-length intervals and returns the upper bound of the interval that contains x. This function is an automated version of `recode()` (see below). See [U] **28 Commands for dealing with categorical variables** for example of use. The algorithm for `autocode()` is

$$\text{if } (x==. \mid n==. \mid x_0==. \mid x_1==. \mid n \leq 0 \mid x_0 \geq x_1)$$
  then return *missing*
  otherwise
    for $i = 1$ to $n - 1$
      $xmap = x_0 + i * (x_1 - x_0)/n$
      if $x \leq xmap$ then return *xmap*
    end
    otherwise
      return x_1

_caller()
  Range: 1.0 to 6.0
  Description: function for use by programmers; returns **version** of the program or session
                that invoked the currently running program; see [R] **version**

cond(x,a,b)
  Domain x: −8e+307 to 8e+307 and missing; 0 ⇒ *false*, otherwise interpreted as *true*
  Domain a: numbers and strings
  Domain b: numbers if a is a number; strings if a is a string
  Range: a and b
  Description: a if x is *true* and b if x is false
                cond(a>2,50,70) = 50 if **a** > 2 or **a** = .
                cond(a>2,"this","that") == **"that"** if **a** $\leq$ 2

e(*name*)
  Domain: names
  Range: real values, strings, and missing
  Description: value of saved result **e(*name*)**;
                see [U] **21.8 Accessing results calculated by other programs**
                e(*name*) = . if the saved result does not exist
                e(*name*) = a numeric value if the saved result is a scalar
                e(*name*) = a numeric value if the saved result is a string (macro) that can be
                          interpreted as a number
                e(*name*) = a string containing the first 80 characters of the saved result otherwise
                Also see [U] **17.8.1 Matrix functions returning matrices**

e(sample)
  Range: 0 and 1
  Description: returns 1 if observation in estimation subsample and 0 otherwise

float(x)
  Domain: −1e+38 to 1e+38
  Range: −1e+38 to 1e+38
  Description: x rounded to **float** precision

Although you may store your numeric variables as **byte**, **int**, **long**, **float**, or **double**, Stata converts all numbers to **double** before performing any calculations. As a consequence, difficulties can arise when numbers that have no finite binary representation are compared.

For example, if the variable **x** is stored as a **float** and contains the value **1.1** (a repeating "decimal" in binary), the expression **x==1.1** will evaluate to *false* because the literal **1.1** is the **double** representation of 1.1, which is different from the **float** representation stored in **x**. (They differ by $2.384 \cdot 10^{-8}$.) The expression **x==float(1.1)** will evaluate to *true* because the **float** function converts the literal **1.1** to its **float** representation before it is compared with **x**. (See [U] **16.10 Precision and problems therein** below for more information.)

group(n)

    Domain: 1 to 8e+307

    Range: integers 1 to $\lceil n \rceil$

    Description: creates categorical variable that divides the data into n as nearly equal-sized subsamples as possible, numbering the first group 1, the second 2, and so on

int(x)

    Domain: −8e+307 to 8e+307

    Range: integers −8e+307 to 8e+307

    Description: the integer obtained by truncating x

                int(5.2) = 5

                int(5.8) = 5

                int(-5.8) = −5

matrix(exp)

    Domain: any valid expression

    Range: evaluation of exp

    Description: restrict name interpretation to scalars and matrices; see scalar() function below

max($x_1,x_2,\ldots,x_n$)

    Domain x_1: −8e+307 to 8e+307; missing

    Domain x_2: −8e+307 to 8e+307; missing

    . . .

    Domain x_k: −8e+307 to 8e+307; missing

    Range: −8e+307 to 8e+307; missing

    Description: maximum value; missing values ignored

                max(2,10,.,7) = 10

                max(.,.,.) = .

min($x_1,x_2,\ldots,x_n$)

    Domain x_1: −8e+307 to 8e+307; missing

    Domain x_2: −8e+307 to 8e+307; missing

    . . .

    Domain x_k: −8e+307 to 8e+307; missing

    Range: −8e+307 to 8e+307; missing

    Description: minimum value; missing values ignored

                min(2,10,.,7) = 2

                min(.,.,.) = .

missing(exp)

    Domain: any string or numeric expression

    Range: 0 and 1

    Description: returns 1 if exp evaluates to missing

Stata has two concepts of missing values: numeric missing value (.) and string missing value (""). missing() returns 1 (meaning true) if the expression evaluates to missing. If x is numeric, missing(x) is equivalent to 'x==.'. If x is string, missing(x) is equivalent to 'x==""'.

recode($x,x_1,x_2,\ldots,x_n$)

    (see next page)

recode($x,x_1,x_2,\ldots,x_n$)
    Domain x: $-8e+307$ to $8e+307$; missing
    Domain x_1: $-8e+307$ to $8e+307$
    Domain x_2: x_1 to $8e+307$
    . . .
    Domain x_n: x_{n-1} to $8e+307$
    Range: $x_1, x_2, \ldots, x_n$; missing
    Description: returns missing if x is missing; x_1 if $x \le x_1$; otherwise x_2 if $x \le x_2$, $\ldots$; otherwise
                 x_n if x is greater than $x_1, x_2, \ldots, x_{n-1}$. Also see [R] **recode** for another
                 style of recode function

r(*name*)
    Domain: names
    Range: real values, strings, and missing
    Description: value of saved result r(*name*);
                 see [U] **21.8 Accessing results calculated by other programs**
                 r(*name*) = . if the saved result does not exist
                 r(*name*) = a numeric value if the saved result is a scalar
                 r(*name*) = a numeric value if the saved result is a string (macro) that can be
                            interpreted as a number
                 r(*name*) = a string containing the first 80 characters of the saved result otherwise;
                 also see [U] **17.8.1 Matrix functions returning matrices**

reldif(x,y)
    Domain x: $-8e+307$ to $8e+307$
    Domain y: $-8e+307$ to $8e+307$
    Range: $-8e+307$ to $8e+307$
    Description: "relative" difference $|x - y|/(|y| + 1)$

replay()
    Range: integers 0 and 1 meaning *false* and *true*
    Description: function for use by programmers writing estimation commands;
                 returns 1 if first nonblank character of local macro `0´ is a comma
                 or if `0´ is empty

return(*name*)
    Domain: names
    Range: real values, strings, and missing
    Description: function for use by programmers; value of to-be-saved result return(*name*);
                 see [U] **21.10 Saving results**
                 return(*name*) = . if the result does not exist
                 return(*name*) = a numeric value if the result is a scalar
                 return(*name*) = a numeric value if the result is a string (macro) that can be
                            interpreted as a number
                 return(*name*) = a string containing the first 80 characters of the result otherwise

round(x,y)
    Domain x: $-8e+307$ to $8e+307$
    Domain x: $-8e+307$ to $8e+307$
    Range: $-8e+307$ to $8e+307$
    Description: round x in units of y

For $y = 1$, this amounts to the closest integer to x; round(5.2,1) is 5 as is round(4.8,1);
round(-5.2,1) is -5 as is round(-4.8,1).

The rounding definition is generalized for $y \neq 1$. With $y = .01$, for instance, x is rounded to two decimal places; `round(sqrt(2),.01)` is 1.41. y may also be larger than 1; `round(28,5)` is 30, which is 28 rounded to the closest multiple of 5. For $y = 0$, the function is defined as returning x unmodified.

s(*name*)
Domain:	names
Range:	real values, strings, and missing
Description:	value of saved result **s**(*name*);
	see [U] **21.8 Accessing results calculated by other programs**
	s(*name*) = . if the saved result does not exist
	s(*name*) = a numeric value if the saved result can be interpreted as a number
	s(*name*) = a string containing the first 80 characters of the saved result otherwise

scalar(*exp*)
Domain:	any valid expression
Range:	evaluation of *exp*
Description:	restrict name interpretation to scalars and matrices

Names in expressions can refer to names of variables in the dataset, names of matrices, or names of scalars. Matrices and scalars can have the same names as variables in the dataset. In the case of conflict, the default is to assume you are referring to the name of the variable in the dataset.

`matrix()` and `scalar()` explicitly state that you are referring to matrices and scalars. `matrix()` and `scalar()` are the same function; scalars and matrices may not have the same names and so cannot be confused. Typing `scalar(x)` makes clear you are referring to the scalar or matrix named **x** and not the variable named **x** should there happen to be a variable of that name.

sign(x)
Domain:	−8e+307 to 8e+307; missing
Range:	−1, 0, 1; missing
Description:	sign of x: -1 if $x < 0$, 0 if $x = 0$, 1 if $x > 0$, and missing if $x = $ missing

sum(x)
Domain:	−8e+307 to 8e+307; missing
Range:	−8e+307 to 8e+307 (and excluding missing)
Description:	running sum of x treating missing values as zero

For example, following the command `generate y=sum(x)`, the jth observation on **y** contains the sum of the first through jth observations on **x**. Also see [R] **egen** for an alternative sum function that produces a constant equal to the overall sum.

16.4   System variables (_variables)

Expressions may also contain _*variables* (pronounced "underscore variables"). These are built-in system variables that are created and updated by Stata. They are called _*variables* because their names all begin with the underscore '_' character.

The _*variables* are

[*eqno*]_**b**[*varname*] (synonym: [*eqno*]_**coef**[*varname*]) contains the value (to machine precision) of the coefficient on *varname* from the most recently estimated model (such as ANOVA, regression, Cox, logit, probit, multinomial logit, and the like). See [U] **16.5 Accessing coefficients and standard errors** below for a complete description.

_**cons** is always equal to the number 1 when used directly and refers to the intercept term when used indirectly, as in _**b**[_**cons**].

_n contains the number of the current observation.

_N contains the total number of observations in the dataset.

_pi contains the value of π to machine precision.

_rc contains the value of the return code from the most recent `capture` command.

[*eqno*] _se[*varname*] contains the value (to machine precision) of the standard error of the coefficient on *varname* from the most recently estimated model (such as ANOVA, regression, Cox, logit, probit, multinomial logit, and the like). See [U] **16.5 Accessing coefficients and standard errors** below for a complete description.

16.5   Accessing coefficients and standard errors

After estimating a model, you can access the coefficients and standard errors and use them in subsequent expressions. You should also see [R] **predict** (and [U] **23 Estimation and post-estimation commands**) for an easier way to obtain predictions, residuals, and the like.

16.5.1   Simple models

Begin by considering estimation methods that yield a single estimated equation with a one-to-one correspondence between coefficients and variables such as `cnreg`, `cox`, `logit`, `ologit`, `oprobit`, `probit`, `regress`, and `tobit`. _b[*varname*] (synonym _coef[*varname*]) contains the coefficient on *varname* and _se[*varname*] contains its standard error, both recorded to machine precision. Thus, _b[age] refers to the calculated coefficient on the `age` variable after typing, say, `regress response age sex` and _se[age] refers to the standard error on the coefficient. _b[_cons] refers to the constant and _se[_cons] to its standard error. Thus, you might type

```
regress response age sex
generate asif = _b[_cons] + _b[age]*age
```

16.5.2   ANOVA models

In ANOVA there is no simple relationship between the coefficients and the variables. For continuous variables in the model, _b[*varname*] refers to the coefficient. This works just as it did in simple models. For categorical variables, you must specify the level as well as the variable. _b[drug[2]] refers to the coefficient on the second level of drug. For interactions, _b[drug[2]*disease[1]] refers to the coefficient on the second level of drug and the first level of disease. Standard errors are obtained similarly using _se[]. Thus, you might type

```
anova outcome sex age drug sex*age drug*age, continuous(age)
gen ageffect = _b[age]*age
replace ageffect = ageffect + _b[sex[1]*age] if sex==1
replace ageffect = ageffect + _b[sex[2]*age] if sex==2
```

16.5.3   Multiple-equation models

The syntax for referring to coefficients and standard errors in multiple-equation models is the same as in the simple-model case except that _b[] and _se[] are preceded by an equation number in square brackets. There are, however, numerous alternatives in how you may type requests. The way you are supposed to type requests is

[*eqno*] _b[*varname*]
[*eqno*] _se[*varname*]

although, of course, you may substitute `_coef[]` for `_b[]`. In fact, you may omit the `_b[]` altogether, and most Stata users do:

$$[eqno] \; [varname]$$

You may also omit the second pair of square brackets:

$$[eqno] \; varname$$

There are two ways to specify the equation number *eqno*: either as an absolute equation number or as an "indirect" equation number. In the absolute form, the number is preceded by a '**#**' sign. Thus, `[#1]displ` refers to the coefficient on `displ` in the first equation (and `[#1]_se[displ]` refers to its standard error). You can even use this form for simple models such as `regress` if you prefer. `regress` estimates a single equation, so `[#1]displ` refers to the coefficient on `displ` just as does `_b[displ]`. Similarly, `[#1]_se[displ]` and `_se[displ]` are equivalent. The logic works both ways—in the multiple equation context, `_b[displ]` refers to the coefficient on `displ` in the first equation and `_se[displ]` refers to its standard error. `_b[`*varname*`]` (`_se[`*varname*`]`) is just another way of saying `[#1]`*varname* (`[#1]_se[`*varname*`]`).

Equations may also be referenced indirectly. `[res]displ` refers to the coefficient on `displ` in the equation named `res`. Equation are often named after the corresponding dependent variable name if there is such a concept in the estimated model, so `[res]displ` might refer to the coefficient on `displ` in the equation for variable `res`.

In the case of multinomial logit (`mlogit`), however, equations are named after the levels of the single dependent categorical variable. In multinomial logit, there is one dependent variable and there is an equation corresponding to each of the outcomes (values taken on) recorded in that variable except for the one that is arbitrarily labeled the base. `[res]displ` would be interpreted as the coefficient on `displ` in the equation corresponding *to the outcome* `res`. If outcome `res` is the base outcome, Stata treats `[res]displ` as zero (and Stata does the same thing for `[res]_se[displ]`).

Continuing with the multinomial logit case, the outcome variable must be numeric, although it need not be an integer. `[res]displ` would only be understood if there were a value label associated with the numeric outcome variable and `res` was one of the labelings. If your data is not labeled, you may refer to the numeric value directly by omitting the '**#**' sign. `[1]displ` refers to the coefficient on `displ` in the equation corresponding to the outcome 1, which may be different from `[#1]displ`. `[1.2]displ` would be the coefficient on `displ` in the equation corresponding to outcome 1.2. `[1.2]_cons` refers to the constant in the equation corresponding to outcome 1.2. `[1.2]_se[_cons]` refers to the standard error on the constant.

Thus, you might type

```
. mlogit outcome displ weight
. gen cont_din1 = [1]displ*displ
```

For every observation in your data, `cont_din1` would contain the coefficient on `displ` in the equation corresponding to the outcome 1 multiplied by `displ`, or the contribution of `displ` in determining outcome 1.

16.6   Accessing results from Stata commands

Most Stata commands—not just estimation commands—save results in a way you can access in subsequent expressions. You do that by referring to `e(`*name*`)`, `r(`*name*`)`, or `s(`*name*`)`:

```
. summarize age
. gen agedev = age-r(mean)
. regress mpg weight
. display "The number of observations used is " e(N)
```

Most commands are categorized r class, meaning they save results in r(). The returned results—such as r(mean)—are available immediately following the command and if you are going to refer to them, you need to refer to them soon because the next command will probably replace what is in r().

e-class commands are Stata's estimation commands—commands that estimate models. Results in e() stick around until the next model is estimated.

s-class commands are parsing commands—commands used by programmers to interpret commands you type. Very few commands save anything in s().

Every command of Stata is designated r, e, or s class, or, if the command saves nothing, n class. r stands for return as in returned results, e stands for estimation as in estimation results, s stands for string and, admittedly, this last acronym is weak; n stands for null.

You can find out what is stored where by looking in the *Saved Results* section for the particular command in the *Reference Manual*. If you know the class of a command—and it is easy enough to guess—you can also see what is stored by typing 'return list', 'estimates list', or 'sreturn list':

See [R] **saved results** and [U] **21.8 Accessing results calculated by other programs**.

16.7   Explicit subscripting

Individual observations on variables can be referenced by subscripting the variables. Explicit subscripts are specified by following a variable name with square brackets that contain an expression. The result of the subscript expression is truncated to an integer, and the value of the variable for the indicated observation is returned. If the value of the subscript expression is less than 1 or greater than _N, a missing value is returned.

16.7.1   Generating lags and leads

When you type something like

. generate y = x

Stata interprets it as if you typed

. generate y = x[_n]

and what that means is that the first observation of y is to be assigned the value from the first observation of x, the second observation of y is to be assigned the value from the second observation on x, and so on. Were you to instead type

. generate y = x[1]

you would set each and every observation of y equal to the first observation on x. If you typed

. generate y = x[2]

you would set each and every observation of y equal to the second observation on x. What would happen if you typed

. generate y = x[0]

Nothing too bad would happen: Stata would merely copy missing value into every observation of y because observation 0 does not exist. Exactly the same thing would happen were you to type

. generate y = x[100]

and you had fewer than 100 observations in your data.

When you type the square brackets, you are specifying explicit subscripts. Explicit subscripting combined with the _variable _n can be used to create lagged values on a variable. The lagged value of a variable x can be obtained by typing

```
. generate xlag = x[_n-1]
```

If you are really interested in lags and leads, you probably have time-series data and then you would be better served by using the time-series operators such as L.x. Time-series operators can be used with varlists and expressions and they are safer because they account for gaps in the data; see [U] **14.4.3 Time-series varlists** and [U] **16.8 Time-series operators**. Even so, it is important that you understand how the above works.

The built-in underscore variable _n is understood by Stata to mean the observation number of the current observation. That is why

```
. generate y = x[_n]
```

results in observation 1 of x being copied to observation 1 of y, and similarly for the rest of the observations. We are considering

```
. generate xlag = x[_n-1]
```

and notice that _n-1 evaluates to the observation number of the previous observation. So for the first observation, _n-1 = 0 and xlag[1] is set to missing. For the second observation, _n-1 = 1 and xlag[2] is set to x[1], and so on.

Similarly, the lead of x can be created by

```
. generate xlead = x[_n+1]
```

In this case, the last observation on the new variable xlead will be *missing* because _n+1 will be greater than _N (_N is the total number of observations in the dataset).

16.7.2 Subscripting within groups

When a command is preceded by the by *varlist*: prefix, subscript expressions and the underscore variables _n and _N are evaluated relative to the subset of the data currently being processed. For example, consider the following (admittedly not very interesting) data:

```
. list

          bvar     oldvar
  1.        1        1.1
  2.        1        2.1
  3.        1        3.1
  4.        2        4.1
  5.        2        5.1
```

To see how _n, _N, and explicit subscripting work, let's create three new variables demonstrating each and then list their values:

```
. generate small_n = _n
. generate big_n = _N
. generate newvar = oldvar[1]
. list

          bvar     oldvar    small_n      big_n     newvar
  1.        1        1.1         1           5        1.1
  2.        1        2.1         2           5        1.1
```

3.	1	3.1	3	5	1.1
4.	2	4.1	4	5	1.1
5.	2	5.1	5	5	1.1

small_n (which is equal to _n) goes from 1 to 5, and big_n (which is equal to _N) is 5. This should not be surprising; there are 5 observations in the data, and _n is supposed to count observations whereas _N is the total number. newvar, which we defined as oldvar[1], is 1.1. Indeed, we see that the first observation on oldvar is 1.1.

Now let's repeat those same three steps, only this time precede each step with the prefix by bvar:. First, we will drop the old values of small_n, big_n, and newvar, so we start fresh:

```
. drop small_n big_n newvar
. sort bvar
. by bvar: generate small_n=_n
-> bvar=         1
-> bvar=         2
. by bvar: generate big_n=_N
-> bvar=         1
-> bvar=         2
. by bvar: generate newvar=oldvar[1]
-> bvar=         1
-> bvar=         2
. list
           bvar      oldvar     small_n      big_n      newvar
  1.          1         1.1           1          3         1.1
  2.          1         2.1           2          3         1.1
  3.          1         3.1           3          3         1.1
  4.          2         4.1           1          2         4.1
  5.          2         5.1           2          2         4.1
```

The results are different. Remember that we claimed that _n and _N are evaluated relative to the subset of data in the by-group. Thus, small_n (_n) goes from 1 to 3 for bvar = 1 and from 1 to 2 for bvar = 2. big_n (_N) is 3 for the first group and 2 for the second. Finally, newvar (oldvar[1]) is 1.1 and 4.1.

▷ Example

You now know enough to do some amazing things.

Suppose you have data on individual states and you have another variable in your data, call it region, that divides the states into the four Census regions. You have a variable x in your data, and you want to make a new variable called avgx to include in your regressions. This new variable is to take on the average value of x for the region in which the state is located. Thus, for California you will have the observation on x and the observation on the average value in the region, avgx. Here's how:

```
. sort region
. by region: generate avgx=sum(x)/_n
. by region: replace avgx=avgx[_N]
```

First, we type sort region to put the data into region order. Next, by region, we generate avgx equal to the running sum of x divided by the number of observations so far. We have, in effect, created the running average of x within region. It is the last observation of this running average, the overall average within the region, that interests us. So, by region, we replace every avgx observation in a region with the last observation within the region, avgx[_N].

Here is what we will see when we type these three commands:

```
. sort region
. by region: generate avgx=sum(x)/_n
-> region= N Cntrl
-> region=      NE
-> region=   South
-> region=    West
. by region: replace avgx=avgx[_N]
-> region= N Cntrl  (11 real changes made)
-> region=      NE  (8 real changes made)
-> region=   South  (15 real changes made)
-> region=    West  (12 real changes made)
```

In our example there are no missing observations on x. If there had been, we would have obtained the wrong answer. When we created the running average, we typed

```
. by region: generate avgx=sum(x)/_n
```

The problem is not with the sum() function. When sum() encounters a missing, it adds zero to the sum. The problem is with _n. Let's assume that the second observation in the first region has recorded a missing for x. When Stata processes the third observation in that region, it will calculate the sum of two elements (remember one is missing) and then divide the sum by 3 when it should be divided by 2. There is an easy solution:

```
. by region: generate avgx=sum(x)/sum(x~=.)
```

Rather than divide by _n, we divide by the total number of nonmissing observations seen on x so far, namely the sum(x~=.).

If our goal were simply to obtain the mean, it could have been more easily accomplished by typing egen avgx=mean(x), by(region); see [R] **egen**. egen, however, is written in Stata and the above is how egen's mean() function works. The general principles are worth understanding.

◁

▷ Example

You have some patient data recording vital signs at various times during an experiment. The variables include **patient**, an id-number or name of the patient; **time**, a variable recording the date or time or epoch of the vital sign reading; and **vital**, a vital sign. You probably have more than one vital sign, but one is enough to illustrate the concept. Each observation in your data represents a patient-time combination.

Let's assume you have 1,000 patients and, for every observation on the same patient, you want to create a new variable called **orig** that records the patient's initial value of this vital sign.

```
. sort patient time
. by patient: generate orig=vital[1]
-> patient=    1
-> patient=    2
-> patient=    3
        .
        .
        .
-> patient= 1000
```

Observe that vital[1] refers not to the first reading on the first patient, but to the first reading on the current patient since we are performing the generate command by patient.

Also, observe that Stata just generated 1,000 lines of output. When we have only a few by-groups, the output was reassuring that things were going well. When we have lots of by-groups, it is obnoxious. There is a solution: We can type `quietly by patient: generate orig=vital[1]`. Putting `quietly` in front of any Stata command suppresses the output that command would otherwise have generated. There is a second gain, too: `generate` will run about ten times faster.

◁

▷ Example

Let's do one more example with this patient data. Suppose we want to create a new dataset from our patient data that records not only the patient's identification, the time of the reading of the first vital sign, and the first vital sign reading itself, but also the time of the reading of the last vital sign and its value. We want one observation per patient. Here's how:

```
. sort patient time
. quietly by patient: generate lasttime=time[_N]
. quietly by patient: generate lastvit=vital[_N]
. quietly by patient: drop if _n~=1
```

◁

16.8   Time-series operators

Time-series operators allow you to refer to the lag of `gnp` by typing `L.gnp`, the second lag by typing `L2.gnp`, etc. There are operators for lead (`F`), difference `D`, and seasonal difference `S` as well.

Time-series operators can be used with varlists and with expressions. See [U] **14.4.3 Time-series varlists** if you have not read it already. This section has to do with using time-series operators in expressions such as with `generate`. You do not have to create new variables; you can use the time-series operated variables directly.

16.8.1   Generating lags and leads

In a time-series context, referring to `L2.gnp` is better than referring to `gnp[_n-2]` because there might be missing observations. Pretend that observation 4 contains data for $t = 25$ and observation 5 data for $t = 27$. `L2.gnp` will still produce correct answers; `L2.gnp` for observation 5 will be the value from observation 4 because the time-series operators look at t to find the relevant observation. The more mechanical `gnp[_n-2]` just goes two observations back and that, in this case, would not produce the desired result.

Time-series operators can be used with varlists or with expressions, so you can type

```
. regress val L.gnp r
```

or you can type

```
. generate gnplagged = L.gnp
. regress val gnplagged
```

Before you can type either one, however, you must use the `tsset` command to tell Stata the identity of the time variable; see [R] **tsset**. Once you have `tsset` the data, anyplace you see an *exp* in a syntax diagram, you may type time-series operated variables, so you can type

```
. summarize r if F.gnp<gnp
```

or

```
. generate grew = 1 if gnp>L.gnp & L.gnp~=.
. replace grew = 0 if grew==. & L.gnp~.
```

or

```
. generate grew = (gnp>L.gnp) if L.gnp~=.
```

16.8.2   Operators within groups

Stata also understands panel or cross-sectional time-series data. For instance, if you

```
. tsset country time
```

You are declaring that have time-series data—the time variable is `time`—and you have that time-series data for separate countries.

Once you have `tsset` both the cross-sectional and time identifiers, you proceed just as you would if you had a simple time series.

```
. generate grew = (gnp>L.gnp) if L.gnp~=.
```

would produce correct results. The `L.` operator will not confuse the observation at the end of one panel with the beginning of the next.

16.9   Label values

(If you have not read [U] **15.6 Dataset, variable, and value labels**, please do so.) You may use labels in an expression in place of the numeric values with which they are associated. To use a label in this way, type the label in double quotes followed by a colon and the name of the value label.

▷ Example

If the value label `yesno` associates the label `yes` with 1 and `no` with 0, then `"yes":yesno` (said aloud as the value of `yes` under `yesno`) is evaluated as 1. If the double-quoted label is not defined in the indicated value label, or if the value label itself is not found, a missing value is returned. Thus, the expression `"maybe":yesno` is evaluated as *missing*.

```
. list

             name      answer
   1.     Sribney          no
   2.      Franks          no
   3.       Hilbe         yes
   4.      DeLeon          no
   5.        Cain          no
   6.      Willis         yes
   7.    Pechacek          no
   8.         Cox          no
   9.      Reimer          no
  10.      Hardin         yes
  11.   Lancaster         yes
  12.     Johnson          no

. list if answer=="yes":yesno

             name      answer
   3.       Hilbe         yes
   6.      Willis         yes
  10.      Hardin         yes
  11.   Lancaster         yes
```

In the above example, the variable `answer` is not a string variable; it is a numeric variable that has the associated value label `yesno`. Since `yesno` associates `yes` with 1 and `no` with 0, we could have typed `list if answer==1` instead of what we did type. We could not have typed `list if answer=="yes"` because `answer` is not a string variable. If we had, we would have received the error message "type mismatch".

◁

16.10   Precision and problems therein

Examine the following short Stata session:

```
. drop _all
. input x y
              x          y
  1. 1 1.1
  2. 2 1.2
  3. 3 1.3
  4. end
. count if x==1
      1
. count if y==1.1
      0
. list
              x          y
  1.          1        1.1
  2.          2        1.2
  3.          3        1.3
```

We created a dataset containing two variables, x and y. The first observation has x equal to 1 and y equal to 1.1. When we asked Stata to `count` the number of times that the variable x took on the value 1, we were told that it occurred once. Yet when we asked Stata to `count` the number of times y took on the value 1.1, we were told zero—meaning that it never occurred. What's gone wrong? When we `list` the data, we see that the first observation has y equal to 1.1.

Despite appearances, Stata has not made a mistake. Stata stores numbers internally in binary, and the number 1.1 has no exact binary representation—that is, there is no finite string of binary digits that is exactly equal to 1.1.

❏ Technical Note

The number 1.1 in binary is 1.0001100110011 . . . , where the period represents the binary point. The problem binary computers have storing numbers like 1/10 is much like the problem we base-10 users have in precisely writing 1/11, which is 0.0909090909 . . . .

❏

The number that appears as 1.1 in the listing above is actually 1.1000000238419, which is off by roughly 2 parts in 10^8. Unless we tell Stata otherwise, it stores all numbers as `floats`, which are also known as *single-precision* or *4-byte reals*. On the other hand, Stata performs all internal calculations in `double`, which is also known as *double-precision* or *8-byte reals*. This is what leads to the difficulty.

In the above example, we compared the number 1.1, stored as a `float`, with the number 1.1 stored as a `double`. The double-precision representation of 1.1 is more accurate than the single-precision representation, but what is important is that it is also different. Those two numbers are not equal.

There are a number of ways around this problem. The problem with 1.1 apparently not equaling 1.1 would never arise if the storage precision and the precision of the internal calculations were the same. Thus, you could store all your data as `doubles`. This takes more computer memory, however, and it is unlikely that (1) your data is really that accurate and (2) the extra digits would meaningfully affect any calculated result even if the data were that accurate.

❏ Technical Note

It is unlikely to affect any calculated result because Stata performs all internal calculations in double precision. This is all rather ironic, since the problem would also not arise if we had designed Stata to use single precision for its internal calculations. Stata would be less accurate, but the problem would have been completely disguised from the user, making this entry unnecessary.

❏

Another solution is to use the `float()` function. `float(x)` rounds x to its `float` representation. If we had typed `count if y==float(1.1)` in the example above, we would have been informed that there is one such value.

16.11   Acknowledgments

We thank George Marsaglia of the Florida State University for providing his KISS (Keep It Simple Stupid) random number generator.

We thank John R. Gleason of Syracuse University for directing our attention to Wichura (1988) for calculating the cumulative normal density accurately, for sharing his experiences about techniques with us, and even providing C code to make the calculations.

16.12   References

Abramowitz, M. and I. A. Stegun, eds. 1968. *Handbook of Mathematical Functions*, 7th printing. Washington, D.C.: National Bureau of Standards.

Haynam, G. E., Z. Govindarajulu, and F. C. Leone. 1970. Tables of the cumulative non-central chi-square distribution. In *Selected Tables in Mathematical Statistics*, vol. 1, ed. H. L. Harter and D. B. Owen, 1–78. Providence, RI: American Mathematical Society.

Marsaglia, G. 1994. Personal communication.

Press, W. H., B. P. Flannery, S. A. Teukolsky, and W. T. Vetterling. 1986. *Numerical Recipes: The Art of Scientific Computing*. Cambridge: Cambridge University Press.

Wichura, M. J. 1988. The percentage points of the normal distribution. *Applied Statistics* 37: 454–477.

17 Matrix expressions

Contents

17.1 Overview

Matrices can be used interactively in Stata and you might do so, for instance, in a teaching situation. The real power of matrices, however, is unleashed when they are used in Stata programs and ado-files to implement other statistical procedures. We do this ourselves and you can, too.

17.1.1 Definition of matrix

Stata's definition of a matrix includes a few details that go beyond the mathematics. To Stata, a matrix is a named entity containing an $r \times c$ ($0 < r \leq$ matsize, $0 < c \leq$ matsize) rectangular array of double-precision numbers, none of which can be missing, and which is bordered by a row and a column of names.

```
. matrix list A
A[3,2]
     c1   c2
r1    1    2
r2    3    4
r3    5    6
```

In this case, we have a 3×2 matrix named A containing elements 1, 2, 3, 4, 5, and 6. Row 1, column 2 (written $A_{1,2}$ in math and A[1,2] in Stata) contains 2. The columns are named c1 and c2 and the rows r1, r2, and r3. These are the default names Stata comes up with when it cannot do better. The names do not play a role in the mathematics, but they are of great help when it comes to labeling the output.

The names are operated on just as the numbers. For instance:

```
. matrix B=A´*A
. matrix list B
symmetric B[2,2]
      c1  c2
c1   35
c2   44  56
```

We defined $\mathbf{B} = \mathbf{A}'\mathbf{A}$. Note that the row and column names of $\mathbf{B}$ are the same. Multiplication is defined for any $a \times b$ and $b \times c$ matrices, the result being $a \times c$. Thus, the row and column names of the result are the row names of the first matrix and the column names of the second matrix. We formed $\mathbf{A}'\mathbf{A}$, using the transpose of $\mathbf{A}$ for the first matrix—which also interchanged the names—and so obtained the names shown.

17.1.2 matsize

Matrices are limited to being no larger than `matsize` $\times$ `matsize`. The default value of `matsize` is 40, but you can reset this using the `set matsize` command; see [R] **matsize**.

The maximum value of `matsize` is 800 and thus matrices are not suitable for holding large amounts of data. This restriction does not prove a limitation because terms that appear in statistical formulae are of the form $(\mathbf{X}'\mathbf{W}\mathbf{Z})$ and Stata provides a command, `matrix accum`, for efficiently forming such matrices; see [U] **17.6 Creating matrices by accumulating data** below.

17.2   Row and column names

Matrix rows and columns always have names. Stata is smart about setting these names when the matrix is created and the matrix commands and operators manipulate these names throughout calculations, with the result that the names typically are set correctly at the conclusion of matrix calculations.

For instance, consider the matrix calculation $\mathbf{b} = (\mathbf{X}'\mathbf{X})^{-1}\mathbf{X}'\mathbf{y}$ performed on real data:

```
. matrix accum XprimeX = weight foreign
. matrix vecaccum yprimeX = mpg weight foreign
. matrix b = syminv(XprimeX)*yprimeX´
. matrix list b
b[3,1]
               mpg
 weight  -.00658789
foreign  -1.6500291
  _cons   41.679702
```

Note that these names were produced without us ever having given a special command to place the names on the result. When we formed matrix `XprimeX`, Stata produced the result

```
. matrix list XprimeX
symmetric XprimeX[3,3]
             weight     foreign       _cons
 weight   7.188e+08
foreign      50950          22
  _cons     223440          22          74
```

`matrix accum` forms $\mathbf{X}'\mathbf{X}$ matrices from data and it sets the row and column names to the variable names used. The names are correct in the sense that, for instance, the (1,1) element is the sum across the observations of squares of `weight` and the (2,1) element is the sum of the product of `weight` and `foreign`.

Similarly, `matrix vecaccum` forms $\mathbf{y'X}$ matrices and it also sets the row and column names to the variable names used, so `matrix vecaccum yprimeX = mpg weight foreign` resulted in

```
. matrix list yprimeX
yprimeX[1,3]
        weight  foreign    _cons
mpg    4493720      545     1576
```

The final step, `matrix b = syminv(XprimeX)*yprimeX´`, manipulated the names and, if you think carefully, you can derive for yourself the rules. `syminv()` (inversion) is much like transposition, so row and column names must be swapped although, in this case, the matrix was symmetric so that amounted to leaving the names as they were. Multiplication amounts to taking the column names of the first matrix and the row names of the second. The final result is

```
. matrix list b
b[3,1]
                mpg
weight   -.00658789
foreign  -1.6500291
 _cons    41.679702
```

and the interpretation is $\mathtt{mpg} = -.00659\,\mathtt{weight} - 1.65\,\mathtt{foreign} + 41.68 + e$.

Researchers have long realized that matrix notation simplifies the description of complex calculations. What they may not have realized is that, corresponding to each mathematical definition of a matrix operator, there is a definition of the operator's effect on the names that can be used to carry the names forward through long and complex matrix calculations.

17.2.1 The purpose of row and column names

Mostly, matrices in Stata are used in programming estimators and mostly, Stata uses row and column names to produce pretty output. For instance, say that we wrote code—interactively or in a program—that produced the following coefficient vector b and covariance matrix V,

```
. matrix list b
b[1,3]
        weight      displ     _cons
y1  -.00656711   .00528078  40.084522

. matrix list V
symmetric V[3,3]
              weight      displ      _cons
weight    1.360e-06
 displ    -.0000103   .00009741
 _cons    -.00207455  .01188356   4.0808455
```

We could now produce standard estimation output by coding two more lines:

```
. estimates post b V
. estimates display
------------------------------------------------------------------------------
             |      Coef.   Std. Err.       z    P>|z|     [95% Conf. Interval]
-------------+----------------------------------------------------------------
      weight |  -.0065671   .0011662    -5.631   0.000    -.0088529   -.0042813
       displ |   .0052808   .0098696     0.535   0.593    -.0140632    .0246248
       _cons |   40.08452    2.02011    19.843   0.000     36.12518    44.04387
------------------------------------------------------------------------------
```

Stata's `estimates` command knew to produce this output because of the row and column names on the coefficient vector and variance matrix. Moreover, in most cases we do nothing special in our code that produces b and V to set the row and column names because, given how matrix names work, they work themselves out.

In addition, sometimes row and column names help us to detect programming errors. Assume we wrote code to produce matrices b and V but made a mistake. Sometimes our mistake will result in the row and column names turning out wrong. Rather than the b vector we previously showed you, we might produce

```
. matrix list b
b[1,3]
           weight         c2      _cons
y1   -.00656711       42.23   40.084522
```

Were we to `post` our estimation results now, Stata would refuse because it can tell by the names that there is a problem:

```
. estimates post b V
name conflict
r(507);
```

Understand, however, that Stata follows the standard rules of matrix algebra; the names are just along for the ride. Matrices are summed by position, meaning a directive to form $C = A + B$ results in $C_{11} = A_{11} + B_{11}$ regardless of the names, and it is not an error to sum matrices with different names:

```
. matrix list a
symmetric a[3,3]
                  mpg      weight       _cons
   mpg         14419
weight       1221120   1.219e+08
 _cons           545       50950          22
. matrix list b
symmetric b[3,3]
             displ        mpg      _cons
displ      3211055
  mpg       227102      22249
_cons        12153       1041         52
. matrix c = a + b
. matrix list c
symmetric c[3,3]
             displ        mpg      _cons
displ      3225474
  mpg      1448222   1.220e+08
_cons        12698       51991         74
```

Matrix row and column names are used to label output; they do not affect how matrix algebra is performed.

17.2.2 Three-part names

Row and column names have three parts: *equation_name*:*ts_operator*.*subname*.

In the examples shown so far, the first two parts have been blank; the row and column names consisted of *subnames* only. This is typical. Run any single-equation model (such as those produced by `regress`, `probit`, `logistic`, etc.), and if you fetch the resulting matrices, you will find they have row and column names of the *subname* form.

Those who work with time-series data will find matrices with row and column names of the form *ts_operator. subname*, such as

```
. matrix list example1

symmetric example1[3,3]
                             L.
              rate         rate         _cons
   rate    3.0952534
 L.rate     .0096504      .00007742
  _cons   -2.8413483     -.01821928    4.8578916
```

We obtained this matrix by running a linear regression on `rate` and `L.rate` and then fetching the covariance matrix. Think of the row and column name `L.rate` no differently than you think of `rate` or, in the previous examples, `r1`, `r2`, `c1`, `c2`, `weight`, and `foreign`.

Equation names are used to label partitioned matrices and, in estimation, occur in the context of multiple equations. Here is a matrix with *equation_names* and *subnames*,

```
. matrix list example2

symmetric example2[5,5]
                          mpg:          mpg:          mpg:       weight:       weight:
                       foreign         displ         _cons       foreign         _cons
   mpg:foreign      1.6483972
     mpg:displ        .004747      .00003876
     mpg:_cons     -1.4266352     -.00905773     2.4341021
weight:foreign     -51.208454     -4.665e-19     15.224135     24997.727
  weight:_cons      15.224135      2.077e-17    -15.224135    -7431.7565     7431.7565
```

and here is an example with all three parts filled in:

```
. matrix list example3

symmetric example3[5,5]
                          val:          val:          val:       weight:       weight:
                                          L.
                          rate          rate         _cons       foreign         _cons
      val:rate      2.2947268
    val:L.rate      .00385216      .0000309
     val:_cons     -1.4533912     -.0072726     2.2583357
weight:foreign     -163.86684     7.796e-17     49.384526     25351.696
  weight:_cons      49.384526    -1.566e-16    -49.384526     -7640.237      7640.237
```

`val:L.rate` is the name of a column just as, in the previous section, `c2` and `foreign` were names.

Let us pretend that this last matrix is the variance matrix produced by a program we wrote and that our program also produced a coefficient vector b:

```
. matrix list b

b[1,5]
                val:          val:          val:       weight:       weight:
                                L.
                rate          rate         _cons       foreign         _cons
  y1       4.5366753     -.00316923     20.68421    -1008.7968     3324.7059
```

Here would be the result of posting and displaying the results:

(continued on next page)

```
. estimates post b example3
. estimates display
```

	Coef.	Std. Err.	z	P>|z|	[95% Conf. Interval]
val						
rate						
--	4.536675	1.514836	2.995	0.003	1.567652	7.505698
L1	-.0031692	.0055591	-0.570	0.569	-.0140648	.0077264
_cons	20.68421	1.502776	13.764	0.000	17.73882	23.6296
weight						
foreign	-1008.797	159.2222	-6.336	0.000	-1320.866	-696.7271
_cons	3324.706	87.40845	38.036	0.000	3153.388	3496.023

The equation names are used to separate one equation from the next.

17.2.3   Setting row and column names

You reset row and column names using the `matrix rownames` and `matrix colnames` commands.

Before resetting the names, list the matrix to verify the names are not set correctly; often they already are. When you enter a matrix by hand, however, the row names are unimaginatively set to r1, r2, ..., and the column names to c1, c2, ...

```
. matrix a = (1,2,3\4,5,6)
. matrix list a
a[2,3]
     c1  c2  c3
r1    1   2   3
r2    4   5   6
```

Regardless of the current row and column names, `matrix rownames` and `matrix colnames` will reset them:

```
. matrix colnames a = foreign alpha _cons
. matrix rownames a = one two
. matrix list a
a[2,3]
      foreign    alpha    _cons
one         1        2        3
two         4        5        6
```

You may set the *ts_operator* as well as the *subname*,

```
. matrix colnames a = foreign l.rate _cons
. matrix list a
a[2,3]
                        L.
      foreign    rate    _cons
one         1       2        3
two         4       5        6
```

and you may set equation names:

```
. matrix colnames a = this:foreign this:l.rate that:_cons
. matrix list a
a[2,3]
          this:    this:    that:
                   L.
        foreign    rate    _cons
  one        1       2        3
  two        4       5        6
```

See [R] **matrix rowname** for more information.

17.2.4   Obtaining row and column names

matrix list displays the matrix with its row and column names. In a programming context, you can fetch the row and column names into a macro using

$$
\begin{aligned}
&\texttt{local} \ldots : \texttt{rowfullnames } \textit{matname} \\
&\texttt{local} \ldots : \texttt{colfullnames } \textit{matname} \\
&\texttt{local} \ldots : \texttt{rownames } \textit{matname} \\
&\texttt{local} \ldots : \texttt{colnames } \textit{matname} \\
&\texttt{local} \ldots : \texttt{roweq } \textit{matname} \\
&\texttt{local} \ldots : \texttt{coleq } \textit{matname}
\end{aligned}
$$

rowfullnames and **colfullnames** return the full names (*equation_name*:*ts_operator*.*subname*) listed one after the other.

rownames and **colnames** omit the equations and return *ts_operator*.*subname*, listed one after the other.

roweq and **coleq** return the equation names, listed one after the other.

See [R] **macro** and [R] **matrix define** for more information.

17.3   Vectors and scalars

Stata does not have vectors as such—they are considered special cases of matrices and are handled by the **matrix** command.

Stata does have scalars, although they are not strictly necessary because they, too, could be handled as special cases. See [R] **scalar** for a description of scalars.

17.4   Inputting matrices by hand

You input matrices using

> **matrix input** *matname* = (...)

or

> **matrix** *matname* = (...)

In either case, you enter the matrices rowwise. You separate one element from the next using commas (,) and one row from the next using backslashes (\). If you omit the word **input** you are using the expression parser to input the matrix:

```
. matrix a = (1,2\3,4)
```

```
. matrix list a
a[2,2]
      c1  c2
r1     1   2
r2     3   4
```

This has the advantage that you can use expressions for any of the elements,

```
. matrix b = (1, 2+3/2 \ cos(_pi), _pi)
. matrix list b
b[2,2]
             c1          c2
r1            1         3.5
r2           -1   3.1415927
```

The disadvantage is that the matrix must be small; say no more than 50 elements (regardless of the value of `matsize`).

`matrix input` has no such restriction, but you may not use subexpressions for the elements:

```
. matrix input c  = (1,2\3,4)
. matrix input d = (1, 2+3/2 \ cos(_pi), _pi)
invalid syntax
r(198);
```

Either way, after inputting the matrix, you will probably want to set the row and column names; see [U] **17.2.3 Setting row and column names** above.

17.5   Accessing matrices created by Stata commands

Some Stata commands—and all estimation commands—leave behind matrices which you can subsequently use. After executing an estimation command, type `estimates list` to see what is available:

```
. probit foreign mpg weight
(output omitted)
. estimates list
scalars:
           e(N)     =  74
           e(ll_0)  =  -45.03320955699139
           e(ll)    =  -26.8441890057987
           e(df_m)  =  2
           e(chi2)  =  36.37804110238538
           e(r2_p)  =  .4039023807124769
macros:
           e(depvar)   :  "foreign"
           e(cmd)      :  "probit"
           e(predict)  :  "probit_p"
           e(chi2type) :  "LR"
matrices:
           e(b)     :  1 x 3
           e(V)     :  3 x 3
functions:
           e(sample)
```

Most estimation commands leave behind e(b) (the coefficient vector) and e(V) (the variance–covariance matrix of the estimator):

```
. matrix list e(b)

e(b)[1,3]
            mpg       weight       _cons
y1  -.10395033   -.00233554    8.275464
```

You can refer to `e(b)` and `e(V)` in any matrix expression:

```
. matrix myb = e(b)

. matrix list myb

myb[1,3]
            mpg       weight       _cons
y1  -.10318674    -.0023264    8.234735

. matrix c = e(b)*syminv(e(V))*e(b)´

. matrix list c

symmetric c[1,1]
            y1
y1  22.440544
```

17.6 Creating matrices by accumulating data

In programming estimators, matrices of the form $\mathbf{X'X}$, $\mathbf{X'Z}$, $\mathbf{X'WX}$, and $\mathbf{X'WZ}$ often occur, where $\mathbf{X}$ and $\mathbf{Z}$ are data matrices. `matrix accum`, `matrix glsaccum`, and `matrix vecaccum` produce such matrices; see [R] **matrix accum**.

We recommend you do not load the data into a matrix and use the expression parser directly to form such matrices, although see [R] **matrix mkmat** if that is your interest. If you think that is your interest, be sure to read the *Technical Note* at the end of [R] **matrix mkmat**. There is much to recommend learning how to use the `matrix accum` commands.

17.7 Matrix operators

You can create new matrices or replace existing matrices by typing

$$\texttt{matrix } \textit{matname} = \textit{matrix\_expression}$$

For instance,

```
. matrix A = syminv(R*V*R´)
. matrix IAR = I(rowsof(A)) - A*R
. matrix beta = b*IAR´ + r*A´
. matrix C = -C´
. matrix D = (A, B \ B´, A)
. matrix E = (A+B)*C´
. matrix S = (S+S´)/2
```

The following operators are provided:

Operator	Symbol
Unary operators	
negation	–
transposition	´
Binary operators	
(lowest precedence)	
row join	\
column join	,
addition	+
subtraction	–
multiplication	*
division by scalar	/
Kronecker product	#
(highest precedence)	

Parentheses may be used to change the order of evaluation.

Note in particular that , and \ are operators; (1,2) creates a 1×2 matrix (vector) and (A,B) creates a rowsof(A) $\times$ colsof(A)+colsof(B) matrix, where rowsof(A) = rowsof(B). (1\2) creates a 2×1 matrix (vector) and (A\B) creates a rowsof(A)+rowsof(B) $\times$ colsof(A) matrix, where colsof(A) = colsof(B). Thus, expressions of the form

```
matrix R = (A,B)*Vinv*(A,B)´
```

are allowed.

17.8   Matrix functions

17.8.1   Matrix functions returning matrices

In addition to the functions listed below, see [R] **matrix svd** for singular value decomposition and [R] **matrix symeigen** for eigenvalues and eigenvectors.

I(n)
   Domain: integer scalars 1 to matsize
   Range: identity matrices
   Description: returns $n \times n$ identity matrix

J(r,c,z)
   Domain r: integer scalars 1 to matsize
   Domain c: integer scalars 1 to matsize
   Domain z: scalars $-8e+307$ to $8e+307$
   Range: $r \times c$ matrices
   Description: $r \times c$ matrix containing elements z

cholesky(M)
   Domain: $n \times n$, positive semidefinite, symmetric matrices
   Range: $n \times n$ lower-triangular matrices
   Description: Cholesky decomposition;
                  if $R = \mathrm{cholesky}(S)$, then $RR = S$

syminv(M)
   Domain: $n \times n$, positive semidefinite, symmetric matrices
   Range: $n \times n$ symmetric matrices
   Description: inverse of symmetric matrix

`inv(`*M*`)`
    Domain: $n \times n$ nonsingular matrices
    Range: $n \times n$ matrices
    Description: inverse of matrix

`sweep(`*M*`,`*i*`)`
    Domain *M*: $n \times n$ matrices
    Domain *i*:   integer scalars 1 to n
    Range: $n \times n$ matrices
    Description: Matrix M with ith row/column swept

`corr(`*M*`)`
    Domain: $n \times n$ symmetric variance matrices
    Range: $n \times n$ symmetric correlation matrices
    Description: correlation matrix of variance matrix

`diag(`*v*`)`
    Domain: $1 \times n$ and $n \times 1$ vectors
    Range: $n \times n$ diagonal matrices
    Description: square, diagonal matrix from row or column vector

`vecdiag(`*M*`)`
    Domain: $n \times n$ matrices
    Range: $1 \times n$ vectors
    Description: row vector containing diagonal of matrix

`e(`*name*`)`
    Domain: names
    Range: strings, scalars, matrices, and missing
    Description: value of saved result `e(`*name*`)`;
                  see [U] **21.8 Accessing results calculated by other programs**
                  `e(`*name*`)` = scalar missing if the saved result does not exist
                  `e(`*name*`)` = specified matrix if the saved result is a matrix
                  `e(`*name*`)` = scalar numeric value if the saved result is a scalar
                  `e(`*name*`)` = scalar numeric value if the saved result is a string (macro)
                                     that can be interpreted as a number
                  `e(`*name*`)` = a string containing the first 80 characters of the saved result otherwise

`r(`*name*`)`
    Domain: names
    Range: strings, scalars, matrices, and missing
    Description: value of saved result `r(`*name*`)`;
                  see [U] **21.8 Accessing results calculated by other programs**
                  `e(`*name*`)` = scalar missing if the saved result does not exist
                  `e(`*name*`)` = specified matrix if the saved result is a matrix
                  `e(`*name*`)` = scalar numeric value if the saved result is a scalar
                  `e(`*name*`)` = scalar numeric value if the saved result is a string (macro)
                                     that can be interpreted as a number
                  `e(`*name*`)` = a string containing the first 80 characters of the saved result otherwise

`return(`*name*`)`
    Domain: names
    Range: strings, scalars, matrices, and missing
    Description: value of to-be-saved result `return(`*name*`)`;
                  see [U] **21.8 Accessing results calculated by other programs**

> return(*name*) = scalar missing if the result does not exist
> return(*name*) = specified matrix if the result is a matrix
> return(*name*) = scalar numeric value if the result is a scalar
> return(*name*) = scalar numeric value if the result is a string (macro)
>   that can be interpreted as a number
> return(*name*) = a string containing the first 80 characters of the result otherwise

get(*systemname*)
    Domain: existing names of system matrices
    Range: matrices
    Description: returns copy of Stata internal system matrix *systemname*
              (this function is included for backwards compatibility with previous versions of Stata)

nullmat(*matname*)
    Domain: matrix names, existing and nonexisting
    Range: matrices including null if *matname* does not exist

nullmat() is for use with the row-join (,) and column-join (\) operators in programming situations. Consider the following code fragment, which is an attempt to create the vector $(1, 2, 3, 4)$:

```
local i = 1
while `i´ <= 4 {
        mat v = (v, `i´)
        local i = `i´ + 1
}
```

The above program will not work because, the first time through the loop, v will not yet exist and thus forming (v, `i´) makes no sense. nullmat() relaxes that restriction:

```
local i = 1
while `i´ <= 4 {
        mat v = (nullmat(v), `i´)
        local i = `i´ + 1
}
```

The nullmat() function informs Stata that you know v might not exist and, in that case, the function row-join is to be generalized. Joining nothing with `i´ results in (`i´). Thus, the first time through the loop, v = (1) is formed. The second time through, v does exist and v = (1, 2) is formed, and so on.

nullmat() can be used only with the , and \ operators.

17.8.2   Matrix functions returning scalars

rowsof(*M*)
    Domain: matrices
    Range: integer scalars 1 to matsize
    Description: number of rows of M

colsof(*M*)
    Domain: matrices
    Range: integer scalars 1 to matsize
    Description: number of columns of M

rownumb(*M*,*s*)
    Domain M: matrices
    Domain s: strings

Range: integer scalars 1 to `matsize` and missing
Description: row number of M associated with row name s;
             returns missing if row cannot be found

`colnumb(`M`,`s`)`
  Domain M: matrices
  Domain s: strings
  Range: integer scalars 1 to `matsize` and missing
  Description: column number of M associated with row name s;
               returns missing if column cannot be found

`trace(`M`)`
  Domain M: $n \times n$ (square) matrices
  Range: scalars −8e+307 to 8e+307
  Description: trace of matrix

`det(`M`)`
  Domain M: $n \times n$ (square) matrices
  Range: scalars −8e+307 to 8e+307
  Description: determinant of matrix

`diag0cnt(`M`)`
  Domain M: $n \times n$ (square) matrices
  Range: integer scalars 0 to n
  Description: number of zeros on diagonal

`mreldif(`X`,`Y`)`
  Domain X: matrices
  Domain Y: matrices with same number of rows and columns as X
  Range: scalars −8e+307 to 8e+307
  Description: maximum relative difference $\max_{i,j} |x_{ij} - y_{ij}|/(|y_{ij}| + 1)$

`el(`s`,`i`,`j`)`
  Domain s: strings containing matrix name
  Domain i: scalars 1 to `matsize`
  Domain j: scalars 1 to `matsize`
  Range: scalars −8e+307 to 8e+307 and missing
  Description: $s[\lfloor i \rfloor, \lfloor j \rfloor]$, the i,j element of the matrix named s
               returns missing if i or j are out of range or matrix s does not exist

17.9 Subscripting

1. In matrix and scalar expressions, you may refer to *matname*`[`r`,`c`]`, where r and c are scalar expressions, to obtain a single element of *matname* as a scalar.

   Examples:
   ```
   matrix A = A / A[1,1]
   gen newvar = oldvar / A[2,2]
   ```

2. In matrix and scalar expressions, you may refer to *matname*`[`s_r`,`s_c`]`, where s_r and s_c are string expressions, to obtain a single element of *matname* as a scalar. The element returned is based on searching the row and column names.

   Examples:
   ```
   matrix B = B / V["mpg:price","mpg:price"]
   generate sdif = dif / sqrt(V["price","price"])
   ```

3. In matrix and scalar expressions, you may mix these two syntaxes and so refer to *matname*$[r, s_c]$ or to *matname*$[s_r, c]$.

Examples:
```
matrix b = b * R[1,"price"]
generate hat = price*b[1,"mpg:price"]
```

4. In matrix expressions, you may refer to *matname*$[r_1..r_2, c_1..c_2]$ to refer to submatrices; r_1, r_2, c_1, and c_2 may be scalar expressions. If r_2 evaluates to missing, it is taken as referring to the last row of *matname*; if c_2 evaluates to missing, it is taken as referring to the last column of *matname*. Thus, *matname*$[r_1..., c_1...]$ is allowed.

Examples:
```
mat S = Z[1..4, 1..4]
mat R = Z[5..., 5...]
```

5. In matrix expressions, you may refer to *matname*$[s_{r1}..s_{r2}, s_{c1}..s_{c2}]$ to refer to submatrices where s_{r1}, s_{r2}, s_{c1}, and s_{c2}, are string expressions. The matrix returned is based on looking up the row and column names.

If the string evaluates to an equation name only, all the rows or columns for the equation are returned.

Examples:
```
mat S = Z["price".."weight", "price".."weight"]
mat L = D["mpg:price".."mpg:weight", "mpg:price".."mpg:weight"]
mat T1 = C["mpg:", "mpg:"]
mat T2 = C["mpg:", "price:"]
```

6. In matrix expressions, any of the above syntaxes may be combined.

Examples:
```
mat T1 = C["mpg:", "price:weight".."price:displ"]
mat T2 = C["mpg:", "price:weight"...]
mat T3 = C["mpg:price", 2..5]
mat T4 = C["mpg:price", 2]
```

7. When defining an element of a matrix, use

$$\texttt{matrix } matname[i,j] = expression$$

where i and j are scalar expressions. The matrix *matname* must already exist.

Example:
```
mat A = J(2,2,0)
mat A[1,2] = sqrt(2)
```

8. To replace a submatrix within a matrix, use the same syntax. If the expression on the right evaluates to a scalar or 1×1 matrix, the element is replaced. If it evaluates to a matrix, the submatrix with top-left element at (i, j) is replaced. The matrix *matname* must already exist.

Example:
```
mat A = J(4,4,0)
mat A[2,2] = C´*C
```

17.10   Using matrices in scalar expressions

Scalar expressions are what are documented as *exp* in the Stata manuals:

```
generate newvar = exp if exp ...
replace  newvar = exp if exp ...
regress ... if exp ...
if exp {... }
while exp {... }
```

Most importantly, scalar expressions occur in `generate` and `replace`, in the `if` *exp* modifier allowed on the end of many commands, and in the `if` and `while` commands for program control.

It is rare that one needs to refer to a matrix in any of these situations except the `if` and `while` commands.

In any case, you may refer to matrices in any of these situations, but the expression cannot require evaluation of matrix expressions returning matrices. Thus, you could refer to `trace(A)` but not to `trace(A+B)`.

It can be difficult to predict when an evaluation of an expression requires evaluating a matrix; even experienced users can be surprised. If you get the error message "matrix operators that return matrices not allowed in this context", r(509), you have encountered such a situation.

The solution, in such cases, is to split the line in two. For instance, one would change

```
if trace(A+B)==0 {
        . . .
}
```

to

```
matrix AplusB = A+B
if trace(AplusB)==0 {
        . . .
}
```

or even to

```
matrix Trace = trace(A+B)
if Trace[1,1]==0 {
        . . .
}
```

18 Printing and preserving output

Contents

18.1 Overview

Stata will record your session into a file. By default, the resulting file—called a log file—contains what you type and what Stata typed in response. The file can be printed or, better, incorporated into documents you create with your word processor.

To start a log:
    Your session is now being recorded
    in file *filename*.`log`.
	. `log using` *filename*

To temporarily stop logging:
    Temporarily stop:
    Resume:
	. `log off`
	. `log on`

To stop logging and close the file:
    You can now print or load *filename*.`log`
    into your word processor.
	. `log close`

Alternative ways to start logging:
    append to an existing log:
    replace an existing log:
	. `log using` *filename*, `append`
	. `log using` *filename*, `replace`

Stata for Windows and Macintosh users:
    The above works, but in addition
    To start a log:
    To temporarily stop logging:
    To resume:
    To stop logging and close the file:
	
	click the **Log** button
	click the **Log** button and choose *Suspend*
	click the **Log** button and choose *Resume*
	click the **Log** button and choose *Close*

An alternative way of logging produces logs containing solely what you typed—logs that, while not containing your results, are sufficient to recreate the session.

To start a what-you-type-only log:	. `log using` *filename*, `noproc`
To stop logging and close the file:	. `log close`
To recreate your session:	. `do` *filename*.`log`

18.1.1 Starting and closing logs

With great foresight, you begin working in Stata and type log using session (or click the **Log** button) before starting your work:

```
. log using session

. use census
(Census Data)

. tabulate reg [freq=pop]

Census      |
region      |      Freq.      Percent        Cum.
------------+-----------------------------------
        NE  |   49135283        21.75       21.75
    N Cntrl |   58865670        26.06       47.81
      South |   74734029        33.08       80.89
       West |   43172490        19.11      100.00
------------+-----------------------------------
      Total |  225907472       100.00

. summarize medage

Variable |     Obs        Mean    Std. Dev.       Min         Max
---------+-------------------------------------------------------
  medage |      50       29.54        1.69       24.20       34.70

. log close
```

There is now a file named session.log on your disk. It is a standard ASCII file. Here is the result of typing the file:

```
. type session.log
. use census
(Census Data)

. tabulate reg [freq=pop]

Census      |
region      |      Freq.      Percent        Cum.
------------+-----------------------------------
        NE  |   49135283        21.75       21.75
    N Cntrl |   58865670        26.06       47.81
      South |   74734029        33.08       80.89
       West |   43172490        19.11      100.00
------------+-----------------------------------
      Total |  225907472       100.00

. summarize medage

Variable |     Obs        Mean    Std. Dev.       Min         Max
---------+-------------------------------------------------------
  medage |      50       29.54        1.69       24.20       34.70

. log close

. _
```

The file is suitable for inclusion into a word processor or printing.

If you intend to exit Stata immediately, you do not have to close the file before exiting. Stata will do that for you.

18.1.2 Appending to an existing log

You previously typed `log using session` and later typed `log close`. In some future Stata session, you type `log using session` again. Here is what happens:

```
. log using session
file session.log already exists
r(602);
```

Stata never lets you accidentally write over a file. You have three choices: (1) choose a different name; (2) append onto the end of the existing file by typing `log using session, append`; or (3) replace the existing file by typing `log using session, replace`. In this last case, you are telling Stata that you know the file already exists and that it is okay to replace it.

Stata for Windows and Macintosh users: All of this applies to you, but you will probably open the log by pressing the **Log** button. Overwriting or appending is handled in the standard way.

18.1.3 Temporarily suspending and resuming logging

Once you are logging your session, you can turn logging on and off. When you turn logging off, Stata temporarily stops recording your session but leaves the log file open. When you turn logging back on, Stata continues to record your session, appending the additional record to the end of the file.

For instance, say the first time something interesting happens, you type `log using results` (or click on **Log** and open `results.log`). You then retype the command that produced the interesting result (or double-click on the command in the Review window, or use the editing key PrevLine to retrieve the command; see [U] **13 Keyboard use**). You now have a hard copy of the interesting result.

You are now reasonably sure that nothing interesting will occur, at least for a while. Rather than type `log close`, however, you type `log off`. (Stata for Windows and Stata for Macintosh users could alternatively click on **Log** and choose *Suspend*.) From now on, nothing goes into the file. The next time something happens, you type `log on` (or click on **Log** and choose *Resume*) and reissue the (interesting) command. After that, you type `log off`. You keep working like this—toggling the log on and off.

18.2 Placing comments in logs

Everything you type and everything Stata types in response goes into the log.

Stata treats lines starting with a '*' as comments and ignores them. Thus, if you are working interactively and wish to make a comment, you can type '*' followed by your comment:

```
. * check that all the spells are completed
.
```

Stata ignores your comment but, if you have a log going, the comment now appears in the file.

18.3 Logging only what you type

Since log files record everything that happens during a session, they tend to become large quickly. The `log on` and `log off` commands are a partial solution. Another solution is to tell Stata to log only what you type but not what the computer types back in response. This method does not give you a record of results, but it does give you a record of what you did. With that record, you can recreate the results whenever you wish. To restrict the log to solely what you type, use the `noproc` option. (Stata for Windows and Stata for Macintosh users: this is a case where you must type the command; the **Log** button cannot access this feature.)

With the same great foresight you demonstrated in our first example, you begin working in Stata but you type `log using session, noproc` before starting to work.

```
. log using session, noproc
. use census
(Census Data)
. tabulate reg [freq=pop]
Census     |
region     |      Freq.      Percent        Cum.
-----------+-----------------------------------
       NE  |   49135283        21.75       21.75
  N Cntrl  |   58865670        26.06       47.81
    South  |   74734029        33.08       80.89
     West  |   43172490        19.11      100.00
-----------+-----------------------------------
    Total  |  225907472       100.00
. summarize medage
Variable |     Obs        Mean    Std. Dev.        Min         Max
---------+-------------------------------------------------------
  medage |      50       29.54        1.69      24.20       34.70
. log close
```

Just as before, Stata creates a file called `session.log`. The contents of that file, however, are different:

```
. type session.log
use census
tabulate reg [freq=pop]
summarize medage
log close

. _
```

Not only does `log using` *filename*, `noproc` create a record of what you have done, but you can easily reexecute the result as a do-file; see [U] **19 Do-files**. If you type `do session.log`, Stata will read and reexecute the commands stored in the file.

❑ Technical Note

`log` can be combined with `#review` (see [U] **13 Keyboard use**) to bail you out when you have not adequately planned ahead. You have been working in front of your computer and you now realize that you have done what you wanted to do. Unfortunately, you are not sure exactly what it is you have done. Did you make a mistake? Could you reproduce the result? Unfortunately, you have not been logging your output. `#review` will allow you to look over what you have done and, combined with `log`, you can still make a record.

Type `log using` *filename*. Type `#review 100`. Stata will list the last 100 commands you gave, or however many it has stored. Since `log` is making a record, that list will also be stored in the file. Finally, type `log close`.

❑

18.4 The log-button alternative (Windows and Macintosh)

You can use the `log` command just as any Stata user can. In addition, all of `log`'s features except for `log, noproc` are available via the **Log** button.

When you open a log via the **Log** button, a Log window appears. You can close the window if you wish—closing the window does not close the log—and you can get it back by clicking on the **Log** button again or, if you use Windows 98/95/NT, clicking on the **Bring Log Window to Front** button.

Even if you do not close the Log window, you will probably want to bring the Results window back to the front. Do this by clicking on the **Results** button. Reviewing the Log window when Stata is idle is one thing, but working with Stata and using the Log window as a substitute for the Results window can be confusing. The Log window is buffered—meaning the output appears in clumps only every so often.

18.5 Printing logs with Stata for Windows and Macintosh

You print a log by pulling down **File** and choosing **Print Log**. You do this while the log is still open.

To print an old log, reopen it for appending, print it, and close it again.

Remember, a log is a plain text (ASCII) file—you can load it into your word processor and print it from there if you wish. Just be sure to use a fixed-width font (e.g., Courier) to print it; otherwise, the columns will not line up.

18.6 Printing logs with Stata for Unix

You can print logs using Unix's lp(1) or lpr(1) commands. To format the log before printing, see [GSU] **fsl**.

You can review logs using Unix's cat(1), more(1), or pg(1) commands or using Stata's `type` command; see [R] **type**.

Remember, a log is a plain ASCII file—you can load it into your editor or word processor.

18.7 References

Cox, N. J. 1994. os13: Using awk and fgrep for selective extraction from Stata log files. *Stata Technical Bulletin* 19: 15–17. Reprinted in *Stata Technical Bulletin Reprints*, vol 4, pp. 78–80.

19 Do-files

19.1 Description

Rather than typing commands at the keyboard, you can create a disk file containing commands and instruct Stata to execute the commands stored in that file. Such files are called *do-files*, since the command that causes them to be executed is do.

A do-file is a standard ASCII text file.

A do-file is executed by Stata when you type do *filename*.

In addition, Stata for Windows 98/95/NT and Stata for Macintosh users can use the built-in do-file editor; see [GSW] **14 Using the do-file editor** or [GSM] **14 Using the do-file editor**.

▷ Example

You can use do-files to create a batch-like environment in which you place all the commands you want to perform in a file and then instruct Stata to do that file. For instance, assume you use your text editor or word processor to create a file called `myjob.do` that contains the three lines

```
──────────────────────────────────────────────────── top of myjob.do ─────────
use census
tabulate reg
summarize mrgrate dvcrate medage if state~="NV"
──────────────────────────────────────────── end of myjob.do ─────────
```

You then enter Stata and instruct Stata to do the file:

```
. do myjob

. use census
(Census data)
```

```
. tabulate reg

Census     |
region     |      Freq.      Percent        Cum.
-----------+-----------------------------------
        NE |          9        18.00       18.00
    N Cntrl |         12        24.00       42.00
     South |         16        32.00       74.00
      West |         13        26.00      100.00
-----------+-----------------------------------
     Total |         50       100.00

. summarize mrgrate dvcrate medage if state~="NV"

Variable |     Obs        Mean    Std. Dev.        Min         Max
---------+--------------------------------------------------------
 mrgrate |      49    .0150694    .0033136    .0103731    .0247333
 dvcrate |      49    .0076565    .0022192    .0039954    .0129728
  medage |      49       29.53        1.71       24.20       34.70

end of do-file

. _
```

You typed only **do myjob** to produce this output. Since you did not specify the file extension, Stata assumed you meant **do myjob.do**; see [U] **14.6 File-naming conventions**.

◁

19.1.1   Version

We recommend that the first line in your do-file declare the Stata release under which you wrote the do-file; **myjob.do** would better read as

```
─────────────────────────────────────── top of myjob.do ───────────
version 6.0
use census
tabulate reg
summarize mrgrate dvcrate medage if state~="NV"
─────────────────────────────────────── end of myjob.do ───────────
```

We admit that we do not always follow our own advice and you will see many examples in this manual that do not include the **version 6.0** line.

If you intend on keeping the do-file, however, you should include the line since it ensures that your do-file will continue to work with future versions of Stata. Stata is under constant development and sometimes things change and in surprising ways.

For instance, in Stata 3.0 a new syntax for specifying the weights was introduced. If you had an old do-file written for Stata 2.1 that analyzed weighted data and did not have **version 2.1** at the top, you would find that today's Stata would flag some of its lines as syntax errors. If you had the **version 2.1** line, it would work just as it used to.

In Stata 4.0, we updated the random number generator **uniform()** —the new one is better in that it has a longer period. If you wrote an old do-file back in the days of Stata 3.1 that made a bootstrap calculation of variance and it did not include **version 3.1** at the top, it would now produce different (but equivalent) results. If you had included the line, it would produce the same results as it used to.

When running an old do-file that includes a **version** statement, you need not worry about setting the version back. Stata automatically restores the previous value of **version** when the do-file completes.

19.1.2   Comments and blank lines in do-files

You may freely include blank lines in your do-file. In the above example, the do-file could just as well have read

```
——————————————————————————————————— top of myjob.do ——————————
version 6.0

use census
tabulate reg
summarize mrgrate dvcrate medage if state~="NV"
—————————————————————————————————— end of myjob.do ——————————
```

You may include comments in do-files in either of two ways:

1. Begin the line with a '*'; Stata ignores such lines.

2. Placing the comment in /* */ delimiters.

Thus, `myjob.do` might read

```
——————————————————————————————————— top of myjob.do ——————————
* a sample analysis job
version 6.0

use census

/* obtain the summary statistics: */
tabulate reg
summarize mrgrate dvcrate medage if state~="NV"
—————————————————————————————————— end of myjob.do ——————————
```

Which style of comment indicator you use is up to you. One advantage of the /* */ method is that it can be put at the end of lines:

```
——————————————————————————————————— top of myjob.do ——————————
* a sample analysis job
version 6.0

use census

tabulate reg                      /* obtain summary statistics */
summarize mrgrate dvcrate medage if state~="NV"
—————————————————————————————————— end of myjob.do ——————————
```

In fact, /* */ can be put anywhere, even in the middle of a line:

```
——————————————————————————————————— top of myjob.do ——————————
* a sample analysis job
version 6.0

use census

tabulate reg                      /* obtain summary statistics */
summarize /* marriage rate */ mrgrate dvcrate medage if state~="NV"
—————————————————————————————————— end of myjob.do ——————————
```

❑ Technical Note

The /* */ comment indicator can be used in do-files only; you may not use it interactively. You can, however, use the '*' comment indicator interactively.

❑

19.1.3   Long lines in do-files

When you use Stata interactively, you press *Enter* to end a line and tell Stata to execute it. If you need to type a line wider than the screen, you just do, letting it wrap or scroll.

You can follow the same procedure in do-files—if your editor or word processor will let you—but you can do better. You can change the end-of-line delimiter to ';' using **#delimit** or you can comment out the line break using **/* */** comment delimiters.

▷ Example

In the following fragment of a do-file, we temporarily change the end-of-line delimiter:

─── fragment of example.do ──────────
```
use mydata
#delimit ;
summarize weight price displ hdroom rep78 length turn gratio
        if substr(company,1,4)=="Ford" |
            substr(company,1,2)=="GM", detail ;
gen byte ford = substr(company,1,4)=="Ford" ;
#delimit cr
gen byte gm = substr(company,1,2)=="GM"
```
── fragment of example.do ─────────

Once we change the line delimiter to semicolon, all lines, even short ones, must end in semicolons. Stata treats carriage returns as no different from blanks. We can change the delimiter back to carriage return by typing **#delimit cr**.

Note that the **#delimit** command is allowed only in do-files—it is not allowed interactively. You need not remember to set the delimiter back to carriage return at the end of a do-file because Stata will reset it automatically.

◁

▷ Example

The other way around long lines is to comment out the carriage return using **/* */** comment brackets. Thus, our code fragment could also read

─── fragment of example.do ──────────
```
use mydata
summarize weight price displ hdroom rep78 length turn gratio /*
    */  if substr(company,1,4)=="Ford" |    /*
    */      substr(company,1,2)=="GM", detail
gen byte ford = substr(company,1,4)=="Ford"
gen byte gm = substr(company,1,2)=="GM"
```
── fragment of example.do ─────────

◁

19.1.4   Error handling in do-files

A do-file completes execution when (1) the end of the file is reached, (2) an **exit** is executed, or (3) an error (nonzero *return code*) occurs. If an error occurs, the remaining commands in the do-file are not executed.

If you press *Break* while executing a do-file, Stata responds as though an error occurred, stopping the do-file. This is because the return code is nonzero; see [U] **11 Error messages and return codes** for an explanation of return codes.

▷ Example

Here is what happens when we start a do-file executing and then press *Break*:

```
. do myjob
. version 6.0
. use census
(Census data)
. tabulate reg
     Census|
     Region|      Freq.      Percent        Cum.
--Break--
r(1);

end of do-file
--Break--
r(1);

. _
```

When we pressed *Break*, Stata responded by typing --Break-- and showed a return code of 1. Stata seemingly repeated itself, typing first "end of do-file" and then --Break-- and the return code of 1 again. Do not worry about the repeated messages. The first message indicates that Stata was stopping the tabulate because you pressed *Break*, and the second message indicates that Stata is stopping the do-file for the same reason.

◁

▷ Example

Let's try our example again, but this time let's introduce an error. We change the file myjob.do to read

```
─────────────────────────────────────── top of myjob.do ───────────
use censas
tabulate reg
summarize mrgrate dvcrate medage if state~="NV"
─────────────────────────────────────── end of myjob.do ───────────
```

Note our subtle typographical error. We typed use censas when we meant use census. We assume that there is no file called censas.dta, so now we have an error. Here is what happens when you instruct Stata to do the file:

```
. do myjob
. version 6.0
. use censas
file censas.dta not found
r(601);

end of do-file
r(601);

. _
```

When Stata was told to **use censas**, it responded with "file censas.dta not found" and a return code of 601. Stata then typed "end of do-file" and repeated the return code of 601. The repeated message phenomenon occurred for the same reason as it did when we pressed *Break* in the previous example. The **use** resulted in a return code of 601, so the do-file itself resulted in the same return code. The important thing to understand is that since there was an error, Stata stopped executing the file.

◁

❏ Technical Note

We can tell Stata to continue executing the file even if there are errors, by typing do *filename*, nostop. Here is the result:

```
. do myjob, nostop

. version 6.0

. use censas
file censas.dta not found
r(601);

. tabulate reg
no variables defined
r(111);

. summarize mrgrate dvcrate medage if state~="NV"
no variables defined
r(111);

end of do-file

. _
```

None of the commands worked because the do-file's first command failed. That is why Stata ordinarily stops. However, if our file contained anything that could work, it would work.

❏

19.1.5   Logging the output of do-files

You log the output of do-files just as you would an interactive session; see [U] **18 Printing and preserving output**.

Many users include the commands to start and stop the logging in the do-file itself:

```
─────────────────────────────────────────── top of myjob.do ───────────
version 6.0
log using myjob, replace
* a sample analysis job
use census
tabulate reg                        /* obtain summary statistics */
summarize mrgrate dvcrate medage if state~="NV"
log close
─────────────────────────────────────────── end of myjob.do ───────────
```

We chose to open with **log using myjob, replace**, the important part being the **replace** option. Had we omitted the option, we could not easily rerun our do-file. If **myjob.log** already existed and **log** were not told that it is okay to replace the file, the do-file would have stopped and instead reported "file myjob.log already exists". We could get around that, of course, by erasing the log file before running the do-file.

19.1.6 Preventing – more – conditions

Assume you are running a do-file and logging the output so that you can look at it later. In that case, Stata's feature of pausing every time the screen is full is just an irritation: It means you have to sit and watch the do-file run so you can clear the `--more--`.

The way around this is to include the line `set more off` in your do-file. Setting more to `off`, as explained in [U] **10 –more– conditions**, prevents Stata from ever issuing a `--more--`.

19.1.7 Calling other do-files

Do-files may call other do-files. For instance, say you wrote `makedata.do` that infiles your data, generates a few variables, and saves `step1.dta`. Say you wrote `anlstep1.do` that performed a little analysis on `step1.dta`. You could then create a third do-file,

```
———————————————————————————————— top of master.do ————————————
      version 6.0
      do makedata
      do anlstep1
—————————————————————————————————— end of master.do ————————————
```

and so, in effect, combine the two do-files.

Do-files may call do-files which, in turn, call do-files, and so on. Stata allows do-files to be nested 32 deep.

Be not confused: `master.do` above could call 1,000 do-files one after the other and still the maximum level of nesting would be only two.

19.2 Ways to run a do-file (Stata for Windows 98/95/NT)

1. You can use the do-file editor to compose, save, load, and execute do-files; see [GSW] **14 Using the do-file editor**. Click on the **do-file editor** button or type `doedit` in the Command window.

2. Alternatively, you can execute do-files by pulling down **File** and choosing **Do...**.

3. Alternatively, you can execute do-files by typing `do` followed by the filename just as we have shown above.

Regardless of the method chosen, if you wish to log the output, you can start the log before executing the do-file or you can include the `log using` and `log close` in your do-file.

4. You can double-click on the do-file to launch Stata and run the do-file. When the do-file completes, Stata will prompt you for the next command just as if you had started Stata the normal way. If you want Stata to exit instead, include `exit, STATA clear` as the last line of your do-file. If you want to log the output, you should include the `log using` and `log close` in your do-file.

5. You can run the do-file in batch mode. See [GSW] **A.9 Executing Stata in background (batch) mode** for details. When you start Stata in this way, Stata will run in the background. When the do-file completes, the Stata icon on the taskbar will flash. You can then click on it to close Stata. If you want to stop the do-file before it completes, click on the Stata icon on the taskbar and Stata will ask you if you want to cancel the job.

When you use Stata in this way, Stata takes the following actions:

a. Stata automatically opens a log. If your do-file is named *xyz*.`do`, the log will be called *xyz*.`log` in the same directory.

b. If your do-file explicitly opens another log, that is okay. You will end up with two copies of the output.

c. Stata ignores --more-- conditions and anything else that would cause the do-file to stop were it running interactively.

19.3   Ways to run a do-file (Stata for Windows 3.1)

1. You can execute do-files by pulling down **File** and choosing **Do...**.

2. Alternatively, you can execute do-files by typing do followed by the filename just as we have shown above.

If you wish to log the output, you can start the log before executing the do-file or you can include the log using and log close in your do-file.

19.4   Ways to run a do-file (Stata for Macintosh)

1. You can use the do-file editor to compose, save, load, and execute do-files; see [GSM] **14 Using the do-file editor**. Click on the **do-file editor** button or type doedit in the Command window.

2. Alternatively, you can execute do-files by pulling down **File** and choosing **Do...**.

3. Alternatively, with Stata running, you can go to the desktop and double-click on the do-file.

4. Alternatively, you can execute do-files by typing do followed by the filename just as we have shown above.

Regardless of the method chosen, if you wish to log the output, you can start the log before executing the do-file or you can include the log using and log close in your do-file.

5. With Stata not running, you can double-click on the do-file to launch Stata and run the do-file. When the do-file completes, Stata will prompt you for the next command just as if you had started Stata the normal way. If you want Stata to exit instead, include exit, STATA clear as the last line of your do-file. If you want to log the output, you should include the log using and log close in your do-file.

19.5   Ways to run a do-file (Stata for Unix)

1. You can execute do-files by typing do followed by the filename just as we have shown above. If you wish to log the output, you can start the log before typing do *filename* or you can include the log using and log close in your do-file.

2. At the Unix prompt, you can type stata do *filename* to start Stata and run the do-file. When the do-file completes, Stata will prompt you for the next command just as if you had started Stata the normal way. If you want Stata to exit instead, include exit, STATA clear as the last line of your do-file. If you want to log the output, you should include the log using and log close in your do-file.

3. At the Unix prompt, you can type 'stata -b do *filename* &' to the Unix prompt. Do this and Stata runs the do-file in the background. If you use this method, Stata takes the following actions:

a. Stata automatically opens a log. If your do-file is named *xyz*.do, the log will be called *xyz*.log in the current directory (the directory from which you issued the stata command).

b. If your do-file explicitly opens another log, that is okay. You will end up with two copies of the output.

c. Stata ignores `--more--` conditions and anything else that would cause the do-file to stop were it running interactively.

Thus, to reiterate, one way to run a do-file in the background is by typing

```
% stata -b do myfile &
```

Another way uses standard redirection:

```
% stata < myfile.do > myfile.log &
```

Warning: If your do-file contains either the `#delimit` command or the comment characters (`/*` at the end of one line and `*/` at the beginning of the next), the second method will not work. We recommend that you use the first method: `stata -b do myfile &`.

Either way, Stata knows it is in the background and ignores `--more--` conditions and anything else that would cause the do-file to stop were it running interactively.

19.6   Programming with do-files

This is an advanced topic and we are going to refer to concepts not yet explained.

19.6.1   Argument passing

Do-files accept arguments just as Stata programs do; this is described in [U] **21 Programming Stata** and [U] **21.4 Program arguments**. In fact, the logic Stata follows when invoking a do-file is the same as when invoking a program: the local macros are saved and new ones are defined. Arguments are stored in the local macros `` `1´ ``, `` `2´ ``, and so on. When the do-file completes, the previous definitions are restored just as with programs.

Thus, if you wanted your do-file to

1. use a dataset of your choosing,

2. tabulate a variable named `reg`,

3. summarize variables `mrgrate` and `dvcrate`,

you could write the do-file

——————————————————————————————————————— top of myxmpl.do ———————
```
use `1´
tabulate reg
summarize mrgrate dvcrate
```
——————————————————————————————————————— end of myxmpl.do ———————

and you could run this do-file by typing, for instance,

```
. do myxmpl census
```
  (*output omitted*)

The first command—`use `1´`—would be interpreted as `use census` because `census` was the first argument you typed after `do myxmpl`.

An even better version of the do-file would read

```
───────────────────────────────────────────────── top of myxmpl.do ───────────
    args varname
    use `varname´
    tabulate reg
    summarize mrgrate dvcrate
──────────────────────────────────────────────────── end of myxmpl.do ─────────
```

The `args` command merely assigns a better name to the argument passed. `args varname` does not verify that what we type following `do myxmpl` is a variable name—we would have to use the `syntax` command if we wanted to do that—but substituting `varname` for `1` does make the code more readable.

If our program were to receive two arguments, we could refer to them as `1` and `2` or we could put an 'args varname other' at the top of our do-file and then refer to `varname` and `other`.

To learn more about argument passing, see [U] **21.4 Program arguments**.

19.6.2 Suppressing output

There is an alternative to typing do *filename*; it is run *filename*. run works in the same way as do except that neither the instructions in the file nor any of the output caused by those instructions is shown on the screen or in the log file.

For instance, using the above `myxmpl.do`, typing `run myxmpl census` results in

```
. run myxmpl census

. _
```

All the instructions were executed, but none of the output was shown.

This is not useful in this case, but if the do-file contained only the definitions of Stata programs—see [U] **21 Programming Stata**—and you merely wanted to load the programs without seeing the code, run would be useful.

20 Ado-files

Contents

20.1   Description

Stata is programmable and even if you never write a Stata program, Stata's programmability is important to you. Many of Stata's features are implemented as Stata programs and new features are being implemented every day for Stata, both by us and by others.

1. You can obtain additions from the *Stata Technical Bulletin*. The additions are distributed on diskette and over the Internet.

2. You can obtain additions from the Stata listserver. There is an active group of users advising each other on how to use Stata and, sometimes, even trading programs. Visit the Stata web site *http://www.stata.com* for instructions on how to subscribe; subscribing to the listserver is free.

3. You may decide to write your own additions to Stata.

This chapter is written for people who want to consume ado-files. All users should read it. If you later decide you want to write ado-files, the technical description for programmers can be found in [U] **21.11 Ado-files**.

20.2   What is an ado-file?

An ado-file defines a Stata command, but not all Stata commands are defined by ado-files.

When you type `summarize` to obtain summary statistics, you are using a command built into Stata.

When you type `ci` to obtain confidence intervals, you are running an ado-file. You probably did not know that because the results of using a built-in command or a program are indistinguishable.

An ado-file is an ASCII text file that contains a Stata program. When you type a command that Stata does not know, it looks in certain places for an ado-file of that name. If Stata finds it, Stata loads and executes it, so it appears to you as if the ado-command is just one more command built into Stata.

We just told you that Stata's `ci` command is implemented as an ado-file. That means that, somewhere, there is a file named `ci.ado`.

Ado-files tend to come with help-files. When you type `help ci` (or pull down **Help** and work your way to `ci`'s manual page), Stata looks for `ci.hlp` just as it looks for `ci.ado` when you use the `ci` command. A help-file is also an ASCII text file that tells Stata's help system what to display.

20.3   How can I tell if a command is built-in or an ado-file?

Use the `which` command to determine if a file is built in or implemented as an ado-file. For instance, `logistic` is an ado-file and here is what happens when you type `which logistic`:

```
. which logistic
c:\stata\ado\base\l\logistic.ado
*! version 3.1.1  04nov1998
```

`logit` is a built-in command:

```
. which logit
ado-file for logit not found
r(111);
```

20.4   Can I look at an ado-file?

Certainly. When you type `which` followed by an ado-command, Stata reports where the file is stored:

```
. which logistic
c:\stata\ado\base\l\logistic.ado
*! version 3.1.1  04nov1998
```

Ado-files are just ASCII text files containing the Stata program, so you can type them (or even look at them in your editor or word processor):

```
. type c:\stata\ado\base\l\logistic.ado
*! version 3.1.1  04nov1998
program define logistic, eclass
        version 6.0
        local options `"Level(integer $S_level)"'
  (output omitted )
end
```

You can also look at the corresponding help-file in raw form if you wish. If there is a help-file, it is stored in the same place as the ado-file:

```
. type c:\stata\ado\base\l\logistic.hlp
.-
help for ^logistic^                                     (manual:  ^[R] logistic^)
.-
  (output omitted )
```

20.5   Where does Stata look for ado-files?

Stata looks for ado-files in seven places which can be categorized three ways:

   I. the official ado-directories, meaning
       1. (UPDATES) the official updates directory and
       2. (BASE) the official base directory;

  II. your personal ado directories, meaning
       3. (SITE) the directory for ado-files your site might have installed,
       4. (STBPLUS) the directory for ado-files you personally might have installed,
       5. (PERSONAL) the directory for ado-files you personally might have written, and
       6. (OLDPLACE) the directory where Stata users used to save their personally written ado-files;

 III. the current directory, meaning
       7. (.) the ado-files you have written just this instant or for just this project.

Where these directories are varies from computer to computer, but Stata's `sysdir` command will tell you where they are on your computer:

```
. sysdir
    STATA:  C:\STATA\
  UPDATES:  C:\STATA\ado\updates\
     BASE:  C:\STATA\ado\base\
     SITE:  C:\STATA\ado\site\
  STBPLUS:  C:\ado\stbplus\
 PERSONAL:  C:\ado\personal\
 OLDPLACE:  C:\ado\
```

20.5.1   Where are the official ado directories?

These are the directories listed as BASE and UPDATES by `sysdir`:

```
. sysdir
    STATA:  C:\STATA\
  UPDATES:  C:\STATA\ado\updates\
     BASE:  C:\STATA\ado\base\
     SITE:  C:\STATA\ado\site\
  STBPLUS:  C:\ado\stbplus\
 PERSONAL:  C:\ado\personal\
 OLDPLACE:  C:\ado\
```

1. BASE contains the ado-files we originally shipped to you.

2. UPDATES contains any updates you might have installed since then. You would install these updates using the `update` command or by pulling down **Help** and choosing **Official Updates**; see [U] **20.8 How do I install STB updates?**.

20.5.2   Where is my personal ado directory?

These are the directories listed as PERSONAL, STBPLUS, SITE, and OLDPLACE by `sysdir`:

```
. sysdir
    STATA:  C:\STATA\
  UPDATES:  C:\STATA\ado\updates\
     BASE:  C:\STATA\ado\base\
     SITE:  C:\STATA\ado\site\
  STBPLUS:  C:\ado\stbplus\
 PERSONAL:  C:\ado\personal\
 OLDPLACE:  C:\ado\
```

1. PERSONAL is for ado-files you personally have written. Store your very private ado-files here; see [U] **20.7 How do I add my own ado-files?**.

2. STBPLUS is for ado-files you personally installed but did not write. Such ado-files are obtained from the STB but they are sometimes found in other places, too. You find and install such files using Stata's **net** command or you pull down **Help** and select **STB and User-written Programs**; see [U] **20.6 How do I install an addition?**.

3. SITE is really the opposite of a personal ado directory—it is a public directory and it corresponds to STBPLUS in a public sort of way. If you are on a networked computer, the site administrator can install ado-files here and all Stata users will then be able to use them just as if each found and installed them in their STBPLUS directory for themselves. Site administrators find and install the ado-files just as you would—using Stata's **net** command, the difference being that they specify an option when they say to install something so that Stata knows to write the files into SITE rather than STBPLUS. Site administrators should see [R] **net**.

4. OLDPLACE is for old Stata users. Prior to Stata 6, all "personal" ado-files, whether personally written or just personally installed, were written in the same directory—OLDPLACE. So that they do not have to go back and rearrange what they have already done, Stata still looks in OLDPLACE.

20.6   How do I install an addition?

Additions come in three flavors:

1. User-written additions which you might find in the STB, etc.
   Discussed here.

2. Ado-files you have written.
   Discussed in [U] **20.7 How do I add my own ado-files?**. If you have an ado-file obtained from the Stata listserver or a friend, treat it as belonging to this case.

3. Official updates provided by StataCorp.
   Discussed in [U] **20.8 How do I install STB updates?**.

User-written additions which you might find in the STB, etc., are obtained over the Internet or from new-format STB diskettes you obtain from StataCorp. Either

1. Pull down **Help** and choose **STB and User-written Programs**. Then either
   a. Click on *http://www.stata.com*.
   b. Or insert the STB diskette and click on *diskette*.

2. Or
   a. Type 'net from http://www.stata.com'.
   b. Or insert the STB diskette and type 'net from a:' (Windows) or 'net from :diskette:' (Macintosh). (Unix users: you have to first copy the diskette; see [GSU] **17 Updating Stata**.)

What to do next should be obvious but, in case it is not, see

Windows:
  Internet: [GSW] **19 Using the Internet**
  diskette: [GSW] **20 Updating Stata**

Macintosh:
  Internet: [GSM] **19 Using the Internet**
  diskette: [GSM] **20 Updating Stata**

Unix:
  Internet: [GSU] **16 Using the Internet**
  diskette: [GSU] **17 Updating Stata**

Also see [U] **32 Using the Internet to keep up to date** and [R] **net**.

For your information, STB stands for *Stata Technical Bulletin*, a journal for Stata users. The journal consists of two parts, a printed copy and a "diskette" containing associated software; see [U] **2.4 The Stata Technical Bulletin** for information on the STB itself. We put "diskette" in quotes because you can obtain the real diskette from us or you can obtain the materials over the Internet. We encourage all users to subscribe to the STB, but if you have Internet access. you can save money by subscribing without diskettes.

Each STB issue and diskette is labeled STB-1, STB-2, and so on, and the STB is published every other month. Each STB "diskette" contains official updates to Stata, which you should always install (see [U] **20.8 How do I install STB updates?** below) along with other materials which you may install if they interest you, and which we just told you how to install.

In the printed copy of the STB you will find that each article—called an insert—is given a letter/number combination. The title of an insert might be "zzz999: Command to jumble data". If there is any software associated with the insert, it will be found on the diskette in a directory with the same letter/number combination: **zzz999**.

20.7 How do I add my own ado-files?

You write a Stata program (see [U] **21 Programming Stata**), store it in a file ending in `.ado`, perhaps write a help-file, and copy everything to the directory `sysdir` lists as PERSONAL:

```
. sysdir
    STATA:  C:\STATA\
  UPDATES:  C:\STATA\ado\updates\
     BASE:  C:\STATA\ado\base\
     SITE:  C:\STATA\ado\site\
  STBPLUS:  C:\ado\stbplus\
 PERSONAL:  C:\ado\personal\
 OLDPLACE:  C:\ado\
```

In this case, we would copy the files to `c:\ado\personal`.

While you are writing your ado-file, it is sometimes convenient to store the pieces in the current directory. Do that if you wish; you can move them to your personal ado directory when the program is debugged.

20.8 How do I install STB updates?

"STB updates" really refers to two types of updates:

1. Official updates which we discuss here, and

2. User-written additions distributed through the STB, which we discussed in [U] **20.6 How do I install an addition?** (you use the `net` command or pull down **Help** and select **STB and User-written Programs**).

These updates and additions go together because of timing; official updates are released at the same time each STB is published and, for those who obtain updates and additions from diskettes, the official updates are placed on the STB diskette along with the user-written additions.

Anyway, to install the official updates, either

1. Pull down **Help** and select **Official Updates**. Then
   a. Click on *http://www.stata.com*.
   b. Or insert the STB diskette and click on *diskette*.

2. Or
   a. Type '`update from http://www.stata.com`'.
   b. Or insert the STB diskette and type '`update from a:`' (Windows) or '`update from :diskette:`' (Macintosh). (Unix users: you have to first copy the diskette; see [GSU] **17 Updating Stata**.)

What to do next should be obvious but, in case it is not, see

> Windows:
> > Internet: [GSW] **19 Using the Internet**
> > diskette: [GSW] **20 Updating Stata**
>
> Macintosh:
> > Internet: [GSM] **19 Using the Internet**
> > diskette: [GSM] **20 Updating Stata**
>
> Unix:
> > Internet: [GSU] **16 Using the Internet**
> > diskette: [GSU] **17 Updating Stata**

Also see [U] **32 Using the Internet to keep up to date** and [R] **net**.

For your information, the official updates include bug fixes and new features. The official updates never change the syntax of an existing command nor change the way Stata works. They add a new feature—which you can use or ignore—or they fix a bug.

Once you have installed the updates, you can enter Stata and type `help whatsnew` (or pull down **Help** and click **What's new**) to learn about what has changed.

21   Programming Stata

Contents

This is an advanced topic. Some Stata users live productive lives without ever programming Stata. After all, you do not need to know how to program Stata to input data, create new variables, and estimate models. On the other hand, programming Stata is not difficult—at least if the problem is not difficult—and Stata's programmability is one of its better features. The real power of Stata is not revealed until you program it.

If you are uncertain whether to read this chapter, we recommend you start reading and then bail out when it gets too arcane for you. You will learn things about Stata that you may find useful even if you never write a Stata program.

For those who want even more, we offer courses over the Internet on Stata programming; see [U] **2.6 NetCourses**.

21.1 Description

When you type a command that Stata does not recognize, the first thing Stata does is look in its memory for a program of that name. If Stata finds it, Stata executes the program.

There is no Stata command named `hello`,

```
. hello
unrecognized command
r(199);
```

but there could be if you defined a program named `hello` and after that, the following might happen when you typed `hello`:

```
. hello
hi there
. _
```

This would happen if, beforehand, you had typed

```
. program define hello
  1. display "hi there"
  2. end
. _
```

So that is the overview of programming. A program is defined by

> program define *progname*
> > *Stata commands*
> end

and it is executed by typing *progname* to Stata's dot prompt.

21.2 Relationship between a program and a do-file

Programs and do-files are not much different to Stata, which is to say, Stata treats a do-file in the same way it treats a program. Below we will discuss argument passing, consuming results from Stata commands, and other topics, but realize that everything we say applies equally to do-files as well as to programs.

The differences between a program and a do-file are

1. You invoke a do-file by typing do *filename*. You invoke a program by simply typing the program's name.

2. Programs must be defined (loaded) before they are used, whereas all that is required to run a do-file is that the file exist. There are ways to make programs load automatically, however, so this difference is not of great importance.

3. When you type do *filename*, Stata displays the commands it is executing and the results. When you type *progname*, Stata shows only the results. The displaying of the underlying commands is suppressed. This is an important difference in outlook: In a do-file, how it does something is as important as what it does. In a program, the how is no longer important. One thinks of it as a new feature of Stata.

Let us now mention some of the similarities:

1. Arguments are passed to programs and do-files in the same way.

2. Programs and do-files are implemented in terms of the same Stata commands, which is to say, any Stata commands.

3. Programs may call other programs. Do-files may call other do-files. Programs may call do-files (rarely happens) and do-files may call programs (often happens). Stata allows programs (and do-files) to be nested up to 32 deep.

Now, here is the interesting thing: *programs are typically defined in do-files* (or in a variant of do-files called ado-files; we will get to that later).

That is, you *can* define a program interactively and that is useful for pedagogical purposes, but in real applications, you compose programs in an editor or word processor and store the definition in a do-file.

You have already seen your first program:

```
program define hello
        display "hi there"
end
```

You *could* type those commands interactively, but if the body of the program were more complicated, that could be inconvenient. So instead, imagine you typed the commands into a do-file:

```
───────────────────────────────────────────────── top of hello.do ───────────
program define hello
        display "hi there"
end
───────────────────────────────────────────────── end of hello.do ───────────
```

Now, returning to Stata, you type

```
. do hello

. program define hello
  1.          display "hi there"
  2. end

.
end of do-file
```

Do you see that typing do hello did nothing but load the program? Typing do hello is the same as typing out the program's definition because that is all the do-file contains. Understand that the do-file was executed, but all the statements in the do-file did was define the program hello. Now that the program is loaded, we could use it interactively:

```
. hello
hi there
```

So that is one way you could use do-files and programs together. If you wanted to create new commands for interactive use, you could

1. Write the command as a **program define** ... **end** in a do-file.

2. **do** the do-file before you use the new command.

3. Use the new command during the rest of the session.

There are more convenient ways to do this that automatically load the definition-providing do-file, but put that aside. The above method would work.

Another way we could use do-files and programs together is to put the definition and the execution together into a do-file:

```
─────────────────────────────────────────────────── top of hello.do ───────────
program define hello
        display "hi there"
end
hello
─────────────────────────────────────────────────── end of hello.do ───────────
```

Here is what would happen if we executed this do-file:

```
. do hello
. program define hello
  1.          display "hi there"
  2. end
. hello
hi there

.
end of do-file
```

Do-files and programs are often used in such combinations. Why? Pretend program **hello** is long and complicated and you have a problem where you need to do it twice. That would be a good reason to write a program. Moreover, you may wish to carry forth this procedure as a step of your analysis and, being cautious, do not want to perform this analysis interactively. You never intended program **hello** to be used interactively—it was just something you needed in the midst of a do-file—so you defined the program and used it there.

Anyway, there are lots of variations on this theme and just understand, few people actually sit in front of Stata and *interactively* type **program define** and then compose a program. They instead do that in front of their editor or word processor. They do that into a do-file. Then they execute the do-file.

There is one other (minor) thing to know: Once a program is defined, Stata will not allow you to redefine it:

```
. program define hello
hello already defined
r(110);
```

Thus, in our most recent do-file that defines and executes **hello**, we could not rerun it in the same Stata session:

```
. do hello
. program define hello
hello already defined
r(110);
end of do-file
r(110);
```

That problem is solved by typing `program drop hello` before redefining it. We could do that interactively or we could modify our do-file:

```
——————————————————— top of hello.do ———————
program drop hello
program define hello
        display "hi there"
end
hello
——————————————————— end of hello.do ———————
```

There is a problem with this solution. We can now rerun our do-file, but the first time we tried to run it in a Stata session, it would now fail:

```
. do hello

. program drop hello
hello not found
r(111);

end of do-file
r(111);
```

The way around this conundrum is

```
——————————————————— top of hello.do ———————
capture program drop hello
program define hello
        display "hi there"
end
hello
——————————————————— end of hello.do ———————
```

`capture` in front of a command makes Stata indifferent whether the command works; see [R] **capture**. In real do-files containing programs, you will often see `capture program drop` followed by `program define`.

You will want to learn about the `program` command itself—see [R] **program**. It has nothing to do with programming, but it manipulates programs. `program` can define programs, drop programs, and show you a directory of programs you have defined.

A program can contain any Stata command, but certain Stata commands are of special interest to program writers; see [U] **21.12 A compendium of useful commands for programmers** below.

21.3 Macros

Before we can undertake programming, we must discuss macros.

Macros are the variables of Stata programs.

A *macro* is a string of characters, called the *macroname*, that stands for another string of characters, called the *macro contents*.

Macros come in two flavors, called local and global macros. We will start with local macros because they are the most commonly used, but there is really nothing to distinguish local from global at this stage.

21.3.1 Local macros

Local macro names are up to 7 (not 8) characters in length.

One sets the contents of a local macro with the `local` command. In fact, we can do this interactively. We will begin by experimenting with macros in this way to learn about them. If we type

```
. local shrtcut "myvar thisvar thatvar"
```

then `` `shrtcut´ `` is a synonym for "`myvar thisvar thatvar`". Note the single quotes around shrtcut. We said that sentence exactly the way we meant to because

if you type	`` `shrtcut´ ``
which is to say	left-single-quote `shrtcut` right-single-quote
Stata hears	`myvar thisvar thatvar`

To access the contents of the macro, we use a left single-quote (located at the upper left on most keyboards), the macro name, and a right single-quote (located under the " on the right side of most keyboards).

The single quotes bracketing the macroname `shrtcut` are called the macro-substitution characters. shrtcut means shrtcut. `` `shrtcut´ `` means `myvar thisvar thatvar`.

So, if you were now to type

```
. list `shrtcut´
```

the effect is exactly as if you typed

```
. list myvar thisvar thatvar
```

Macros can be used literally anywhere. For instance, if we also defined

```
. local cmd "list"
```

then we could type

```
. `cmd´ `shrtcut´
```

to mean `list myvar thisvar thatvar`.

As another example, consider the definitions

```
. local prefix "my"
. local suffix "var"
```

Then

```
. `cmd´ `prefix´`suffix´
```

would mean `list myvar`.

21.3.2 Global macros

Let's put aside why Stata has two kinds of macros—local and global—and focus right now on how global macros work.

The first difference is that global macros are allowed to have names that are up to 8 (not 7) characters in length.

The second difference is that you set the contents of a global macro using the `global` rather than the `local` command:

```
. global shrtcut "alpha beta"
```

You obtain the contents of a global macro by prefixing its name with a dollar sign: `$shrtcut` is equivalent to "alpha beta".

In the previous section, we defined a local macro named `shrtcut`. That is a different macro. `` `shrtcut´ `` is still "myvar thisvar thatvar"

Local and global macros may have the same names, but even if they do, they are unrelated and still distinguishable.

At this stage the only thing to know about global macros is that they are just like local macros except that you set their contents with `global` rather than `local` and that you substitute their contents by prefixing them with a `$` rather than enclosing them in `` ` ´ ``.

21.3.3   The difference between local and global macros

The difference between local and global macros is that local macros are private and global macros are public.

Pretend you have written a program

```
program define myprog
        code using local macro alpha
end
```

The local macro `alpha` in `myprog` is private in that no other program can modify or even look at `alpha`'s contents. To make this point absolutely clear, assume your program looks like this:

```
program define myprog
        code using local macro alpha
        mysub
        more code using local macro alpha
end
program define mysub
        code using local macro alpha
end
```

Note that `myprog` calls `mysub` and that both programs use a local macro named `alpha`. Even so, the local macros in each program are different. `mysub`'s `alpha` macro may contain one thing, but that has nothing to do with what `myprog`'s `alpha` contains. Even when `mysub` begins execution, its `alpha` macro is different from `myprog`'s. It is not that `mysub`'s inherits `myprog`'s `alpha` macro contents but is then free to change it. It is that `myprog`'s `alpha` and `mysub`'s `alpha` are completely different things.

When you write a program using local macros, you need not worry that some other program has been written using local macros with the same names. Local macros are just that: local to your program.

Global macros, on the other hand, are known to all programs. If both `myprog` and `mysub` use global macro `beta`, they are using the same macro. Whatever are the contents of `$beta` when `mysub` is invoked, those are the contents when `mysub` begins execution and, whatever are the contents of `$beta` when `mysub` completes, those are the contents when `myprog` regains control.

21.3.4   Macros and expressions

From now on we are going to use local and global macros according to whichever is convenient; you understand that whatever is said about one applies to the other.

Consider the definitions

```
. local one 2+2
. local two = 2+2
```

(which we could just as well have illustrated using the `global` command). In any case, note the equal sign in the second macro definition and the lack of the equal sign in the first. Formally, the first ought to be typed

```
. local one "2+2"
```

but Stata does not mind if we omit the double quotes in the `local` (`global`) statement.

`local one 2+2` (with or without double quotes) copies the string 2+2 into the macro named `one`.

`local two = 2+2` evaluates the expression 2+2, producing 4, and stores 4 in the macro named `two`.

That is, you type

```
local macname contents
```

if you want to copy *contents* to *macname* and you type

```
local macname = expression
```

if you want to evaluate *expression* and store the result in *macname*.

In the second form, *expression* can be numeric or string. 2+2 is a numeric expression. As an example of a string expression,

```
. local res = substr("this",1,2) + "at"
```

stores `that` in `res`.

Since the expression can be either numeric or string, what is the difference between the following statements?

```
. local a "example"
. local b = "example"
```

Both statements store `example` in their respective macros. The first does it by a simple copy operation whereas the second evaluates the expression `"example"`, which is a string expression because of the double quotes, and which, in this case, evaluates to itself.

There is, however, a difference. Stata's expression parser is limited to handling strings of 80 characters. If the string is longer than 80 characters, it is truncated.

The copy operation of the first syntax is not limited—it can copy up to the maximum length of a macro, which is currently 18,623 (1,000 for Small Stata).

As a programmer, the 80-character limit for string expressions may seem limiting, but it turns out it is not, due to another feature discussed in [U] **21.3.6 Extended macro functions** below.

Put that aside. There are some other issues of using macros and expressions that look a little strange to programmers from other languages the first time they see them. For instance, pretend the macro `` `i' `` contains 5. How would you increment it so that it contained $5 + 1 = 6$? The command is

```
local i = `i' + 1
```

Do you see why the single quotes are on the right but not the left? Remember, `` `i´ `` refers to the contents of the local macro named i which, we just said, is 5. Thus, after expansion, the line reads

```
local i = 5 + 1
```

which is the desired result.

Now consider another local macro `` `answ´ `` which might contain **yes** or might contain **no**. In a program that was supposed to do something different based on **answ**'s content, you would code

```
if "`answ´" == "yes" {
      . . .
}
else {
      . . .
}
```

Note the odd-looking `` "`answ´" `` and now think about the line after substitution. The line either reads

```
if "yes" == "yes" {
```

or it reads

```
if "no" == "yes" {
```

either of which is the desired result. Had we omitted the double quotes, the line would have read

```
if no == "yes" {
```

(assuming `` `answ´ `` contains **no**) and that is not at all the desired result. As the line reads now, **no** would not be a macro but would be interpreted as a variable in the data.

The key to all of this is to think of the line after substitution.

21.3.5 Double quotes

Double quotes are used to enclose strings: **"yes"**, **"no"**, **"my dir\my file"**, `` "`answ´" `` (meaning the contents of local macro **answ**, treated as a string), and so on. Double quotes are used with macros,

```
local a "example"
if "`answ´" == "yes" {
      . . .
}
```

and double quotes are used by lots of Stata commands,

```
. regress lnwage age ed if sex=="female"
. gen outa = outcome if drug=="A"
. use "person file"
```

In fact, Stata has two sets of double-quote characters of which **""** is one. The other is `` `"""´ `` and they work the same way as **""**:

```
. regress lnwage age ed if sex==`"female"´
. gen outa = outcome if drug==`"A"´
. use `"person file"´
```

No rational user would use `` `"""` `` (called compound double quotes) instead of `""` (called simple double quotes), but smart programmers do use them:

```
local a `"example"'
if `"`answ'"' == `"yes"' {
        ...
}
```

Why is `` `"example'" `` better than `"example"`, `` `"`answ'"' `` better than `"answ"`, and `` `"yes"' `` better than `"yes"`? The answer is that only `` `"`answ'"' `` is better than `"`answ'"`; `` `"example"' `` and `` `"yes"' `` are no better—and no worse—than `"example"` and `"yes"`.

`` `"`answ'"' `` is better than `"`answ'"` because the macro **answ** might itself contain (simple or compound) double quotes. The really great thing about compound double quotes is that they nest. Pretend `` `answ' `` contained the string "I "think" so". Then

Stata would find	`if "`answ'"=="yes"`
confusing because it would expand to	`if "I "think" so"=="yes"`
Stata would not find	`if `"`answ'"'==`"yes"'`
confusing because it would expand to	`if `"I "think" so"'==`"yes"'`

Open and close double quote in the simple form look the same; open quote is `"` and so is close quote. Open and close double quote in the compound form are distinguishable; open quote is `` `" `` and close quote is `"'` and so Stata can pair the close with the corresponding open double quote. `` `"I "think" so"' `` is easy for Stata to understand whereas `"I "think" so"` is a hopeless mishmash. (If you disagree, consider what `"A"B"C"` might mean. Is it the quoted string A"B"C or is it quoted string A followed by B followed by quoted string C?)

Since Stata can distinguish open from close quotes, even nested compound double quotes are understandable: `` `"I `"think"' so"' ``. (What does `"A"B"C"` mean? It means either `` `"A`"B"'C"' `` or it means `` `"A"`B'"C"' ``.)

Yes, compound double quotes make you think your vision is stuttering, especially when combined with the macro substitution `` ` ' `` characters. That is why we rarely use them, even when writing programs. You do not have to use exclusively one or the other style of quotes. It is perfectly acceptable to code

```
local a "example"
if `"`answ'"' == "yes" {
        ...
}
```

using compound double quotes where it might be necessary (`` `"`answ'"' ``) and using simple double quotes in other places (such as `"yes"`). It is also acceptable to use simple double quotes around macros (e.g., `"`answ'"`) if you are certain that the macros themselves do not contain double quotes or you do not much care what happens if they do.

There are instances where careful programmers should use compound double quotes, however. Later you will learn that Stata's **syntax** command interprets standard Stata syntax and so makes writing programs that understand things like

```
. myprog mpg weight if index(make,"VW")~=0
```

easy. The way **syntax** works—we are getting ahead of ourselves—the if *exp* typed by the user is placed in the local macro **if**. Thus, `` `if' `` will contain "if index(make,"VW")~=0" in this case. Now say you are at a point in your program and want to know whether the user specified an if *exp*. It would be natural to code

```
if `"`if´"´ ~= "" {
            /* the if exp was specified */
            . . .
}
else {
            /* it was not */
            . . .
}
```

Note that we used compound double quotes around the macro `` `if´ ``. The local macro `` `if´ `` might contain double quotes, so we used compound double quotes around it.

21.3.6 Extended macro functions

In addition to allowing =*exp*, `local` and `global` provide what are known as *extended functions*. The use of an extended function is denoted by a colon (:) following the macro name. For instance,

```
local lbl : variable label myvar
```

copies the variable label associated with `myvar` to the local macro `lbl`. Thus, if `myvar` had the variable label `My variable`, the macro `lbl` would now contain "My variable" too. If `myvar` had no variable label, the macro `lbl` would contain nothing. The full list of extended functions is

$\big\{$ global | local $\big\}$ *macroname* : char $\big\{$ *varname* | _dta $\big\}$
$\big\{$ global | local $\big\}$ *macroname* : colfullnames *matrixname*
$\big\{$ global | local $\big\}$ *macroname* : colnames *matrixname*
$\big\{$ global | local $\big\}$ *macroname* : coleq *matrixname*
$\big\{$ global | local $\big\}$ *macroname* : data label
$\big\{$ global | local $\big\}$ *macroname* : display *display_directives*
$\big\{$ global | local $\big\}$ *macroname* : environment *envvar*   Unix only
$\big\{$ global | local $\big\}$ *macroname* : format *varname*
$\big\{$ global | local $\big\}$ *macroname* : label *valuelabelname* # $[\#]$
$\big\{$ global | local $\big\}$ *macroname* : label (*varname*) # $[\#]$
$\big\{$ global | local $\big\}$ *macroname* : label *valuelabelname* maxlength
$\big\{$ global | local $\big\}$ *macroname* : label (*varname*) maxlength
$\big\{$ global | local $\big\}$ *macroname* : log [on]
$\big\{$ global | local $\big\}$ *macroname* : piece # # of "*string*"
$\big\{$ global | local $\big\}$ *macroname* : rowfullnames *matrixname*
$\big\{$ global | local $\big\}$ *macroname* : rownames *matrixname*
$\big\{$ global | local $\big\}$ *macroname* : roweq *matrixname*
$\big\{$ global | local $\big\}$ *macroname* : set adosize
$\big\{$ global | local $\big\}$ *macroname* : set beep
$\big\{$ global | local $\big\}$ *macroname* : set display linesize
$\big\{$ global | local $\big\}$ *macroname* : set display pagesize
$\big\{$ global | local $\big\}$ *macroname* : set graphics
$\big\{$ global | local $\big\}$ *macroname* : set level
$\big\{$ global | local $\big\}$ *macroname* : set log linesize
$\big\{$ global | local $\big\}$ *macroname* : set log pagesize
$\big\{$ global | local $\big\}$ *macroname* : set matsize
$\big\{$ global | local $\big\}$ *macroname* : set more
$\big\{$ global | local $\big\}$ *macroname* : set rmsg
$\big\{$ global | local $\big\}$ *macroname* : set textsize
$\big\{$ global | local $\big\}$ *macroname* : set trace
$\big\{$ global | local $\big\}$ *macroname* : set type

$\{$ global | local $\}$ *macroname* : set virtual
$\{$ global | local $\}$ *macroname* : sortedby
$\{$ global | local $\}$ *macroname* : subinstr $\{$local|global$\}$ *macname* "*string*" "*string*" $\big[$,
all word count($\{$local|global$\}$ *macname*) $\big]$
$\{$ global | local $\}$ *macroname* : sysdir *dir*
$\{$ global | local $\}$ *macroname* : tempvar
$\{$ global | local $\}$ *macroname* : tempfile
$\{$ global | local $\}$ *macroname* : type *varname*
$\{$ global | local $\}$ *macroname* : value label *varname*
$\{$ global | local $\}$ *macroname* : variable label *varname*
$\{$ global | local $\}$ *macroname* : word count *string*
$\{$ global | local $\}$ *macroname* : word # of *string*

See [R] **macro** for details and examples.

char returns the list of all defined characteristics for *varname* or _dta; see [U] **15.8 Characteristics**.

colfullname, colnames, and coleq return lists of matrix column names; see [U] **17.2.4 Obtaining row and column names**.

data label returns the dataset label; see [U] **15.6.1 Dataset labels**.

display fills in the *macroname* with the *display_directive*. This is, in effect, Stata's display command—see [R] **display for programmers**—with its output redirected to a macro. display used in this way does not allow in *color*, _newline, _continue, or _request(*macname*).

environment, allowed only under Unix, imports the contents of the Unix environment variable *envvar* into the Stata macro.

format returns the display format associated with *varname*.

label looks up # in *valuelabelname* and returns the label value of # if *valuelabelname* exists and # is recorded in it, or it returns # otherwise. If a second number is specified, the result is trimmed to being no more than that many characters in length.

If (*varname*) (note parentheses) is specified rather than *valuelabelname*, the logic is the same except that the value label for variable *varname*, if any, is used.

maxlength specifies that, rather than looking up a number in a value label, label is return the maximum length of the labelings. For instance, if value label yesno mapped 0 to no and 1 to yes, then its maxlength would be 3 because yes is the longest labeling and it has three characters.

log returns the name of the log file if one is open, or nothing otherwise. log on returns "on" if a log file is open and logging is on; otherwise, it returns "off".

piece $\#_i$ $\#_l$ of "*string*" returns the *i*th piece of *string* with length not to exceed *l*. The pieces are formed at word breaks if that is possible. Say `stub` contained a potentially long string and you needed to display it in 20 columns on multiple lines. "local x: piece 1 20 of "`stub`"" would produce the first line of text. "local x: piece 2 20 of "`stub`"" would produce the second line, and so on until piece returned "" (nothing) in x. See [R] **macro**.

rowfullname, rownames, and roweq return lists of matrix row names; see [U] **17.2.4 Obtaining row and column names**.

set ... returns the current value of the set parameter.

sortedby fills in the *macroname* with the list of variables that the dataset is sorted by, or leaves the macro undefined if the dataset is not sorted.

subinstr {local|global} *macname* "*string$_{from}$*" "*string$_{to}$*" changes the first occurrence of "*string$_{from}$*" to "*string$_{to}$*" in *macname*.

Option `all` specifies that all occurrences are to be changed.

Option `word` specifies that substitutions are allowed only on space-separated tokens (words). In this case, tokens at the beginning or end also count as being space separated.

The `count()` option places the number of substitutions that occur into the named macro.

sysdir *dir* substitutes directory names for the codenames. Different people have Stata installed in different places. There are seven directories that, in system maintenance, are important; they are

Codename	directory
STATA	directory where Stata is installed
UPDATES	directory where Stata's official ado-file updates go
BASE	directory containing base-level official ado-files
SITE	directory where personal, site-level ado-files go
STBPLUS	directory where user-written ado-files go
PERSONAL	directory where personal ado-files go
OLDPLACE	(for backwards compatibility with Stata 5)

For instance, coding "`local x: sysdir PERSONAL`" would place in `` `x´ `` the name of the directory containing the user's personal ado-files. `` `x´ `` will end in the appropriate directory separator for the operating system, so you could subsequently refer to `` `x´`fname´ `` if `` `fname´ `` contained, say, "`myfile.ado`".

tempvar returns a name that can be used as a temporary variable.

tempfile returns a name that can be used as a temporary file.

type returns the storage type for *varname*.

value label returns the name of the value label associated with the variable; see [U] **15.6.3 Value labels**.

variable label returns the variable label associated with the variable; see [U] **15.6.2 Variable labels**.

word count returns the number of words in the string.

word # of returns the #th word of string.

21.3.7 Advanced local macro manipulation

This section is really an aside to help test your understanding of local macro substitution. The tricky examples illustrated below rarely, but sometimes, occur in real programs.

We already mentioned in [U] **21.3.4 Macros and expressions** incrementing macros:

```
local i = `i´ + 1
```

Now pretend you had macros `x1`, `x2`, `x3`, and so on. Obviously, `` `x1´ `` refers to the contents of `x1`, `` `x2´ `` to the contents of `x2`, etc.

What does `` `x`i´´ `` refer to? Pretend `` `i´ `` contains 6. The rule is to expand the inside first:

`` `x`i´´ ``	expands to	`` `x6´ ``
`` `x6´ ``	expands to	the contents of local macro x6

So there you have a vector of macros.

We have already shown adjoining expansions: `` `alpha'`beta' `` expands to `myvar` if `` `alpha' `` contains `my` and `` `beta' `` contains `var`.

The only other thing to know is that Stata does not mind if you reference a nonexisting macro. A nonexisting macro is treated as a macro with no contents. If local macro `gamma` does not exist, then

$$`gamma' \text{ expands to}$$

which is to say, nothing. It is not an error. Thus, `` `alpha'`gamma'`beta' `` still expands to `myvar`.

Correspondingly, you clear a local macro by setting its contents to nothing:

> local *macname*
> or   local *macname* `""`
> or   local *macname* = `""`

21.3.8   Advanced global macro manipulation

We continue with our aside to test your understanding of macro substitution, this time with global macros.

In [U] **21.3.4 Macros and expressions** we mentioned incrementing local macros,

```
local i = `i' + 1
```

the corresponding command for global macros is

```
global i = $i + 1
```

although the construct never arises in practice. Global macros are rarely used and, when they are used, are typically used for communication between programs. You should never use a global macro where a local macro would suffice.

Things like `$x$i` are expanded sequentially. If `$x` contained `this` and `$i` 6, then `$x$i` expands to `this6`. If `$x` was undefined then `$x$i` contains just 6 because undefined global macros, just like undefined local macros, are treated as containing nothing.

You can nest macro expansion by including braces, so assuming `$i` contains 6, `${x$i}` expands to `${x6}` which expands to the contents of `$x6` (or nothing if `$x6` is undefined).

You can mix global and local macros. Assume local macro j contains 7. Then `` ${x`j'} `` expands to the contents of `$x7`.

You also use braces to force the contents of global macros to run up against the succeeding text. For instance, assume the macro `drive` contains "b:". Were `drive` a local macro, you could type

> `` `drive'myfile.dta ``

to obtain `b:myfile.dta`. Since `drive` is a global macro, however, you must type

> `${drive}myfile.dta`

You could not type

> `$drive myfile.dta`

because that would expand to `b: myfile.dta`. You could not type

> `$drivemyfile.dta`

because that would expand to (surprise) `ile.dta`. Global macros are allowed to have names up to 8-characters long. Given no other information, Stata ate the first 8 characters as the name of the macro, forming macro `$drivemyf`. `$drivemyf` is presumably undefined, meaning it evaluates to nothing, leaving `ile.dta`.

Because Stata uses `$` to mark global-macro expansion, printing a real `$` is sometimes tricky. To display the string $22.15 using the `display` command, you are supposed to type `display "\$22.15"` although you can get away with `display "$22.15"` because Stata is rather smart. Stata would not be smart about `display "$this"` if you really wanted to display $this and not the contents of the macro `this`. You would have to type `display "\$this"`.

Real dollar signs can also be placed into the contents of macros, thus postponing substitution. First, let's understand what happens when we do not postpone substitution; consider the following definitions:

```
global baseset "myvar thatvar"
global bigset "$baseset thisvar"
```

Then `$bigset` is equivalent to "`myvar thatvar thisvar`". Now say we redefine the macro `baseset`:

```
global baseset "myvar thatvar othvar"
```

The definition of `bigset` has not changed—it is still equivalent to "`myvar thatvar thisvar`". It has not changed because `bigset` used the definition of `baseset` that was current at the time it was defined. `bigset` no longer knows that its contents are supposed to have any relation to `baseset`.

Instead, let us assume we had defined `bigset` as

```
global bigset "\$baseset thisvar"
```

at the outset. Then `$bigset` is equivalent to "`$baseset thisvar`" which in turn is equivalent to "`myvar thatvar othvar thisvar`". Since `bigset` explicitly depends upon `baseset`, anytime we change the definition of `baseset`, we will automatically change the definition of `bigset` as well.

21.3.9   Constructing Windows filenames using macros

Stata uses the \ character to flag its parser not to expand macros.

Windows uses the \ character as the directory path separator.

Mostly, there is no problem. However, if you are writing a program that contains a Windows path in macro `path` and a filename in `fname`, do *not* assemble the final result as

```
`path'\`fname'
```

because Stata will interpret the \ as an instruction to not expand `` `fname' ``. Instead, assemble the final result as

```
`path'/`fname'
```

Stata understands / as a directory separator on all platforms.

21.3.10   Built-in global system macros

Global macros that begin with the characters 'S_' are called *system macros* and some are predefined for you.

`$S_ADO`
    Path over which Stata looks for ado-files; see [U] **20 Ado-files**.

$S_DATE

Contains the current date expressed as an 11-character string in the format

dd mon yyyy

For instance, 3 March 1996 is recorded as " 3 Mar 1996".

$S_FLAVOR

Contains Intercooled or Small.

$S_FN

Contains filename last specified with use or save. The contents of this macro are used by describe to state the source of the data and by save when the user does not state the name of the file. S_FN is undefined (equivalent to null string) at the start of a Stata session and is made undefined when a drop results in no remaining data or when the user executes collapse, or any other Stata command that reforms the existing dataset. If you write a program that reforms the data, you should clear $S_FN by coding "global S_FN" in your program.

$S_FNDATE

S_FNDATE contains the date and time the file corresponding to the filename stored in $S_FN was last saved. The contents of this macro are used by describe to display the date and time.

$S_level

Records the default significance level, in percent, for significance tests and confidence intervals; see [R] **level**.

$S_MACH

Contains a string describing the hardware. The current possibilities are Convex, DEC ALPHA, DEC RISC, HP-9000, IBM RS/6000, Macintosh, PC, Power Macintosh, and Sun SPARC, although you should not depend on the completeness of this list.

$S_MODE

This macro is set to "BATCH" on Unix and Windows 98/95/NT computers when Stata is invoked in background (batch) mode, meaning that there is no console. Windows users see [GSW] **A.9 Executing Stata in background (batch) mode** and Unix users see [GSU] **A.8 Executing Stata in background (batch) mode**.

$S_OS

Contains DOS, MacOS, OS/2, Unix, or Windows, although you should not depend on the completeness of this list.

$S_OSDTL

Contains added information identifying the operating system. For instance, S_OSDTL contains 3.1 if the operating system is Windows 3.1 and contains nothing if the operating system is Windows 98/95/NT.

$S_TIME

Contains the current time expressed as an 8-character string in the format *hh:mm:ss* on a 24-hour clock. For instance, 14 seconds after 3:38 p.m. would be recorded as 15:38:14.

User-written programs may examine and change the contents of system macros. Typing macro drop _all does not, however, eliminate system macros and the contents of system macros such as S_DATE and S_TIME cannot be changed.

21.3.11   Referencing characteristics

Characteristics—see [U] **15.8 Characteristics**—are like macros associated with variables. They have names of the form *varname*[*charname*]—such as mpg[comment]—and you quote their names just as you do macro names to obtain their contents:

To substitute the value of *varname*[*charname*], type   `` `varname[charname]` ``

for example   `` `mpg[comment]` ``

You set the contents using the char command:

$$\text{char } varname[charname] \ [[\text{"}]text[\text{"}]]$$

This is similar to the local and global commands except that there is no =*exp* variation. You clear a characteristic by setting its contents to nothing just as you would with a macro:

Type   char *varname*[*charname*]

or   char *varname*[*charname*] ""

What is unique about characteristics is that they are saved with the data, meaning their contents survive from one session to the next, and they are associated with variables in the data, so if the user ever drops a variable, the associated characteristics disappear, too. (In addition, there is _dta[*charname*] that is associated with the data but not with any variable in particular.)

All the standard rules apply: characteristics may be referenced by quotation in any context and all that happens is that the characteristic's contents are substituted for the quoted characteristic name. As with macros, referencing a nonexistent characteristic is not an error; it merely substitutes to nothing.

21.4   Program arguments

When you invoke a program or do-file, what you type following that are the arguments. For instance, if you have a program called xyz and type

```
. xyz mpg weight
```

then mpg weight are the program's arguments, mpg being the first argument and weight being the second.

Program arguments are passed to programs via local macros:

Macro	Contents
`` `0` ``	what the user typed exactly as the user typed it odd spacing, double quotes, and all
`` `1` ``	going back; the first argument (first word of `` `0` ``)
`` `2` ``	the second argument (second word of `` `0` ``)
`` `3` ``	the third argument (third word of `` `0` ``)
...	...
`` `*` ``	the arguments `` `1` ``, `` `2` ``, `` `3` ``, ..., listed one after the other and with a single blank in between; similar to but different from `` `0` `` because odd spacing and double quotes are gone

That is, what the user types is passed to you in three different ways:

1. It is passed in `` `0´ `` exactly as the user typed it, meaning quotes, odd spacing, and all.

2. It is passed in `` `1´ ``, `` `2´ ``, ..., broken out into arguments on the basis of blanks (but with quotes used to force binding; we will get to that).

3. It is passed in `` `*´ `` as "`` `1´ `2´ `3´ ``...", which is a sort of crudely cleaned up version of `` `0´ ``.

It is not anticipated that you will use all three forms in one program.

We recommend you ignore `` `*´ ``, at least for receiving arguments; it is included so that old Stata programs continue to work.

Operating directly with `` `0´ `` takes considerable programming sophistication, although Stata's `syntax` command makes interpreting `` `0´ `` according to standard Stata syntax easy. That will be covered in [U] **21.4.4 Parsing standard Stata syntax** below.

The easiest way to receive arguments, however, is to deal with the positional macros `` `1´ ``, `` `2´ ``, ...

At the start of this section we imagined an **xyz** program invoked by typing "xyz mpg weight". In that case, `` `1´ `` would contain **mpg**, `` `2´ `` would contain **weight**, and `` `3´ `` would contain nothing.

Let's write a program to report the correlation between two variables. Of course, Stata already has a command that can do this—**correlate**—and in fact we will implement our program in terms of **correlate**. It is silly, but all we want to accomplish right now is to show how Stata passes arguments to a program.

Here is our program:

```
program define xyz
        correlate `1´ `2´
end
```

Once the program is defined, we can try it:

```
. use auto
(1978 Automobile Data)

. xyz mpg weight
(obs=74)

             |      mpg    weight
    ---------+------------------
         mpg|   1.0000
      weight|  -0.8072    1.0000
```

See how this works? We typed 'xyz mpg weight', that invoked our **xyz** program with `` `1´ `` being **mpg** and `` `2´ `` being **weight**, our program gave the command 'correlate `` `1´ `2´ ``', and that expanded to 'correlate **mpg weight**'.

Stylistically, this is not a good example of the use of positional arguments but realistically, there is nothing wrong with it. The stylistic problem is that if **xyz** is really to report the correlation between two variables, it ought to allow standard Stata syntax and really, that is not a difficult thing to do. Realistically, the program works.

Positional arguments, however, play an important role even for programmers who care about style. When we write a subroutine—a program to be called by another program and not intended for direct human use—we often pass information using positional arguments.

Stata forms the positional arguments `` `1´ ``, `` `2´ ``, ..., by taking what the user typed following the command (or do-file), parsing it on white space with double quotes used to force binding, and finally strips the quotes. What that means is that the arguments are formed on the basis of words but double-quoted strings are kept together as a single argument but with the quotes removed.

Let's create a program to illustrate these concepts. Although one does not normally define programs interactively, this program is short enough that we will:

```
. program define listargs
  1.   display "The 1st argument you typed is:  `1´"
  2.   display "The 2nd argument you typed is:  `2´"
  3.   display "The 3rd argument you typed is:  `3´"
  4.   display "The 4th argument you typed is:  `4´"
  5.   end
```

The `display` command simply types the double-quoted string following it; see [R] **display**.

Let's try our program:

```
. listargs
The 1st argument you typed is:
The 2nd argument you typed is:
The 3rd argument you typed is:
The 4th argument you typed is:
```

We type `listargs` and the result shows us what we already know—we typed nothing after the word `listargs`. There are no arguments. Let's try it again, this time adding `this is a test`:

```
. listargs this is a test
The 1st argument you typed is:  this
The 2nd argument you typed is:  is
The 3rd argument you typed is:  a
The 4th argument you typed is:  test
```

We learn that the first argument is 'this', the second is 'is', and so on. Blanks always separate arguments. You can, however, override this feature by placing double quotes around what you type:

```
. listargs "this is a test"
The 1st argument you typed is:  this is a test
The 2nd argument you typed is:
The 3rd argument you typed is:
The 4th argument you typed is:
```

This time we typed only *one* argument, 'this is a test'. When we place double quotes around what we type, Stata interprets whatever we type inside the quotes to be a single argument. In this case, `1´ contains 'this is a test' (and note, the double quotes were removed).

We can use double quotes more than once:

```
. listargs "this is" "a test"
The 1st argument you typed is:  this is
The 2nd argument you typed is:  a test
The 3rd argument you typed is:
The 4th argument you typed is:
```

The first argument is 'this is' and the second argument is 'a test'.

21.4.1   Renaming positional arguments

Positional arguments can be renamed: in your code you do not have to refer to `1´, `2´, `3´, ..., you can instead refer to more meaningful names such as n, a, and b or numb, alpha, and beta, or whatever else you find convenient. You want to do this because programs coded in terms of `1´, `2´, ... are hard to read and therefore more likely to contain errors.

You obtain better-named positional argument using the **args** command:

```
program define progname
        args argnames
        ...
end
```

For instance, if your program was to receive four positional arguments and you wanted to call them **varname**, **n**, **oldval**, and **newval**, you would code

```
program define progname
        args varname n oldval newval
        ...
end
```

varname, **n**, **oldval**, and **newval** become new local macros and all **args** does is copy `` `1´ ``, `` `2´ ``, ..., to them. It does not change `` `1´ ``, `` `2´ ``, etc.—you can still refer to the numbered macros if you wish—and it does not verify that your program receives the right number of arguments. If our example above were invoked with just two arguments, then `` `oldval´ `` and `` `newval´ `` would contain nothing. If it were invoked with five arguments, the fifth argument would still be out there, stored in local macro `` `5´ ``.

Let's make a command to create a dataset containing n observations on **x** ranging from a to b. Such a command would be useful, for instance, if we wanted to graph some complicated mathematical function and experiment with different ranges. It would be convenient if we can type the range of **x** over which we wish to make the graph rather than concocting the range by hand. (In fact, Stata already has such a command—**range**—but it will be instructive to write our own.)

Before writing this program, we had better know how to proceed, so here is how, in Stata, you could create a dataset containing n observations with **x** ranging between a to b:

1. **drop _all** to clear whatever data is in memory.

2. **set obs** n to make a dataset of n observations on no variables; if n were 100, we would type **set obs 100**.

3. **gen x = (_n-1)/($n-1$)*(b-a)+a** because the built-in variable **_n** is 1 in the first observation, 2 in the second, and so on; see [U] **16.4 System variables (_variables)**.

So, the first version of our program might read:

```
program define rng /* arguments are n a b */
        drop _all
        set obs `1´
        generate x = (_n-1)/(_N-1)*(`3´-`2´)+`2´
end
```

The above is just a direct translation of what we just said. `` `1´ `` corresponds to n, `` `2´ `` corresponds to a, and `` `3´ `` corresponds to b. This program, however, would be far more understandable if we changed it to read

```
program define rng
        args n a b
        drop _all
        set obs `n´
        generate x = (_n-1)/(_N-1)*(`b´-`a´)+`a´
end
```

21.4.2   Shifting through positional arguments

Some programs contain k arguments, where k varies, but it does not much matter because the same thing is done to each argument. summarize is an example of a program like that: type summarize mpg to obtain summary statistics on mpg and type summarize mpg weight to obtain first summary statistics on mpg and then summary statistics on weight.

One way to code such programs it to shift through the arguments:

```
program define ...
        while "`1´" ~= "" {
                        logic stated in terms of `1´
                        macro shift
                }
end
```

macro shift shifts `1´, `2´, `3´, ..., one to the left: what was `1´ disappears, what was `2´ becomes `1´, what was `3´ becomes `2´, and so on.

The outside while loop continues the process until macro `1´ contains nothing.

❑ Technical Note

You can combine **args** and shifting through an unknown number of positional arguments. For instance, say you were writing a subroutine that was to receive (1) **varname**, the name of some variable, (2) **n**, which is some sort of count, and (3) at least one and maybe 20 variable names. Perhaps you are to sum the variables, divide by **n** and store the result in the first variable. What the program does is irrelevant; here is how we could receive the arguments:

```
program define progname
        args varname n
        mac shift 2
        while "`1´" ~= "" {
                        ...
                        mac shift
                }
end
```

❑

❑ Technical Note

`*´, the result of listing the contents of the numbered macros one after the other with a single blank in between, changes with macro shift. Say your program received a list of variables and that the first variable had the interpretation of the dependent variable and the rest were independent variables. You want to save the first variable name in `lhsvar´ and all the rest in `rhsvars´. You could code

```
program define progname
        local lhsvar "`1´"
        macro shift 1
        local rhsvars "`*´"
        ...
end
```

Getting ahead of ourselves, it sometimes happens that a single macro contains a list of variables and you want to split the contents of the macro in two. Perhaps `varlist´ is the result of a syntax command (see [U] **21.4.4 Parsing standard Stata syntax**) and you now wish to split `varlist´ into `lhsvar´ and `rhsvars´. tokenize will reset the numbered macros:

```
program define progname
        ...
        tokenize `varlist´
        local lhsvar "`1´"
        macro shift 1
        local rhsvars "`*´"
        ...
end
```

❏

21.4.3   Incrementing through positional arguments

Another way to code the repeat-the-same-process problem for each argument is

```
program define ...
        local i 1
        while "``i´´" ~= "" {
                logic stated in terms of ``i´´
                local i = `i´ + 1
        }
end
```

Note the tricky construction ``i´´ which then itself had to be placed in double quotes for the while loop; see [U] **21.3.7 Advanced local macro manipulation**.

Whether you use the shift or increment logic makes no difference.

❏ Technical Note

The increment logic has the advantage when you need to pass through the list more than once because the shift logic shifts the list away. Say you were writing a subroutine that was to receive (1) **newvar**, a new variable you are to create followed by k numbers. You need to first count the number (and so derive k) and then, knowing k, pass through the list again.

```
program define progname
        args newvar
        mac shift                   /* shift away newvar */
                                    /* count the number of arguments */
        local k 1
        while "``k´´" ~= "" {
                local k = `k´ + 1
        }
                                    /* now pass through again */
        local i 1
        while `i´ <= `k´ {
                code in terms of ``i´´ and `k´
                local i = `i´ + 1
        }
end
```

❏

21.4.4 Parsing standard Stata syntax

Let us now switch to `` `0' `` from the positional arguments `` `1' ``, `` `2' ``, ...

You can parse `` `0' `` (what the user typed) according to standard Stata syntax with a single command. Remember that standard Stata syntax is

$$\big[\text{by } varlist\text{:}\big] \ command \ \big[varlist\big] \ \big[=exp\big] \ \big[\text{using } filename\big] \ \big[\text{if } exp\big] \ \big[\text{in } range\big] \ \big[weight\big] \ \big[\text{, } options\big]$$

See [U] **14 Language syntax**.

syntax parses standard syntax. You type out the syntax diagram in your program and then **syntax** looks at `` `0' `` (it knows to look there) and compares what the user typed to what you are willing to accept. Then one of two things happens: either **syntax** stores the pieces in an easy-for-you-to-process way or, if what the user typed does not match what you specified, **syntax** issues the appropriate error message and stops your program.

Consider a program that is to take two or more variable names along with an optional if *exp* and in *range*. The program would read

```
program define ...
        syntax varlist(min=2) [if] [in]
        ...
    end
```

You will have to read [R] **syntax** to learn how to specify the syntactical elements, but the command is certainly readable and it will not be long until you are guessing correctly about how to fill it in. And yes, the square brackets really do mean optional and you just use them with **syntax** in the natural way.

Understand, that one command is the entire parsing process. In this case, if what the user typed matches "two-or-more variables and an optional if and in", **syntax** defines new local macros:

`` `varlist' ``	the two variable names
`` `if' ``	the if *exp* specified by the user (or nothing)
`` `in' ``	the in *range* specified by the user (or nothing)

To see that this works, experiment with the following program

```
program define tryit
        syntax varlist(min=2) [if] [in]
        display "varlist now contains |`varlist'|"
        display `"if now contains |`if'|"'
        display "in now contains |`in'|"
    end
```

Here is a little bit of experimentation:

```
. tryit mpg weight
varlist now contains |mpg weight|
if now contains ||
in now contains ||
. tryit mpg weight displ if foreign==1
varlist now contains |mpg weight displ|
if now contains |if foreign==1|
in now contains ||
. tryit mpg wei in 1/10
varlist now contains |mpg weight|
if now contains ||
in now contains |in 1/10|
. tryit mpg
too few variables specified
r(102);
```

Note that in our third try we abbreviated the weight variable as `wei` yet, after parsing, `syntax` unabbreviated the variable for us.

If what this program were next going to do was step through the variables in the varlist, the positional macros `` `1´ ``, `` `2´ ``, ..., can be reset by coding

```
tokenize `varlist´
```

see [R] **tokenize**. This step resets `` `1´ `` to be the first word of `` `varlist´ ``, `` `2´ `` to be the second word, and so on (if there is a so on).

21.4.5 Parsing immediate commands

Immediate commands are described in [U] **22 Immediate commands**—they take numbers as arguments. By convention, when you name immediate commands, you should always make the last letter 'i'. Assume `mycmdi` takes as arguments two numbers, the first number of which must be a positive integer, and also allows the options `alpha` and `beta`. The basic structure is

```
program define mycmdi
        gettoken n 0 : 0, parse(" ,")        /* get first number */
        gettoken x 0 : 0, parse(" ,")        /* get second number */
        confirm integer number `1´           /* verify first is number */
        confirm number `2´                   /* verify second is number */

        if `n´<=0 { error 2001 }             /* check that n is positive     */
        place any other checks here

        syntax [, Alpha Beta]                /* parse remaining syntax */
        make calculation and display output
end
```

See [R] **gettoken**.

21.4.6 Parsing nonstandard syntax

If you wish to interpret nonstandard syntax and positional arguments are not adequate for you, you know that you face a formidable programming task. The key to the solution is the `gettoken` command.

`gettoken` has the ability to pull a single token from the front of a macro according to the parsing characters you specify and, optionally, to define another macro or redefine the initial macro to contain the remaining (unparsed) characters. That is,

Say `` `0´ `` contains	"this is what the user typed"
After `gettoken`	
new macro `` `token´ `` could contain	"this"
and `` `0´ `` could still contain	"this is what the user typed"
or	
new macro `` `token´ `` could contain	"this"
and new macro `` `rest´ `` could contain	" is what the user typed"
and `` `0´ `` could still contain	"this is what the user typed"
or	
new macro `` `token´ `` could contain	"this"
and `` `0´ `` could contain	" is what the user typed"

A simplified syntax of `gettoken` is

gettoken *macname*$_1$ [*macname*$_2$] : *macname*$_3$ [, **parse**("*string*") **quote** **match**(*macname*$_4$)]

where $macname_1$, $macname_2$, $macname_3$, and $macname_4$ are the names of local macros. (There is a way to work with global macros but, in practice, that is seldom necessary; see [R] **gettoken**.)

gettoken pulls the first token from $macname_3$, stores it in $macname_1$, and if $macname_2$ is specified, stores the remaining characters from $macname_3$ in $macname_2$. Any of $macname_1$, $macname_2$, and $macname_3$ may be the same macro. Typically, gettoken is coded

$$\text{gettoken } macname_1 \quad : \text{ 0 } [, \text{ options}]$$
$$\text{gettoken } macname_1 \text{ 0 } : \text{ 0 } [, \text{ options}]$$

since `0´ is the macro containing what the user typed. The first coding is used for token lookahead, should that be necessary, and the second is used for committing to taking the token.

gettoken's options are

parse("*string*")	for specifying parsing characters the default is parse(" "), meaning parse on white space it is common to specify parse(`"" "´), meaning parse on white space and double quote (`"" "´ is the string double-quote-space in compound double quotes)
quotes	to specify that outer double quotes are *not* to be stripped
match(*macname₄*)	to bind on parentheses and square brackets $macname_4$ will be set to contain "(", "[", or nothing depending on whether $macname_1$ was bound on parentheses, brackets, or match() turned out to be irrelevant $macname_1$ will have the outside parentheses or brackets removed

gettoken binds on double quotes whenever a (simple or compound) double quote is encountered at the beginning of $macname_3$. Specifying parse(`"" "´) ensures that double-quoted strings are isolated.

quote specifies that double quotes are not to be removed from the source in defining the token. For instance, in parsing '"this is" a test", the next token is "this is" if quote is not specified and is '"this is"' if quote is specified.

match() specifies that parenthesis and square brackets are to be matched in defining tokens. The outside level of parentheses or brackets are stripped. In parsing "(2+3)/2", the next token is "2+3" if match() is specified. In practice, match() might be used with expressions, but it is more likely to be used to isolate bound varlists and time-series varlists.

21.5   Scalars and matrices

In addition to macros, scalars, and matrices are provided for programmers; see [U] **17 Matrix expressions**, [R] **scalar** and [R] **matrix**.

As far as scalar calculations go, you can use macros or scalars. Remember, macros can hold numbers. Stata's scalars are, however, slightly faster and a little more accurate than macros. The speed issue is so slight as to be nearly immeasurable. As for accuracy, macros are accurate to a minimum of 12 decimal digits and scalars are accurate to roughly 16 decimal digits. Which you use makes little difference except in iterative calculations.

21.6   Temporarily destroying the data in memory

It is sometimes necessary to modify the user's data to accomplish a particular task. A well-behaved program, however, ensures that the user's data is always restored. The `preserve` command makes this easy:

```
code before the data needs changing
preserve
code that changes data freely
```

When you give the `preserve` command, Stata makes a copy of the user's data on disk. When your program terminates—no matter how—Stata restores the data and erases the temporary file. `preserve` is described in [R] **preserve**.

21.7   Temporary objects

If you write a substantial program, it will invariably require use of temporary variables in the data, or temporary scalars, or temporary matrices, or temporary files. By temporary, it is meant that these objects are necessary while the program is making its calculations and, once the program completes, they can be discarded.

Stata provides three commands to deal with this: `tempvar` creates names for variables in the dataset, `tempname` creates names for scalars and matrices, and `tempfile` creates names for files. All are described in [R] **macro** and all have the same syntax:

{ `tempvar` | `tempname` | `tempfile` } *macname* [*macname* ...]

The commands create local macros containing names you may use.

21.7.1   Temporary variables

Say that, in the process of making a calculation, you need to add variables `sum_y` and `sum_z` to the data. One possible code fragment is

```
prior code
tempvar sum_y
gen `sum_y´ = etc.
tempvar sum_z
gen `sum_z´ = etc.
code continues
```

Or you may obtain both temporary variable names in a single call:

```
prior code
tempvar sum_y sum_z
gen `sum_y´ = etc.
gen `sum_z´ = etc.
code continues
```

It is not necessary that you explicitly `drop` `sum_y´ and `sum_z´ when you are finished, although you may if you wish. Stata will automatically drop any variables with names assigned by `tempvar`. After issuing the `tempvar` command, it is important to remember always to refer to the names with the enclosing quotes, which signifies macro expansion. Thus, after typing `tempvar sum_y`, the one case where you do not put single quotes around the name, refer thereafter to the variable `sum_y´ with quotes. `tempvar` does not create a temporary variable: it creates a name that we may subsequently use as a temporary variable and stores that name in the local macro whose name you provide.

A full description of `tempvar` can be found in [R] **macro**.

21.7.2 Temporary scalars and matrices

tempname works just like tempvar. For instance, a piece of your code might read:

```
tempname YXX XXinv
matrix accum `YXX´ = price weight mpg
matrix `XXinv´ = syminv(`YXX´[2..., 2...])
tempname b
matrix `b´ = `XXinv´*`YXX´[1..., 1]
```

The above code solves for the coefficients of a regression on **price** on **weight** and **mpg**; see [U] **17 Matrix expressions** and [R] **matrix** for more information on the matrix commands.

As with temporary variables, temporary scalars and matrices are automatically dropped at the conclusion of your program.

21.7.3 Temporary files

In cases where you ordinarily might think you need temporary files, you may not because of Stata's ability to preserve and automatically restore the data in memory; see [U] **21.6 Temporarily destroying the data in memory** above.

For more complicated programs, Stata does provide temporary files. A code fragment might read:

```
preserve                        /* save original data */
tempfile males females
keep if sex==1
save `males´
restore, preserve               /* get back original data */
keep if sex==0
save `females´
```

As with temporary variables, scalars, and matrices, it is not necessary to delete the temporary files when you are through with them; Stata automatically erases them when your program ends.

21.8   Accessing results calculated by other programs

Stata commands that report results also save the results where they can be subsequently used by other commands or programs. This is documented in the "Saved Results" section of the particular command in the *Reference Manual*. Commands save results in one of three places:

1. r-class commands such as **summarize** save their result in **r()**; most commands are r class.

2. e-class commands such as **regress** save their results in **e()**; e-class commands are Stata's model estimation commands.

3. s-class commands (there are no good examples) save their results in **s()**; this is a rarely used class programmers sometimes find useful to help parse input.

Commands that do not save results are called n-class commands. More correctly, these commands require you to state where the result is to be saved as in **generate** *newvar* = ...

▷ Example

You wish to write a program to calculate the standard error of the mean, which is given by the formula $\sqrt{s^2/n}$, where s^2 is the calculated variance. (You could obtain this statistic using the ci command, but we will pretend that is not true.) You look at [R] **summarize** and learn that the mean is stored in r(mean), the variance in r(Var), and the number of observations in r(N). With that knowledge, you write the following program:

```
program define meanse
        quietly summarize `1´
        display "      mean = " r(mean)
        display "SE of mean = " sqrt(r(Var)/r(N))
end
```

The result of executing this program is

```
. meanse mpg
      mean = 21.297297
SE of mean = .67255109
```

◁

If you run an r-class command and type 'return list' or run an e-class command and type 'estimates list', Stata will summarize what was saved:

```
. regress mpg weight displ
(output omitted)

. estimates list
scalars:
        e(N)       =  74
        e(df_m)    =  2
        e(df_r)    =  71
        e(F)       =  66.78504752026512
        e(r2)      =  .6529306984682527
        e(rmse)    =  3.456061765708281
        e(mss)     =  1595.409691543724
        e(rss)     =  848.0497679157355
        e(r2_a)    =  .6431540984251049
        e(ll)      =  -195.2397979466294
        e(ll_0)    =  -234.3943376482347

macros:
        e(depvar)   :  "mpg"
        e(cmd)      :  "regress"
        e(predict)  :  "regres_p"
        e(model)    :  "ols"

matrices:
        e(b)        :  1 x 3
        e(V)        :  3 x 3

functions:
        e(sample)

. summarize mpg if foreign
    Variable |     Obs        Mean    Std. Dev.      Min         Max
---------+-----------------------------------------------------------
       mpg |      22    24.77273    6.611187         14          41
```

```
. return list
scalars:
            r(N)        =    22
            r(sum_w)    =    22
            r(mean)     =    24.77272727272727
            r(Var)      =    43.70779220779221
            r(min)      =    14
            r(max)      =    41
            r(sum)      =    545
```

In the example above we ran **regress** followed by **summarize**. As a result, e(N) records the number of observations used by **regress** (equal to 74) and r(N) records the number of observations used by **summarize** (equal to 22). r(N) and e(N) are separate things.

Were we now to run another r-class command—say **tabulate**—the contents of r() would change but those in e() would remain unchanged. You might therefore think that were we then to run another e-class command—say **probit**—the contents of e() would change but r() would remain unchanged. While it is true that e() results remain in place until the next e-class command is executed, do not depend on r() remaining unchanged. if an e-class or n-class command were to use an r-class command as a subroutine, that would cause r() to change. Anyway, most commands are r-class, so the contents of r() change frequently.

❏ Technical Note

It is therefore of great importance that you access results stored in r() immediately after the command that sets them. If you needed the mean and variance of the variable `1´ for subsequent calculation, do *not* code

```
summarize `1´
...
... r(mean) ... r(Var) ...
```

instead, code

```
summarize `1´
local mean = r(mean)
local var = r(Var)
...
... `mean´ ... `var´ ...
```

or code

```
tempname mean var
summarize `1´
scalar `mean´ = r(mean)
scalar `var´ = r(Var)
...
... `mean´ ... `var´ ...
```

❏

Saved results, be they in r() or e(), come in three flavors: scalars, macros, and matrices. If you look back at the **estimates list** and **return list** output, you will see that **regress** saves examples of all three whereas **summarize** saves just scalars. (**regress** also saves the "function" e(sample) and so do all the other e-class commands; see [U] **23.4 Specifying the estimation subsample**.)

Regardless of the flavor of e(*name*) or r(*name*), you can just refer to e(*name*) or r(*name*). That was the rule we gave in [U] **16.6 Accessing results from Stata commands** and that rule is sufficient to get most users by. There is, however, another way to refer to saved results. Rather than referring to r(*name*) and e(*name*), you can embed the reference in macro substitution characters

`` ` `` `´` to produce `` `r(``*name*`)´`` and `` `e(``*name*`)´``. The result is the same as macro substitution; the saved result is evaluated and then the evaluation is substituted:

```
. display "You can refer to " e(cmd) " or to `e(cmd)´"
You can refer to regress or to regress
```

This means, for instance, that typing `` `e(cmd)´`` is the same as typing `regress` because `e(cmd)` contains "`regress`":

```
. `e(cmd)´
    Source |       SS       df       MS              Number of obs =      74
-----------+-----------------------------           F(  2,     71) =   66.79
     Model |  1595.40969      2  797.704846          Prob > F       =  0.0000
```
(remaining output omitted)

In the `estimates list`, `e(cmd)` was listed as being a macro, and when you place a macro's name in single quotes, the macro's contents are substituted, so this is hardly a surprise.

What is surprising is that you can do this with scalar and even matrix saved results. `e(N)` is a scalar equal to 74 and may be used as such in any expression such as "`display e(mss)/e(N)`" or "`local meanss = e(mss)/e(N)`". `` `e(N)´`` substitutes to the string "74" and may be used in any context whatsoever such as "`local val`e(N)´ = e(N)`" (which would create a macro named `val74`). The rules for referring to saved results are

1. You may refer to `r(`*name*`)` or `e(`*name*`)` without single quotes in any expression and only in an expression. (Referring to s-class `s(`*name*`)` without single quotes is not allowed.)

    1.1 If *name* does not exist, missing value (.) is returned; it is not an error to refer to a nonexisting saved result.

    1.2 If *name* is a scalar, the full double-precision value of *name* is returned.

    1.3 If *name* is a macro, it is examined to determine whether its contents can be interpreted as a number. If so, the number is returned; otherwise the first 80 characters of *name* are returned.

    1.4 If *name* is a matrix, the full *matrix* is returned.

2. You may refer to `` `r(``*name*`)´``, `` `e(``*name*`)´``, or `` `s(``*name*`)´``—note the presence of quotes indicating macro substitution—in any context whatsoever.

    2.1 If *name* does not exist, nothing is substituted; it is not an error to refer to a nonexisting saved result. The resulting line is as if you had never typed `` `r(``*name*`)´``, `` `e(``*name*`)´``, or `` `s(``*name*`)´``.

    2.2 If *name* is a scalar, a string representation of the number accurate to no less than 12 digits is substituted.

    2.3 If *name* is a macro, the full contents (up to 18,623 characters for Intercooled and 1,000 characters for Small Stata) are substituted.

    2.4 If *name* is a matrix, the word `matrix` is substituted.

In general, you should refer to scalar and matrix saved results without quotes—`r(`*name*`)` and `e(`*name*`)`—and macro saved results with quotes—`` `r(``*name*`)´``, `` `e(``*name*`)´``, and `` `s(``*name*`)´``—but it is sometimes convenient to switch. For instance, say returned result `r(example)` contains the number of time periods patients are observed and assume that `r(example)` was saved as a macro and not a scalar. One could still refer to `r(example)` without the quotes in an expression context and obtain the expected result. It would have made more sense for the programmer to have stored `r(example)` as a scalar, but really it would not matter and you as a user would not even have to be cognizant of how the saved result was stored.

Switching the other way is sometimes useful, too. Say that returned result `r(N)` is a scalar and contains the number of observations used. You now want to use some other command that has an option `n(#)` which specifies the number of observations used. You could not type `n(r(N))` because the syntax diagram says that option `n()` expects its argument to be a literal number. You could type `n(`r(N)´)`.

21.9   Accessing results calculated by estimation commands

Estimation results are saved in `e()` and you access them in the same way you access any saved result; see [U] **21.8 Accessing results calculated by other programs** above. In summary,

1. Estimation commands—`regress`, `logistic`, etc.—save results in `e()`.

2. Estimation commands save their name in `e(cmd)`. For instance, `regress` saves "regress" and `poisson` saves "poisson" in `e(cmd)`.

3. Estimation commands save the number of observations used in `e(N)` and they identify the estimation subsample by setting `e(sample)`. You could type, for instance, "`summarize if e(sample)`" to obtain summary statistics on the observations used by the estimator.

4. Estimation commands save the entire coefficient vector and variance–covariance matrix of the estimators in `e(b)` and `e(V)`. These are matrices and may be manipulated as you would any other matrix:

```
. matrix list e(b)

e(b)[1,3]
        weight        displ       _cons
y1  -.00656711    .00528078    40.084522

. mat y = e(b)*e(V)*e(b)´
. mat list y

symmetric y[1,1]
          y1
y1  6556.982
```

5. Estimation commands set `_b[`*name*`]` and `_se[`*name*`]` as convenient ways to use coefficients and their standard errors in expressions; see [U] **16.5 Accessing coefficients and standard errors**.

6. Estimation commands may set other `e()` scalars, macros, or matrices containing additional information. This is documented in the "Saved Results" section of the particular command in the command reference.

▷ Example

If you are writing a command for use after `regress`, early in your code you should include the following:

```
if "`e(cmd)´" ~= "regress" {
        error 301
}
```

This is how you verify that the estimation results that are stored have been set by `regress` and not some other estimation command. Error 301 is Stata's "last estimates not found" error.

◁

21.10   Saving results

If your program calculates something, it should save the results of the calculation so that other programs can access them. This way your program can not only be used interactively, it can be used as a subroutine for other commands.

Saving results is easy:

1. On the `program define` line, specify the option `rclass`, `eclass`, or `sclass` according to whether you intend to return results in `r()`, `e()`, or `s()`.

2. Code

`return scalar` *name* = *exp*	(same syntax as `scalar` without the `return`)
`return local` *name* ...	(same syntax as `local` without the `return`)
`return matrix` *name matname*	(moves *matname* to `r(`*name*`)`)

    to save results in `r()`.

3. Code

`estimates scalar` *name* = *exp*	(same syntax as `scalar` without the `estimates`)
`estimates local` *name* ...	(same syntax as `local` without the `estimates`)
`estimates matrix` *name matname*	(moves *matname* to `e(`*name*`)`)

    to save results in `e()`. (You do not save the coefficient vector and variance matrix `e(b)` and `e(V)` in this way, however; you use `estimates post`.)

4. Code

`sreturn local` *name* ...	(same syntax as `local` without the `sreturn`)

    to save results in `s()`. (The s class has only macros.)

A program must be exclusively r, e, or s class.

21.10.1   Saving results in r()

In [U] **21.8 Accessing results calculated by other programs** we showed an example that reported the mean and standard error of the mean. A better version would save in `r()` the results of its calculations and read

```
program define meanse, rclass
        quietly summarize `1´
        local mean = r(mean)
        local sem  = sqrt(r(Var)/r(N))
        display "      mean = " `mean´
        display "SE of mean = " `sem´
        return scalar mean = `mean´
        return scalar se = `sem´
end
```

Running `meanse` now sets `r(mean)` and `r(se)`:

```
. meanse mpg
      mean = 21.297297
SE of mean = .67255109

. return list

scalars:
        r(se)      =   .6725510870764975
        r(mean)    =   21.2972972972973
```

In this modification, we added option `rclass` to the `program define` statement and we added two `return` commands to the end of the program.

Although we placed the `return` statements at the end of the program, they may be placed at the point of calculation if that is more convenient. A more concise version of this program would read,

```
program define meanse, rclass
        quietly summarize `1´
        return scalar mean = r(mean)
        return scalar se = sqrt(r(Var)/r(N))
        display "     mean = " return(mean)
        display "SE of mean = " return(se)
end
```

The `return()` function is just like the `r()` function except that `return()` refers to the results that this program *will* return rather than to the saved results that currently *are* returned (which in this case are due to `summarize`). That is, when you code the `return` command, the result is not immediately posted to `r()`. Rather, Stata holds onto the result in `return()` until your program concludes and then it copies the contents of `return()` to `r()`. While your program is active, you may use the `return()` function to access results you have already "returned". (`return()` works just like `r()` works after your program returns, meaning you may code `return()´` to perform macro substitution.)

21.10.2 Saving results in e()

Saving in `e()` is in some ways similar to saving in `r()`: you add the `eclass` option to the `program define` statement and then you use `estimates ...` just as you used `return ...` to store results. There are, however, some significant differences:

1. Unlike `r()`, estimation results are saved in `e()` the instant you issue an `estimates scalar`, `estimates local`, or `estimates matrix` command. This is because estimation results can consume considerable memory and Stata does not want to have multiple copies of the results floating around. That means you must be more organized and post your results at the end of your program.

2. There will come a point in your code where you will have your estimates and be ready to begin posting. The first step is to clear the previous estimates, set the coefficient vector `e(b)` and corresponding variance matrix `e(V)`, and to set the estimation-sample function `e(sample)`. How you do this depends on how you obtained your estimates:

    2.1 If you obtained your estimates using Stata's likelihood maximizer `ml`, this is automatically handled for you; skip to step 3.

    2.2 If you obtained estimates by "stealing" an already existing estimator, `e(b)`, `e(V)`, and `e(sample)` already exist and you do not want to clear them; skip to step 3.

    2.3 If you wrote your own code from start to finish, you use the `estimates post` command; see [R] **estimates**. You will code something like "estimates post `b´ `V´`, esample(`touse´)" where `b´ is the name of the coefficient vector, `V´ is the name of the corresponding variance matrix, and `touse´ is the name of a variable containing 1 if the observation was used and 0 if it was ignored. `estimates post` clears the previous estimates and moves the coefficient vector, variance matrix, and variable into `e(b)`, `e(V)`, and `e(sample)`.

    2.4 A variation on (2.3) is when you use an already existing estimator to produce the estimates but do not want all the other `e()` things stored by the estimator. In that case, you code

```
tempvar touse
tempname b V
matrix `b´ = e(b)
matrix `V´ = e(V)
qui gen byte `touse´ = e(sample)
estimates post `b´ `V´, esample(`touse´)
```

3. You now save anything else in `e()` that you wish using the `estimates scalar`, `estimates local`, or `estimates matrix` commands.

4. You code `estimates local cmd "cmdname"`. Stata does not consider estimation results complete until this is posted, and Stata considers the results to be complete when this is posted, so you must remember to do this and to do this last. If you set `e(cmd)` too early and the user pressed *Break*, Stata would consider your estimates complete when they are not.

Say you wish to write the estimation command with syntax

> myest *depvar var*$_1$ *var*$_2$ [if *exp*] [in *range*], *optset1 optset2*

where *optset1* affects how results are displayed and *optset2* affects the estimation results themselves. One important characteristic of estimation commands is that, when typed without arguments, they redisplay the previous estimation results. The outline is

```
program define myest, eclass
        local options "optset1"
        if replay() {
                if "e(cmd)"~="myest" {
                        error 301                    /* last estimates not found */
                }
                syntax [, `options´]
        }
        else {
                syntax varlist [if] [in] [, `options´ optset2]
                marksample touse
Code either contains this,
                tempnames b V
                commands for performing estimation
                assume produces `b´ and `V´
                estimates post `b´ `V´, esample(`touse´)
                estimates local depvar "`depv´"
or this,
                ml model ... if `touse´ ...
and regardless, concludes,
                perhaps other estimates commands appear here
                estimates local cmd "myest"
        }
                                        /* (re)display results ...       */
code typically reads
        code to output header above coefficient table
        estimates display               /* displays coefficient table    */
or
        ml display                      /* displays header and coef. table */
end
```

Here is a list of the commonly saved `e()` results. Of course, you may create any `e()` results that you wish.

`e(cmd)` (macro)
    Name of the estimation command.

`e(clustvar)` (macro)
    Name of the cluster variable, if any.

e(N_clust) (scalar)
  Number of clusters, if any.

e(depvar) (macro)
  Name(s) of the dependent variable(s).

e(ll) (scalar)
  Log-likelihood value, if relevant.

e(ll_0) (scalar)
  Log-likelihood value for constant-only model, if relevant.

e(F) (scalar)
  Test of the model against the constant-only model, if relevant, and if nonasymptotic results.

e(chi2) (scalar)
  Test of the model against the constant-only model, if relevant, and if asymptotic results.

e(chi2type) (macro)
  LR or Wald or other depending on how e(chi2) was performed.

e(df_m) (scalar)
  Model degrees of freedom.

e(N) (scalar)
  Number of observations.

e(r2_p) (scalar)
  Value of the pseudo-R^2 if it is calculated. (If a "real" R^2 is calculated as it would be in linear regression, it is stored in (scalar) e(r2).)

e(df_r) (scalar)
  "Denominator" degrees of freedom if estimates nonasymptotic.

e(vcetype) (macro)
  Text to appear above standard errors in estimation output; typically "Robust" or "".

e(wtype) and e(wexp) (macros)
  If weighted estimation was performed, e(wtype) contains the weight type (fweight, pweight, etc.) and e(wexp) contains the weighting expression.

e(predict) (macro)
  Name of the command that predict is to use; if this is blank, predict uses the default _predict.

e(b) and e(V) (matrices)
  The coefficient vector and corresponding variance matrix. Saved when you coded estimates post.

e(sample) (function)
  This function was defined by matrix post's esample() option if you specified it. You specified a variable containing 1 if you used an observation and 0 otherwise. estimates post stole the variable and created e(sample) from it.

21.10.3 Saving results in s()

S is a strange class because, whereas the other classes allow scalars, macros, and matrices, s allows only macros.

S is seldom used, too.

S is for subroutines that you might write to assist in parsing the user's input prior to evaluating any user-supplied expressions.

So think of s as standing for strange, seldomly used subroutines.

Here is the problem s solves: Say you create a non-standard syntax for some command so that you have to parse through it yourself. The syntax is so complicated that you want to create subroutines to bite off pieces of it and then return information to your main routine. Assume that, in your syntax, are expressions that the user might type. Now pretend that one of the expressions the user types is, say, `r(mean)/sqrt(r(Var))`—perhaps the user is consuming results left behind by `summarize`.

If, in your parsing step, you call subroutines and they return results in `r()`, you will wipe out `r(mean)` and `r(Var)` before you ever get around to seeing them, much less evaluating them.

So you must be careful to leave `r()` intact until your parsing is complete; you must use no r-class commands and any subroutines you write must not touch `r()`.

The way to do this is to use s-class subroutines because s-class routines return results in `s()` rather than `r()`. S class provides macros only because that is all you need to solve parsing problems.

To create an s-class routine, specify the `sclass` option on the `program define` line and then use `sreturn local` to return results.

s-class results are posted to `s()` at the instant you issue the `sreturn()` command, so you must organize your results. In addition, `s()` is never automatically cleared, so occasionally coding `sreturn clear` at appropriate points in your code is a good idea.

Very few programs need s-class subroutines.

21.11 Ado-files

Ado-files were introduced in [U] **20 Ado-files**.

When a user types '*gobbledegook*', Stata first asks itself if *gobbledegook* is one of its built-in commands. If so, the command is executed. Otherwise, it asks itself if *gobbledegook* is a defined program. If so, the program is executed. Otherwise, Stata looks in various directories for *gobbledegook*.ado. If there is no such file, the process ends with the "unrecognized command" error.

If Stata finds the file, it quietly issues to itself the command '`run` *gobbledegook*.ado' (specifying the path explicitly). If that runs without error, Stata asks itself again if *gobbledegook* is a defined program. If not, Stata issues the "unrecognized command" error. (In this case, somebody wrote a bad ado-file.) If the program is defined, as it should be, Stata executes it.

Thus, you can arrange for programs you write to be loaded automatically. For instance, if you were to create `hello.ado` containing

─────────────────────────────────────── top of hello.do ───────────

```
program define hello
        display "hi there"
end
```

─────────────────────────────────────── end of hello.do ───────────

and store the file in your current directory or your personal directory (we will tell you where that is momentarily), you could type `hello` and be greeted by a reassuring

```
. hello
hi there
```

You could, at that point, think of `hello` as just a part of Stata.

There are two places for putting your personal ado files. One is the current directory, and that is a good choice when the ado-file is unique to a project and you will want to use it only when you are in that directory, and the other place is your *personal ado directory*, which is probably something like `c:\ado\personal` if you use Windows, `~/ado/personal` if you use Unix, and `~:ado:personal` if you use a Macintosh. We are guessing.

> To find your personal ado directory,
> enter Stata and type   . personal

❏ Technical Note

Stata looks in various directories for ado-files, defined by the global macro S_ADO, which contains

UPDATES;BASE;SITE;.;PERSONAL;STBPLUS;OLDPLACE

The words in capital letters are codenames for directories and the mapping from codenames to directories can be obtained by typing the `sysdir` command. Here is what `sysdir` shows on one particular Windows computer:

```
. sysdir
   STATA:  D:\STATA\
 UPDATES:  D:\STATA\ado\updates\
    BASE:  D:\STATA\ado\base\
    SITE:  D:\STATA\ado\site\
 STBPLUS:  C:\ado\stbplus\
PERSONAL:  C:\ado\personal\
OLDPLACE:  C:\ado\
```

Even if you use Windows, your mapping might be different because it all depends on where you installed Stata. That is the point of the codenames. They make it possible to refer to directories according to their logical purposes rather than their physical location.

Global macro S_ADO is the search path, so in looking for an ado-file, Stata first looks in UPDATES, then in BASE, and so on, until it finds the file. Actually, Stata not only looks in UPDATES, it also takes the first letter of the ado-file it is looking for and looks in the lettered subdirectory. Say Stata was looking for `gobbledegook.ado`. Stata would look up UPDATES (D:\STATA\ado\updates in our example) and, if the file were not found there, it would look in the g subdirectory of UPDATES (D:\STATA\ado\updates\g) before moving on to looking in BASE, whereupon it would follow the same rules.

Why the extra complication? We distribute literally hundreds of ado-files with Stata and some operating systems have difficulty dealing with so many files in the same directory. All operating systems experience at least a performance degradation. To prevent this, the ado directory we ship is split 27 ways (letters *a–z* and underscore). Thus, the Stata command `ci`, which is implemented as an ado-file, can be found in the subdirectory c of BASE.

If you write ado-files, you can structure your personal ado directory this way, too, but there is no reason to do so until you have more than, say, 250 files in a single directory.

❏

❏ Technical Note

After finding and running *gobbledegook*.`ado`, Stata calculates the total size of all programs that it has automatically loaded. If this exceeds `adosize` (see [R] **sysdir**) Stata begins discarding the oldest automatically loaded programs until the total is less than `adosize`. Oldest here is measured by the time last used, not the time loaded. This discarding saves memory and does not affect you since any program that was automatically loaded could be automatically loaded again should the need arise.

It does, however, affect performance. Loading the program takes time and you will again have to wait if you use one of the previously loaded-and-discarded programs. Increasing `adosize` reduces this possibility, but at the cost of memory. The `set adosize` command allows you to change this parameter; see [R] **sysdir**. The default value of `adosize` is 65 for Intercooled Stata and 15 for Small Stata. A value of 40 for `adosize` means that up to 40K can be allocated to autoloaded programs. Experimentation has shown that this is a good number—increasing it does not improve performance much.

❏

21.11.1 Version

We recommend that the first line following `program define` in your ado-file declare the Stata release under which you wrote the program; `hello.ado` would better read as

```
———————————————————————————————————————— top of hello.do ————
program define hello
        version 6.0
        display "hi there"
end
———————————————————————————————————————— end of hello.do ————
```

We introduced the concept of version in [U] **19.1.1 Version**. In regular do-files, we recommend the `version` line appear as the first line of the do-file. For ado-files, the line appears after the `program define` because the loading of the ado-file is one step and the execution of the program another. It is when Stata executes the program defined in the ado-file that we want to stipulate the interpretation of the commands.

The appearance of the `version` line is of more importance in ado-files than in do-files because (1) ado-files have longer lives than do-files and so it is more likely that you will use an ado-file with a later release and (2) ado-files tend to use more of Stata's features and so increase the probability that, if a change to Stata is made, the change will affect them.

21.11.2 Comments and long lines in ado-files

Comments in ado-files are handled the same way as in do-files: enclose the text in `/*` comment `*/` brackets or
`*` begin the line with an asterisk; see [U] **19.1.2 Comments and blank lines in do-files**.

Logical lines longer than physical lines are also handled as they are in do-files: either you change the delimiter to semicolon `;`
or you comment out the newline `/*`
`*/` character; see [U] **19.1.2 Comments and blank lines in do-files**.

21.11.3   Debugging ado-files

Debugging ado-files is just a little tricky because it is Stata and not you that is in control of when the ado-file is loaded.

Assume you wanted to change `hello` to say "Hi, Mary". Assume that your editor is called `vi` and that you are in the habit of calling your editor from Stata, so you do this

```
. !vi hello.ado
```

and make the obvious change to the program. Equivalently, you can pretend that you are using a windowed operating system and you jump out of Stata—leaving it running—and modify the `hello.ado` file. Anyway, you change `hello.ado` to read

```
───────────────────────────────────── top of hello.do ─────────
program define hello
        version 6.0
        display "hi, Mary"
end
───────────────────────────────────── end of hello.do ─────────
```

Back in Stata, you try it:

```
. hello
hi there
```

What just happened is that Stata ran the old copy of `hello`—the copy it still has in its memory. Stata wants to be fast about executing ado-files, so when it loads one, it keeps it around a while—waiting for memory to get short—before clearing it from its memory. Naturally, Stata can drop `hello` anytime because it can always reload it from disk.

So, you changed the copy on disk but Stata still has the old copy loaded into memory. You type `discard` to tell Stata to forget these automatically loaded things and so force itself to get new copies of the ado-files from disk:

```
. discard
. hello
hi, Mary
```

Understand, the only reason you had to type `discard` is because you changed the ado-file while Stata was running. Had you exited Stata and returned later to use `hello`, the `discard` would not have been necessary because Stata forgets things between sessions anyway.

21.11.4   Local subroutines

A single ado-file can contain more than one `program define`, but if it does, the other programs defined in the ado-file are assumed to be subroutines of the main program. For example:

```
───────────────────────────────────── top of decoy.ado ─────────
program define decoy
        ...
        duck ...
        ...
end
program define duck
        ...
end
───────────────────────────────────── end of decoy.ado ─────────
```

duck is considered a local subroutine of decoy. Even after decoy.ado were loaded, if you were to type duck, you would be told "unrecognized command". To emphasize what local means, assume you have also written an ado-file named duck.ado:

```
                                                          ── top of duck.ado ──
        program define duck
                ...
        end
                                                          ── end of duck.ado ──
```

Even so, when decoy called duck, it would be the program duck defined in decoy.ado that was called. To further emphasize what local means, assume that duck.ado contains

```
                                                          ── top of decoy.ado ──
        program define decoy
                ...
                manic ...
                ...
                duck ...
                ...
        end
        program define duck
                ...
        end
                                                          ── end of decoy.ado ──
```

and that manic.ado contained

```
                                                          ── top of manic.ado ──
        program define manic
                ...
                duck ...
                ...
        end
                                                          ── end of manic.ado ──
```

Here is what would happen when you executed decoy:

1. decoy in decoy.ado would begin execution. decoy calls manic.

2. manic in manic.ado would begin execution. manic calls duck

3. duck in duck.ado (yes) would begin execution. duck would do whatever and return.

4. manic regains control and eventually returns.

5. decoy is back in control. decoy calls duck.

6. duck in decoy.ado would execute, complete, and return.

7. decoy would regain control and return.

Note well, when manic called duck, it was the global ado-file duck.ado that was executed yet when decoy called duck, it was the local program duck that was executed.

Stata does not find this confusing and neither should you.

21.11.5   Development of a sample ado-command

Below we demonstrate how one creates a new Stata command. We will program an influence measure for use with linear regression. It is an interesting statistic in its own right, but even if you are not interested in linear regression and influence measures, the focus here is on programming, not on the particular statistic chosen.

Hadi (1992) presents a measure of influence (see [R] **hadimvo**) in linear regression defined

$$H_i^2 = \frac{k}{(1 - h_{ii})} \frac{d_i^2}{1 - d_i^2} + \frac{h_{ii}}{1 - h_{ii}}$$

where k is the number of estimated coefficients; $d_i^2 = e_i^2/e'e$ and e_i is the ith residual; and h_{ii} is the ith diagonal element of the hat matrix. The ingredients of this formula are all available through Stata and so, after estimating a regression, one can easily calculate H_i^2. For instance, one might type

```
. regress mpg weight displ
. predict hii if e(sample), hat
. predict ei if e(sample), resid
. gen eTe = sum(ei*ei)
. gen di2 = (ei*ei)/eTe[_N]
. gen Hi = (3/(1-hii))*(di2/(1-di2)) + hii/(1-hii)
```

The number 3 in the formula for `Hi` represents k, the number of estimated parameters (which is an intercept plus coefficients on **weight** and **displ**).

❑ Technical Note

Do you understand why this works? `predict` can create h_{ii} and e_i, the trick is in getting $e'e$—the sum of the squared e_i's. Stata's `sum()` function creates a running sum. The first observation of `eTe` thus contains e_1^2; the second, $e_1^2 + e_2^2$; the third $e_1^2 + e_2^2 + e_3^2$; and so on. The last observation, then, contains $\sum_{i=1}^{N} e_i^2$, which is $e'e$. (We specified if e(sample) on our `predict` commands to restrict calculations to the estimation subsamples, so `hii` and `eii` might have missing values, but that does not matter because `sum()` treats missing values as contributing zero to the sum.) We use Stata's explicit subscripting feature and then refer to `eTe[_N]`, the last observation. (See [U] **16.3 Functions** and [U] **16.7 Explicit subscripting**.) After that, we plug into the formula to obtain the result.

❑

Assuming we often wanted this influence measure, it would be easier and less prone to error if we canned this calculation in a program. Our first draft of the program reflects exactly what we would have typed interactively:

```
───────────────────────────────────────────── top of hinflu.ado, version 1 ─────────
program define hinflu
        version 6.0
        predict hii if e(sample), hat
        predict ei if e(sample), resid
        gen eTe = sum(ei*ei)
        gen di2 = (ei*ei)/eTe[_N]
        gen Hi = (3/(1-hii))*(di2/(1-di2)) + hii/(1-hii)
        drop hii ei eTe di2
end
───────────────────────────────────────────── end of hinflu.ado, version 1 ─────────
```

All we have done is enter what we would have typed into a file, bracketing it with **program define hinflu**—meaning we decided to call our new command **hinflu**—and **end**. Since our command is to be called `hinflu`, the file must be named `hinflu.ado` and it must be stored in either the current directory or our personal ado directory (see [U] **20.5.2 Where is my personal ado directory?**).

That done, when we type `hinflu`, Stata will be able to find it, load it, and execute it. In addition to copying the interactive lines into a program, we added the line 'drop hii ...' to eliminate the working variables we had to create along the way.

So now we can interactively type

```
. regress mpg weight displ
. hinflu
```

and add the new variable Hi to our data.

Our program is not general. It is suitable for use after estimating a regression model on two and only two independent variables because we coded a 3 in the formula for k. Stata statistical commands like **regress** store information about the problem and answer in e(). Looking under "Saved Results" in [R] **regress**, we find that e(df_m) contains the model degrees of freedom, which is $k - 1$ assuming the model has an intercept. Thus, the second draft of our program reads

```
                                                  —— top of hinflu.ado, version 2 ————
    program define hinflu
            version 6.0
            predict hii if e(sample), hat
            predict ei if e(sample), resid
            gen eTe = sum(ei*ei)
            gen di2 = (ei*ei)/eTe[_N]
            gen Hi = ((e(df_m)+1)/(1-hii))*(di2/(1-di2)) + hii/(1-hii)
            drop hii ei eTe di2
    end
                                                  —— end of hinflu.ado, version 2 ————
```

In the formula for Hi, we substituted (e(df_m)+1) for the literal number 3.

Returning to the substance of our problem, **regress** also saves the residual sum of squares in e(rss), so the calculation of eTe is not really necessary:

```
                                                  —— top of hinflu.ado, version 3 ————
    program define hinflu
            version 6.0
            predict hii if e(sample), hat
            predict ei if e(sample), resid
            gen di2 = (ei*ei)/e(rss)            /* changed this version */
            gen Hi = ((e(df_m)+1)/(1-hii))*(di2/(1-di2)) + hii/(1-hii)
            drop hii ei di2
    end
                                                  —— end of hinflu.ado, version 3 ————
```

Our program is now shorter and faster and it is completely general. This program is probably good enough for most users; if you were implementing this for solely your own occasional use, you could stop right here. The program does, however, have the following deficiencies:

1. When we use it with data with missing values, the answer is correct but we see messages about the number of missing values generated. (These messages appear when the program is generating the working variables.)

2. We cannot control the name of the variable being produced—it is always called Hi. Moreover, when Hi already exists (say from a previous regression), we get an error message about Hi already existing. We then have to **drop** the old Hi and type **hinflu** again.

3. If we have created any variables named hii, ei, or di2, we also get an error about the variable already existing and the program refuses to run.

Fixing these problems is not difficult. The fix for problem 1 is exceedingly easy; we embed the entire program in a **quietly** block:

```
                                                    ─── top of hinflu.ado, version 4 ───────
      program define hinflu
             version 6.0
             quietly {                              /* new this version */
                    predict hii if e(sample), hat
                    predict ei if e(sample), resid
                    gen di2 = (ei*ei)/e(rss)
                    gen Hi = ((e(df_m)+1)/(1-hii))*(di2/(1-di2)) + hii/(1-hii)
                    drop hii ei di2
             }                                      /* new this version */
      end
                                                    ─── end of hinflu.ado, version 4 ───────
```

The output for the commands between the `quietly` { and } is now suppressed—the result is the same as if we had put `quietly` in front of each command.

Solving problem 2—that the resulting variable is always called `Hi`—requires use of the `syntax` command. Let's put that off and deal with problem 3—that the working variables have nice names like `hii`, `ei`, and `di2` and so prevent users from using those names in their data.

One solution would be to change the nice names to unlikely names. We could change `Hi` to `MyHiVaR`—that would not guarantee the prevention of a conflict, but it would certainly make it unlikely. It would also make our program difficult to read, an important consideration should we want to change it in the future. There is a better solution. Stata's `tempvar` command (see [U] **21.7.1 Temporary variables**) places names into local macros that are guaranteed to be unique:

```
                                                    ─── top of hinflu.ado, version 5 ───────
      program define hinflu
             version 6.0
             tempvar hii ei di2            /* new this version              */
             quietly {
                    predict `hii´ if e(sample), hat  /* changed, as are other lines */
                    predict `ei´ if e(sample), resid
                    gen `di2´ = (`ei´*`ei´)/e(rss)
                    gen Hi = ((e(df_m)+1)/(1-`hii´))*(`di2´/(1-`di2´)) + /*
                            */ `hii´/(1-`hii´)
             }
      end
                                                    ─── end of hinflu.ado, version 5 ───────
```

At the top of our program, we declare the temporary variables. (We can do it outside or inside the `quietly`—it makes no difference—and we do not have to do it at the top or even all at once; we could declare them as we need them, but at the top is prettiest.) When we refer to a temporary variable, we do not refer directly to it (such as by typing `hii`), we refer to it indirectly by typing open and close single quotes around the name (`` `hii´ ``). And at the end of our program, we no longer bother to `drop` the temporary variables—temporary variables are dropped automatically by Stata when a program concludes.

❏ Technical Note

Why do we type single quotes around the names? `tempvar` creates local macros containing the real temporary variable names. `hii` in our program is now a local macro and `` `hii´ `` refers to the contents of the local macro, which is the variable's actual name.

❏

We now have an excellent program—its only fault is that we cannot specify the name of the new variable to be created. Here is the solution to that problem:

```
────────────────────────────────── top of hinflu.ado, version 6 ──────────
program define hinflu
        version 6.0
        syntax newvarname                              /* new this version   */
        tempvar hii ei di2
        quietly {
                predict `hii´ if e(sample), hat
                predict `ei´ if e(sample), resid
                gen `di2´ = (`ei´*`ei´)/e(rss)
                gen `typlist´ `varlist´     /* changed this version */  /*
                */  ((e(df_m)+1)/(1-`hii´))*(`di2´/(1-`di2´)) +  /*
                */  `hii´/(1-`hii´)
        }
end
────────────────────────────────── end of hinflu.ado, version 6 ──────────
```

It took a change to one line and the addition of another to obtain the solution. This magic all happens because of syntax (see [U] **21.4.4 Parsing standard Stata syntax** above).

'syntax newvarname' specifies that one new variable name must be specified (had we typed 'syntax [newvarname]' the new varname would have been optional; had we typed 'syntax newvarlist' the user would have been required to specify at least one new variable and allowed to specify more). In any case, syntax compares what the user types to what is allowed. If what the user types does not match what we have declared, syntax will issue the appropriate error message and stop our program. If it does match, our program will continue and what the user typed will be broken out and stored in local macros for us. In the case of a newvarname, the new name typed by the user is placed in the local macro varlist and the type of the variable (float, double, ...) is placed in typlist (even if the user did not specify a storage type).

This is now an excellent program. There is, however, one more improvement we could make. hinflu is intended to be used sometime after regress. How do we know the user is not misusing our program and executing it after, say, logistic? e(cmd) will tell us the name of the last estimation command; see [U] **21.9 Accessing results calculated by estimation commands** and [U] **21.10.2 Saving results in e()** above. We should change our program to read

```
────────────────────────────────── top of hinflu.ado, version 7 ──────────
program define hinflu
        version 6.0
        if "`e(cmd)´"~="regress" { error 301 }    /* new this version */
        syntax newvarname
        tempvar hii ei di2
        quietly {
                predict `hii´ if e(sample), hat
                predict `ei´ if e(sample), resid
                gen `di2´ = (`ei´*`ei´)/e(rss)
                gen `typlist´ `varlist´ = /*
                */  ((e(df_m)+1)/(1-`hii´))*(`di2´/(1-`di2´)) + /*
                */  `hii´/(1-`hii´)
        }
end
────────────────────────────────── end of hinflu.ado, version 7 ──────────
```

The error command (see [R] **error**) issues one of Stata's prerecorded error messages and stops our program. Error 301 is "last estimates not found"; see [R] **error messages**. (Try typing error 301 at the console.)

In any case, this is a perfect program.

❏ Technical Note

You do not have to go to all the trouble we did to program the Hadi measure of influence or any other statistic that appeals to you. Whereas version 1 was not really an acceptable solution—it was too specialized—version 2 was acceptable. Version 3 was better, and version 4 better yet, but the improvements were of less and less importance.

Putting aside the details of Stata's language, you should understand that final versions of programs do not just happen—they are the results of drafts that have been subject to refinement. How much refinement should depend on how often and who will be using the program. In this sense, the "official" ado-files that come with Stata are poor examples. They have been subject to substantial refinement because they will be used by strangers with no knowledge of how the code works. When writing programs for yourself, you may want to stop refining at an earlier draft.

❏

21.11.6   Writing on-line help

When you write an ado-file, you should also write a help file to go with it. This file is a standard ASCII text file, named *command*.hlp, that you place in the same directory as your ado-file *command*.ado. This way when the user (who may be you at a future date) types `help` followed by the name of your new command (or pulls down **Help**), they will see something better than "help for ... not found".

You can obtain examples of help files by examining the .hlp files in the official ado directory; type "`sysdir`" and look in the lettered subdirectories of the directory defined as BASE:

```
. sysdir
    STATA:  C:\STATA\
  UPDATES:  C:\STATA\ado\updates\
     BASE:  C:\STATA\ado\base\
     SITE:  C:\STATA\ado\site\
  STBPLUS:  C:\ado\stbplus\
 PERSONAL:  C:\ado\personal\
 OLDPLACE:  C:\ado\
```

In this case, you would find examples of .hlp files in the a, b, ..., subdirectories of C:\STATA\ado\base.

Inside help files, two characters have been assigned special meanings. Placing either carets (^) or at signs (@) around text highlights the text. The difference is that @ also signifies to windowed operating systems a hypertext link—that the word should be highlighted and is clickable and, if clicked on, help for that word should be presented.

Thus, if a line in your .hlp file reads

```
Also see ^help^ for the @finishup@ command.
```

when the user reads your .hlp entry, he or she will see

Also see **help** for the **finishup** command.

For Stata for Windows and Stata for Macintosh users, **finishup** would appear in the hypertext-help color, indicating that they could click on the word to obtain more help.

Within the @ symbols, you can specify that words displayed in the hypertext-help color link to a file with a different name. For example, if a line in your .hlp file reads

```
Also see ^help^ for the @finishup!new finishup command@.
```

a user reading your `.hlp` entry will see

Also see **help** for the **new finishup command**.

For Stata for Windows and Stata for Macintosh users, **new finishup command** would appear in the hypertext-help color. Clicking on any of these words would display the **finishup** help file.

The ^ and @ do not count in spacing. In

```
^finishup^ has the unique ability to
```

the 'f' of finishup will be displayed in column 1.

If you want to display the real ^ or @ characters, you type ^^ or @@.

❑ Technical Note

When you include `@this@` in your help file, the user sees **this** and, should the user click on it, file `this.hlp` would be invoked.

`@...@` has a second syntax: `@filename!text to display@`

If you coded `@this:click here@` the user would see **click here** and, should the user click on it, file `this.hlp` would be invoked. Be careful with this: Unix users do not have the ability to click and neither do Windows and Macintosh users when they access the help via the `help` command versus the **Help** pulldown.

`@...@` has a third syntax: `@cmd:args!text to display@`

where allowed for `cmd:args` are

`browser:`*url*	invoke user's browser pointing to *url*
`search:`*keywords*	Like pulling down **Help**, selecting **Search. . .**, and entering *keywords*
`back:`	equivalent to clicking **Back** button on Help window
`net:from` *url*	displays output from `net from` *url*
`netfrom:`	pops up dialog box requesting "*Site URL*", then displays output from `net from` *specified_url*
`net:describe` *pkgname*	displays output from `net describe` *pkgname*
`net:install` *pkgname*	performs `net install` *pkgname*
`net:get` *pkgname*	performs `net get` *pkgname*
`ado:dir` [*pkgid*]	displays output from `ado dir`
`ado:describe` [*pkgid*]	displays output from `ado describe`
`adosearch:`	pops up dialog box requesting "Search for:", then displays output from `ado dir, find(`*search_for_entered*`)`

❑

A period in the first column of a `.hlp` file has special meaning. '.-' means display 79 dashes. The beginning of the official `brier.hlp` file reads

```
.-
help for ^brier^                                          (manual:  [R] brier)
.-
```

and when the user types **help brier**, he or she will see

```
-------------------------------------------------------------------------------
help for brier                                            (manual:  [R] brier)
-------------------------------------------------------------------------------
```

Another dot command is '.h *name*', which must be not only in the first column but also on the first line of the .hlp file. It means "go to". For instance, the official ereg.hlp file reads, in its entirety,

```
.h weibull
```

This tells the help system to reissue the command help weibull to itself. Thus, if the user types help ereg, it is just as if the user typed help weibull. (Similarly, if the Windows user clicks on **ereg**, it is just as if the user clicked on **weibull**.)

Why the feature? Sometimes commands are related and it is logical to document them together. The help for both weibull and ereg is in weibull.hlp.

Another .hlp feature concerns blank lines. Some editors make it difficult to add blank lines, especially at the end of files. Thus, Stata treats a period followed by nothing (.) in column one as a blank line. You can use real blank lines or these marked-blank lines in your file. If you use real blank lines, be sure that they have no white-space characters prior to the carriage return or line feed. Such white-space characters confuse the help line-counting mechanism.

❏ Technical Note

Users will find it easier to understand your programs if you document them in the same way as we document ours. We offer the following guidelines:

1. The first line should contain a row of dashes; '.-' will provide them.

2. The second line should contain

    ```
    help for ^yourcmd^
    ```
           (ultimate source of information)

    Highlight *yourcmd* by placing carets (^) around it. In the official Stata files, the ultimate source of information is the printed manual, so we reference it. When a new command has been published in the STB, we place the STB issue and insert number there; for example, (STB-20: sg25). For your files, we suggest putting your name.

    The line should be 79 characters long (before insertion of carets).

    The line should mention all commands documented here, so if there are multiple commands, list them separated by commas. Use more than one line if necessary.

3. The next line should contain a '.-' row of dashes.

4. Include one blank line. Include the title; do not highlight it. On the following line, include a line of dashes underlining the title and include one blank line following that:

    ```
    ( blank line )
    Title
    -----
    ( blank line )
    ```

5. Include syntax diagrams. Syntax diagrams are not centered; they are indented one tab stop (8 characters), although less indentation is acceptable. In the syntax diagram, highlight minimum acceptable abbreviations and other things that must be typed. Do not highlight terms that stand for things the user substitutes, such as *varlist*.

6. Include the heading "Description", meaning two blank lines, the subtitle, a line of dashes underlining it, and another blank line. In general, the format for a heading is

    ```
    ( blank line )
    ( blank line )
    Heading
    -------
    ( blank line )
    ```

7. Provide a short description of what the command does. Do not burden the user with details yet. Assume the user is at the point of asking whether this is what he or she is looking for.

8. If your commands have options, include the heading "Options" followed by a description of each option, preferably in the order they appear in the syntax diagram. Option paragraphs are reverse indented: the first line is flush left and begins with the option's name; subsequent lines are indented four spaces. One blank line separates paragraphs. If an option requires more than one paragraph, subsequent paragraphs begin with lines that are also indented four spaces.

9. Optionally include the heading "Remarks" and whatever lengthy discussion you feel necessary. Stata's official on-line help files omit this because the discussions appear in the manual. Stata's official help files for features added between releases (obtained from the STB or the Stata web site, etc.), however, include this because the appropriate STB issue may not be as accessible as the manuals. This choice is yours.

10. Include the heading "Examples" and provide some examples. Nothing communicates better than providing something beyond theoretical discussion. Examples seldom need lengthy or even any explanation.

11. Optionally include the heading "Author" and your identity. Exercise caution. If you include a telephone number, expect it to ring. An email address may be more appropriate. In email addresses, be sure to use @@ instead of @ so Stata will not think that it is a hypertext link.

12. Optionally include the heading "References" and any printed references. Stata's official help files seldom do this because they are in the printed manual.

13. Include the heading "Also see" and provide cross references to related on-line help.

We also warn that it is easy to use too much highlighting. Use highlighting only to indicate what is to be typed as is—what would be shown in a `typewriter` typeface were the documentation printed in this manual.

❏

❏ Technical Note

If you write a collection of programs, you need to somehow index the programs so that users (and you) can find the command they want. We do that with our `contents.hlp` entry. You should create a similar kind of entry. We suggest you call your private entry `user.hlp` in your personal ado directory; see [U] **20.5.2 Where is my personal ado directory?**. This way, to review what you have added, you can type `help user`.

We suggest Unix users at large sites also add `site.hlp` to the SITE directory (typically `/usr/local/ado`, but type "sysdir" to be sure). Then you can type `help site` for a list of the commands available site-wide.

❏

21.12   A compendium of useful commands for programmers

You can use literally any Stata command in your programs and ado-files. In addition, some commands are intended solely for use by Stata programmers. You should read

Basics

[R] **program**	Define and manipulate programs
[R] **macro**	Macro definition and manipulation

Parsing and program arguments

[R] **syntax**	Parse Stata syntax
[R] **gettoken**	Low-level parsing
[R] **confirm**	Argument verification

Program control

[R] **version**	version control
[R] **if**	if programming command
[R] **while**	Looping
[R] **error**	Display generic error message and exit
[R] **exit**	Exit from a program or do-file
[R] **capture**	Capture return code

Console output

[R] **display (for programmers)**	Display strings and values of scalar expressions
[R] **tabdisp**	Display tables

Commonly used programming commands

[R] **#delimit**	Change delimiter
[R] **exit (for programmers)**	Exit from a program or do-file
[R] **quietly**	Quietly and noisily perform Stata command
[R] **mark**	Mark observations for inclusion
[R] **matrix**	Introduction to matrix commands
[R] **preserve**	Preserve and restore data
[R] **scalar**	Scalar variables

Debugging

[R] **beep**	Make computer emit a beep
[R] **pause**	Program debugging command
[R] **program**	set trace debugging command

Advanced programming commands

[R] **break**	Suppress Break key
[R] **char**	Characteristics
[R] **estimates**	Post and redisplay estimation results
[R] **ml**	Maximum-likelihood estimation
[R] **more**	Pause until key is depressed
[R] **postfile**	Save results in Stata dataset
[R] **tokenize**	Divide strings into tokens
[R] **unab**	Unabbreviate variable list
[R] **window**	Programming menus, dialogs, and windows

21.13 References

Hadi, A. S. 1992. A new measure of overall potential influence in linear regression. *Computational Statistics and Data Analysis* 14: 1–27.

22 Immediate commands

22.1 Overview

An *immediate* command is a command that obtains data not from the data stored in memory but from numbers typed as arguments. Immediate commands, in effect, turn Stata into a glorified hand calculator.

There are many instances where you may not have the data, but you do know something about the data, and what you know is adequate to perform statistical tests. For instance, you do not have to have individual-level data to obtain the standard error of the mean and thereby a confidence interval if you know the mean, standard deviation, and number of observations. There are other instances where you may actually have the data and you could enter it and perform the test, but it would be easier if you could just ask for the statistic based on a summary. For instance, you flip a coin 10 times and it comes up heads twice. You could enter a 10 observation dataset with two 1s (standing for heads) and eight 0s (meaning tails).

Immediate commands are meant to solve those problems. Immediate commands have the following properties:

1. They never disturb the data in memory. You can perform an immediate calculation as an aside and your data remains unchanged.

2. The syntax for all is the same, the command name followed by numbers which are the summary statistics from which the statistic is calculated. Almost always the numbers are summary statistics and the order in which they are specified is in some sense "natural".

3. Immediate commands all end in the letter i, although the converse is not true. In most cases, if there is an immediate command, there is a nonimmediate form also, that is, a form that works on the data in memory. For every statistical command in Stata, we have included an immediate form if it is reasonable to assume that you might know the requisite summary statistics without being in possession of the underlying data, and typing those statistics is not absurdly burdensome.

4. Immediate commands are documented along with their nonimmediate cousins. Thus, if you want to obtain a confidence interval, whether it be from summary data with an immediate command or using the data in memory, use the table of contents or index to discover that [R] **ci** discusses confidence intervals. There, you learn that `ci` calculates confidence intervals using the data in memory and `cii` does the same with the data specified immediately following the command.

22.1.1 Examples

▷ Example

Taking the example of confidence intervals, professional papers often publish the mean, standard deviation, and number of observations for variables used in the analysis. Those statistics are sufficient for calculating a confidence interval. If we know that the mean mileage rating of cars in some sample is 24, the standard deviation 6, and that there are 97 cars in the sample:

```
. cii 97 24 6
Variable |      Obs         Mean     Std. Err.      [95% Conf. Interval]
---------+-------------------------------------------------------------
         |       97           24     .6092077        22.79073    25.20927
```

We learn that the mean's standard error is 0.61 and its 95% confidence interval is $[22.8, 25.2]$. To obtain this we typed `cii` (the immediate form of the `ci` command) followed by the number of observations, the mean, and the standard deviation. We knew the order in which to specify the numbers because we had read [R] **ci**.

We could use the immediate form of the `ttest` command to test the hypothesis that the true mean is 22:

```
. ttesti 97 24 6 22
One-sample t test
-----------------------------------------------------------------------------
         |      Obs     Mean    Std. Err.    Std. Dev.    [95% Conf. Interval]
---------+-------------------------------------------------------------------
      x  |       97       24    .6092077            6     22.79073    25.20927
-----------------------------------------------------------------------------
Degrees of freedom: 96
                              Ho: mean(x) = 22
   Ha: mean < 22              Ha: mean ~= 22              Ha: mean > 22
    t =   3.2830               t =   3.2830               t =   3.2830
 P < t =   0.9993           P > |t| =   0.0014         P > t =   0.0007
```

The first three numbers were as we specified in the `cii` command. `ttesti` requires a fourth number, which is the constant against which the mean is being tested; see [R] **ttest**.

◁

▷ Example

We mentioned flipping a coin 10 times and having it come up heads twice. The 99% confidence interval can also be obtained from `ci`:

```
. cii 10 2, level(99)
                                               -- Binomial Exact --
Variable |      Obs         Mean     Std. Err.    [99% Conf. Interval]
---------+-------------------------------------------------------------
         |       10           .2     .1264911      .0109375         .65
```

In the previous example, we specified `cii` with three numbers following it; in this example, we specify 2. Immediate commands often determine what to do by the number of arguments following the command. With two arguments, `ci` assumes we are specifying the number of trials and successes from a binomial experiment; see [R] **ci**.

The immediate form of the `bitest` command performs exact hypothesis testing:

```
. bitesti 10 2 .5
          N   Observed k   Expected k   Assumed p   Observed p
      -----------------------------------------------------------
         10            2            5     0.50000      0.20000
      Pr(k >= 2)               = 0.989258  (one-sided test)
      Pr(k <= 2)               = 0.054688  (one-sided test)
      Pr(k <= 2 or k >= 8) = 0.109375  (two-sided test)
```

For a full explanation of this output, see [R] **bitest**.

◁

▷ Example

Stata's `tabulate` command makes tables and calculates various measures of association. The immediate form, `tabi`, does the same, but you specify the contents of the table following the command:

```
. tabi 5 10 \ 2 14
            | col
      row |        1         2 |    Total
    -----------+----------------------+----------
        1 |        5        10 |       15
        2 |        2        14 |       16
    -----------+----------------------+----------
    Total |        7        24 |       31
              Fisher's exact =              0.220
      1-sided Fisher's exact =              0.170
```

The `tabi` command is slightly different from most immediate commands in that it uses '\' to indicate where one row ends and another begins.

◁

22.1.2   A list of the immediate commands

Command	Reference	Description
bitesti	[R] **bitest**	Binomial probability test
cii	[R] **ci**	Confidence intervals for means, proportions, and counts
cci	[R] **epitab**	Tables for epidemiologists
csi		
iri		
mcci		
prtesti	[R] **prtest**	One and two-sample tests of proportions
sampsi	[R] **sampsi**	Sample size and power determination
sdtesti	[R] **sdtest**	Variance comparison tests
symmi	[R] **symmetry**	Symmetry and marginal homogeneity tests
tabi	[R] **tabulate**	One- and two-way tables of frequencies
ttesti	[R] **ttest**	Mean comparison tests

22.2 The display command

display is not really an immediate command, but it can be used to perform as a hand-calculator.

```
. display 2+5
7
. display sqrt(2+sqrt(3^2-4*2*-2))/(2*3)
.44095855
```

See [R] **display**.

23 Estimation and post-estimation commands

Contents

23.1   All estimation commands work the same way

All Stata commands that estimate statistical models—commands such as `regress`, `logit`, `sureg`, and so on—work the same way. Most single-equation commands have similar syntax

$$command\ varlist\ \left[weight\right]\ \left[\texttt{if}\ exp\right]\ \left[\texttt{in}\ range\right]\ \left[,\ options\right]$$

as do most multiple-equation commands:

$$command\ (varlist)\ (varlist)\ \dots\ (varlist)\ \left[weight\right]\ \left[\texttt{if}\ exp\right]\ \left[\texttt{in}\ range\right]\ \left[,\ options\right]$$

Adopt a loose definition of single and multiple equation in interpreting this. For instance, `heckman` is a two-equation system mathematically speaking yet we categorize it, syntactically, with single-equation commands because most researchers think of it as a linear regression with an adjustment for the censoring.

In any case, the important thing is that most estimation commands have one or the other of these two syntaxes.

In single-equation commands, the first variable in the *varlist* is the dependent variable and the remaining are the independent variables. There can be variations. For instance, `anova` allows you to specify *terms* in addition to variables for the independent variables, but the foregoing is generally correct.

More importantly, all estimation commands—whether single or multiple equation—share the following features:

1. You can use the features of Stata's syntax to specify the estimation sample; you do not have to make a special dataset.

2. You can, at any time, review the last estimates by typing the estimation command without arguments. After estimating a regression model with `regress`, for instance, you can see the last estimates again by typing `regress` by itself. You do not have to do this immediately—any number of commands can occur in between the estimation and the replaying and, in fact, you can even replay the last estimates after the data has changed or you have dropped the data altogether. Stata never forgets (unless you type `discard`; see [R] **discard**).

3. All estimation commands display confidence intervals for the coefficients and allow the `level()` option to indicate the width of the intervals. The default is `level(95)`, meaning 95% confidence intervals. You can reset the default with `set level`; see [R] **level**.

4. You can obtain the variance–covariance matrix of the estimators (VCE), presented as either a correlation matrix or a covariance matrix, using `vce` at any time after estimation. (You can also obtain the estimated coefficients and covariance matrix as vectors and matrices and manipulate them with Stata's matrix capabilities; see [U] **17.5 Accessing matrices created by Stata commands**.)

5. You can obtain predictions, residuals, influence statistics, and the like, either for the data on which you just estimated or for some other data, using `predict`.

6. You can refer to coefficients and standard errors in expressions (such as with `generate`); see [U] **16.5 Accessing coefficients and standard errors**.

7. You can perform tests on the estimated parameters using `test` (Wald test of linear hypotheses), `nltest` (Wald test of nonlinear hypotheses), and `lrtest` (likelihood-ratio tests). You can also obtain point estimates and confidence intervals for linear combinations of the estimated parameters using `lincom`.

8. You can hold estimates, perform other estimation commands, and then restore the prior estimates. This is of particular interest to programmers; see [R] **estimates**.

Eventually it will also be true that

9. You can obtain the Huber/White/robust alternate estimate of variance by specifying the `robust` option and you can relax the assumption of independence of the observations by also specifying the `cluster()` option.

Right now there are a few estimation commands that do not allow `robust`.

23.2   Standard syntax

The syntax for `regress`, or any of the estimation commands, is the same as that for all single-equation commands. Most importantly, you can combine `if` *exp* and `in` *range* with the estimation command, and some estimation commands also allow `by` *varlist*:.

▷ Example

You have data on 74 automobiles, recording the mileage rating (mpg), weight (weight) and whether the car is domestic or foreign-produced (foreign). You can estimate a linear regression model of mpg on weight and weightsq, using just the foreign-made automobiles, by typing

```
. regress mpg weight weightsq if foreign

      Source |       SS       df       MS                  Number of obs =      22
-------------+------------------------------              F(  2,    19) =    8.31
       Model |  428.256889       2  214.128444            Prob > F      =  0.0026
    Residual |  489.606747      19  25.7687762            R-squared     =  0.4666
-------------+------------------------------              Adj R-squared =  0.4104
       Total |  917.863636      21  43.7077922            Root MSE      =  5.0763

---------------------------------------------------------------------------------
         mpg |      Coef.   Std. Err.       t    P>|t|     [95% Conf. Interval]
-------------+-------------------------------------------------------------------
      weight |  -.0132182   .0275711     -0.479   0.637    -.0709252    .0444888
    weightsq |   5.50e-07   5.41e-06      0.102   0.920    -.0000108    .0000119
       _cons |   52.33775   34.1539       1.532   0.142    -19.14719    123.8227
---------------------------------------------------------------------------------
```

You can run separate regressions for the domestic and foreign-produced automobiles using the by varlist: prefix:

```
. by foreign: regress mpg weight weightsq

-> foreign=Domestic
      Source |       SS       df       MS                  Number of obs =      52
-------------+------------------------------              F(  2,    49) =   91.64
       Model |  905.395466       2  452.697733            Prob > F      =  0.0000
    Residual |  242.046842      49  4.93973146            R-squared     =  0.7890
-------------+------------------------------              Adj R-squared =  0.7804
       Total |  1147.44231      51  22.4988688            Root MSE      =  2.2226

---------------------------------------------------------------------------------
         mpg |      Coef.   Std. Err.       t    P>|t|     [95% Conf. Interval]
-------------+-------------------------------------------------------------------
      weight |  -.0131718   .0032307     -4.077   0.000    -.0196642   -.0066794
    weightsq |   1.11e-06   4.95e-07      2.249   0.029     1.19e-07    2.11e-06
       _cons |   50.74551   5.162014      9.830   0.000     40.37205    61.11896
---------------------------------------------------------------------------------

-> foreign= Foreign
      Source |       SS       df       MS                  Number of obs =      22
-------------+------------------------------              F(  2,    19) =    8.31
       Model |  428.256889       2  214.128444            Prob > F      =  0.0026
    Residual |  489.606747      19  25.7687762            R-squared     =  0.4666
-------------+------------------------------              Adj R-squared =  0.4104
       Total |  917.863636      21  43.7077922            Root MSE      =  5.0763

---------------------------------------------------------------------------------
         mpg |      Coef.   Std. Err.       t    P>|t|     [95% Conf. Interval]
-------------+-------------------------------------------------------------------
      weight |  -.0132182   .0275711     -0.479   0.637    -.0709252    .0444888
    weightsq |   5.50e-07   5.41e-06      0.102   0.920    -.0000108    .0000119
       _cons |   52.33775   34.1539       1.532   0.142    -19.14719    123.8227
---------------------------------------------------------------------------------
```

Although all estimation commands do allow if *exp* and in *range*, not all allow the by *varlist*: prefix. In any case, the duration of Stata's memory is limited: it remembers only the *last* set of estimates. This means that if you were to use any of the other features which we describe below, they would use the last regression estimated, which right now is mpg on weight and weightsq for the foreign subsample.

◁

23.3   Replaying prior results

When you type an estimation command without arguments, it redisplays prior results.

▷ Example

To estimate a regression of `mpg` on the variables `weight` and `displ`, you type

```
. regress mpg weight displ
  Source |       SS       df       MS                  Number of obs =      74
---------+------------------------------                F(  2,    71) =   66.78
   Model | 1595.40969       2  797.704846              Prob > F      =  0.0000
Residual | 848.049768      71  11.9443629              R-squared     =  0.6529
---------+------------------------------                Adj R-squared =  0.6432
   Total | 2443.45946      73  33.4720474              Root MSE      =  3.4561

------------------------------------------------------------------------------
     mpg |      Coef.   Std. Err.       t     P>|t|     [95% Conf. Interval]
---------+--------------------------------------------------------------------
  weight |  -.0065671   .0011662     -5.631   0.000    -.0088925   -.0042417
   displ |   .0052808   .0098696      0.535   0.594    -.0143986    .0249602
   _cons |   40.08452    2.02011     19.843   0.000     36.05654    44.11251
------------------------------------------------------------------------------
```

You now go on to do other things, summarizing data, listing observations, performing hypothesis tests, or anything else. If you decide you want to see the last set of estimates again, type the estimation command without arguments.

```
. regress
  Source |       SS       df       MS                  Number of obs =      74
---------+------------------------------                F(  2,    71) =   66.78
   Model | 1595.40969       2  797.704846              Prob > F      =  0.0000
Residual | 848.049768      71  11.9443629              R-squared     =  0.6529
---------+------------------------------                Adj R-squared =  0.6432
   Total | 2443.45946      73  33.4720474              Root MSE      =  3.4561

------------------------------------------------------------------------------
     mpg |      Coef.   Std. Err.       t     P>|t|     [95% Conf. Interval]
---------+--------------------------------------------------------------------
  weight |  -.0065671   .0011662     -5.631   0.000    -.0088925   -.0042417
   displ |   .0052808   .0098696      0.535   0.594    -.0143986    .0249602
   _cons |   40.08452    2.02011     19.843   0.000     36.05654    44.11251
------------------------------------------------------------------------------
```

This feature works with *every* estimation command, so you could just as well have done it with, say, `cox` or `logit`.

◁

23.4   Specifying the estimation subsample

Once an estimation command has been run, Stata remembers the estimation subsample and you can use the modifier `if e(sample)` on the end of Stata commands. The term estimation subsample refers to the set of observations used to produce the last estimation results. That might turn out to be all the observations (as it was in the above example) or a subset of the observations:

```
. generate exc = rep78==5 if rep78~=.
(5 missing values generated)
. regress mpg weight exc if foreign
```

```
  Source |       SS       df       MS              Number of obs =      21
---------+------------------------------           F( 2,   18) =   10.21
   Model | 423.317154      2  211.658577           Prob > F      =  0.0011
Residual | 372.96856      18  20.7204756           R-squared     =  0.5316
---------+------------------------------           Adj R-squared =  0.4796
   Total | 796.285714     20  39.8142857           Root MSE      =   4.552

------------------------------------------------------------------------------
     mpg |      Coef.   Std. Err.       t    P>|t|     [95% Conf. Interval]
---------+--------------------------------------------------------------------
  weight |  -.0131402   .0029684    -4.427   0.000    -.0193765    -.0069038
     exc |   5.052676    2.13492     2.367   0.029     .5673764     9.537977
   _cons |   52.86088   6.540147     8.083   0.000     39.12054     66.60122
------------------------------------------------------------------------------

. summarize mpg weight exc if e(sample)
Variable |     Obs        Mean    Std. Dev.       Min        Max
---------+-----------------------------------------------------
     mpg |      21    25.28571    6.309856        17         41
  weight |      21    2263.333    364.7099      1760       3170
     exc |      21   .4285714    .5070926         0          1
```

Note that 21 observations were used in the above regression and we subsequently obtained the means for those same 21 observations by typing 'summarize ... if e(sample)'. There are two reasons observations were dropped: we specified if foreign when we ran the regression and there were observations for which exc was missing. The reason does not matter; e(sample) is true if the observation was used and false otherwise.

You can use if e(sample) on the end of any Stata command that allows an if *exp*.

23.5 Specifying the width of confidence intervals

You specify the width of the confidence intervals for the coefficients using the level() option, and you can specify the width at estimation or when you play back the results.

▷ Example

To obtain narrower, 90% confidence intervals when we estimate the model, we type

```
. regress mpg weight level, level(90)
  Source |       SS       df       MS              Number of obs =      74
---------+------------------------------           F( 2,   71) =   66.78
   Model | 1595.40969      2  797.704846           Prob > F      =  0.0000
Residual | 848.049768     71  11.9443629           R-squared     =  0.6529
---------+------------------------------           Adj R-squared =  0.6432
   Total | 2443.45946     73  33.4720474           Root MSE      =  3.4561

------------------------------------------------------------------------------
     mpg |      Coef.   Std. Err.       t    P>|t|     [90% Conf. Interval]
---------+--------------------------------------------------------------------
  weight |  -.0065671   .0011662    -5.631   0.000    -.0085108    -.0046234
   displ |   .0052808   .0098696     0.535   0.594    -.0111679     .0217294
   _cons |   40.08452    2.02011    19.843   0.000     36.71781     43.45124
------------------------------------------------------------------------------
```

Were we to subsequently type regress, without arguments, 95% confidence intervals would be reported. Had we initially estimated the model with 95% confidence intervals, we could later type 'regress, level(90)' to redisplay results with 90% confidence intervals.

In addition, we could type set level 90 to make 90% intervals our default; see [R] **level**.

◁

23.6   Obtaining the variance–covariance matrix

Typing `vce` displays the variance–covariance matrix of the estimators after estimation.

▷ Example

Continuing with our regression of `mpg` on variables `weight` and `displ` example, we have previously typed `regress mpg weight displ`. The full variance–covariance matrix of the estimators can be displayed at any time after estimation:

```
. vce

         |   weight     displ     _cons
---------+-----------------------------
  weight|   1.4e-06
   displ|  -.00001   .000097
   _cons|  -.002075  .011884   4.08085
```

`vce` with the `corr` option will present this matrix as a correlation matrix:

```
. vce, corr

         |   weight     displ     _cons
---------+-----------------------------
  weight|   1.0000
   displ|  -0.8949   1.0000
   _cons|  -0.8806   0.5960    1.0000
```

See [R] **vce**.

In addition, Stata's matrix commands understand `e(V)` to be a reference to the matrix:

```
. matrix list e(V)

symmetric e(V)[3,3]
              weight        displ        _cons
weight   1.360e-06
 displ    -.0000103    .00009741
 _cons    -.00207455   .01188356    4.0808455

. mat Vinv = syminv(e(V))

. mat list Vinv

symmetric Vinv[3,3]
              weight        displ        _cons
weight   60175851
 displ    4081161.2   292709.46
 _cons    18706.732   1222.3339    6.1953911
```

See [U] **17.5 Accessing matrices created by Stata commands**.

◁

23.7   Obtaining predicted values

Our discussion below, while cast in terms of predicted values, applies equally to all the statistics generated by `predict`; see [R] **predict**.

When Stata estimates a model, be it regression or whatever, it internally saves the results, which includes the estimated coefficients and the variable names. The `predict` command allows you to use that information.

▷ Example

Let's estimate a linear regression of `mpg` on `weight` and `weightsq`:

```
. regress mpg weight weightsq

     Source |       SS       df       MS              Number of obs =      74
------------+------------------------------           F(  2,    71) =   72.80
      Model | 1642.52197       2  821.260986          Prob > F      =  0.0000
   Residual |  800.937487      71  11.2808097          R-squared     =  0.6722
------------+------------------------------           Adj R-squared =  0.6630
      Total | 2443.45946       73  33.4720474          Root MSE      =  3.3587

------------------------------------------------------------------------------
        mpg |     Coef.   Std. Err.       t    P>|t|     [95% Conf. Interval]
------------+-----------------------------------------------------------------
     weight | -.0141581   .0038835     -3.646   0.001    -.0219016   -.0064145
   weightsq |  1.32e-06   6.26e-07      2.116   0.038     7.67e-08    2.57e-06
      _cons |  51.18308   5.767884      8.874   0.000     39.68225    62.68392
------------------------------------------------------------------------------
```

After estimating the regression, `predict` is defined to be

$$-.0141581\text{weight} + 1.32 \cdot 10^{-6}\text{weightsq} + 51.18308$$

(Actually, it is more precise, because the coefficients are internally stored at much higher precision than shown in the output.) Thus, we can create a new variable—call it `fitted`—equal to the prediction by typing `predict fitted` and then use `graph` to make a display of the fitted and actual values separately for domestic and foreign automobiles:

```
. predict fitted
(option xb assumed; fitted values)
. graph mpg fitted weight, by(foreign) total c(.l) s(oi) sort
```

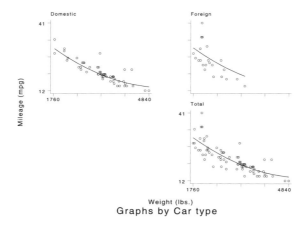

Graphs by Car type

`predict` can calculate much more than just predicted values. In the case of `predict` after linear regression `predict` can calculate residuals, standardized residuals, studentized residuals, influence statistics, and the list goes on and on. In any case, what is to be calculated is specified via an option, so if we wanted the residuals stored in new variable `r`, we would type

```
. predict r, resid
```

What options may be specified following `predict` vary according to the estimation command previously used; the `predict` options are documented along with the estimation command. If we wanted to know all the things `predict` can do for us following `regress`, we would see [R] **regress**.

◁

23.7.1 predict can be used after any estimation command

The use of `predict` is not limited to linear regression.

▷ Example

You estimate a logistic regression model of whether a car is manufactured outside the United States based on its weight and mileage rating using either the `logistic` or the `logit` command; see [R] **logistic** and [R] **logit**. We will use `logit`.

```
. logit foreign weight mpg

Iteration 0:   log likelihood =  -45.03321
Iteration 1:   log likelihood = -29.898968
Iteration 2:   log likelihood = -27.495771
Iteration 3:   log likelihood = -27.184006
Iteration 4:   log likelihood = -27.175166
Iteration 5:   log likelihood = -27.175156

Logit estimates                                Number of obs   =         74
                                               LR chi2(2)      =      35.72
                                               Prob > chi2     =     0.0000
Log likelihood = -27.175156                    Pseudo R2       =     0.3966

------------------------------------------------------------------------------
   foreign |      Coef.   Std. Err.       z    P>|z|     [95% Conf. Interval]
-----------+------------------------------------------------------------------
    weight |  -.0039067   .0010116    -3.862   0.000    -.0058894    -.001924
       mpg |  -.1685869   .0919174    -1.834   0.067    -.3487418     .011568
     _cons |   13.70837   4.518707     3.034   0.002     4.851864    22.56487
------------------------------------------------------------------------------
```

After `logit`, `predict` without options calculates the probability of a positive outcome (we learned that by looking at [R] **logit**). To obtain the predicted probabilities that each car is manufactured outside the U.S.:

```
. predict probhat
(option p assumed; Pr(foreign))

. summarize probhat

Variable |     Obs        Mean   Std. Dev.       Min        Max
---------+-----------------------------------------------------
 probhat |      74    .2972973   .3052979    .000729   .8980594

. list in 1/5

        make                 mpg   weight   foreign    probhat
  1. AMC Concord              22    2,930   Domestic   .1904363
  2. AMC Pacer                17    3,350   Domestic   .0957767
  3. AMC Spirit               22    2,640   Domestic   .4220815
  4. Buick Century            20    3,250   Domestic   .0862625
  5. Buick Electra            15    4,080   Domestic   .0084948
```

◁

23.7.2   Making in-sample predictions

predict does not retrieve a vector of prerecorded values—it calculates the predictions based on the recorded coefficients and the data currently in memory. In the above examples, when we have typed things like

```
. predict phat
```

predict has filled in the prediction everywhere it could be calculated.

Sometimes we may have more data in memory than was used by the estimation command, either because we explicitly ignored some of the observations by specifying an if *exp* with the estimation command, or because there are missing values. In such cases, if we want to restrict the calculation to the estimation subsample, we would do that in the usual way, by adding if e(sample) to the end of the command:

```
. predict phat if e(sample)
```

23.7.3   Making out-of-sample predictions

Because predict makes its calculations based on the recorded coefficients and the data in memory, predict can do more than calculate predicted values for the data on which the sample was estimated—it can make out-of-sample predictions as well.

If we estimated our model on a subset of the observations, we could then predict the outcome for all the observations:

```
. logit foreign weight mpg if rep78>3
. predict pall
```

If we do not specify if e(sample) at the end of the predict command, predict calculates the predictions for all observations possible.

In fact, because predict works from the estimated model, we can use predict with *any* dataset that contains the necessary variables.

▷ Example

Continuing with our previous logit example, assume you have a second dataset containing the mpg and weight of a different sample of cars. You have just estimated your model and now continue:

```
. use otherdat, clear
(Different cars)

. predict probhat                 Stata remembers previous model
(option p assumed; Pr(foreign))

. summarize probhat foreign
```

Variable	Obs	Mean	Std. Dev.	Min	Max
probhat	12	.2505068	.3187104	.0084948	.8920776
foreign	12	.1666667	.3892495	0	1

◁

▷ Example

There are numerous ways to obtain out-of-sample predictions. Above, we estimated on one dataset and then used another. If our first dataset had contained both sets of cars, marked say by the variable `difcars` being 0 if from the first sample and 1 if from the second, we could type

```
. logit foreign weight mpg if difcars==0
same output as above appears
. predict probhat
(option p assumed; Pr(foreign))
. summarize probhat foreign if difcars==1
same output as directly above appears
```

If we had just a small number of additional cars, we could even input them after estimation. Assume our data once again contains only the first sample of cars, and assume that we are interested in an additional sample of only 2 rather than 12 cars, we could type

```
. logit foreign weight mpg
same output as above appears
. input
                    make       mpg    weight    foreign
75. "Merc. Zephyr" 20 2830 0              we type in our new data
76. "VW Dasher" 23 2160 1
77. end
. predict probhat                    obtain all the predictions
(option p assumed; Pr(foreign))
. list in -2/l                       and list the new ones
                    make       mpg    weight   foreign    probhat
75.        Merc. Zephyr          20     2830   Domestic   .3275397
76.          VW Dasher           23     2160    Foreign   .8009743
```

◁

23.8 Accessing estimated coefficients

You can access coefficients and standard errors after estimation by referring to _b[*name*] and _se[*name*]; see [U] **16.5 Accessing coefficients and standard errors**.

▷ Example

Let us return to linear regression. You are doing a study of earnings of men and women at a particular company. In addition to each person's earnings, you have information on their educational attainment and tenure with the company. You do the following:

```
. generate femten = female*tenure
. generate femed = female*ed
. regress lnearn ed tenure female femed femten
output appears
```

You now wish to predict everyone's income as if they were male and then compare these as-if earnings with the actual earnings:

```
. generate asif = _b[_cons] + _b[ed]*ed + _b[tenure]*tenure
```

◁

▷ Example

    You are analyzing the mileage of automobiles and are using a slightly more sophisticated model than any we have used so far. As we have previously, you assume `mpg` is a function of `weight` and `weightsq`, but you also add the interaction of `foreign` multiplied by `weight` (called `fweight`), the car's gear ratio (`gratio`), and `foreign` multiplied by `gratio` (`fgratio`).

```
. regress mpg weight weightsq fweight gratio fgratio

    Source |       SS       df       MS                Number of obs =      74
-----------+------------------------------             F(  5,    68) =   33.44
     Model | 1737.05293       5  347.410585            Prob > F      =  0.0000
  Residual | 706.406534      68  10.3883314            R-squared     =  0.7109
-----------+------------------------------             Adj R-squared =  0.6896
     Total | 2443.45946      73  33.4720474            Root MSE      =  3.2231

-------------------------------------------------------------------------------
       mpg |      Coef.   Std. Err.       t    P>|t|     [95% Conf. Interval]
-----------+-------------------------------------------------------------------
    weight |  -.0118517   .0045136    -2.626   0.011    -.0208584    -.002845
  weightsq |   9.81e-07   7.04e-07     1.392   0.168    -4.25e-07    2.39e-06
   fweight |  -.0032241   .0015577    -2.070   0.042    -.0063326   -.0001157
    gratio |   1.159741   1.553418     0.746   0.458    -1.940057    4.259539
   fgratio |   1.597462   1.205313     1.325   0.189    -.8077035    4.002627
     _cons |   44.61644   8.387943     5.319   0.000     27.87856    61.35432
-------------------------------------------------------------------------------
```

Unless you are experienced in both regression technology and automobile technology, you may find it difficult to interpret this regression. Putting aside issues of statistical significance, you find that mileage decreases with a car's weight but increases with the square of weight; decreases even more rapidly with weight for foreign cars; increases with higher gear ratio and increases even more rapidly with higher gear ratio in foreign cars.

    Thus, do foreign cars yield better or worse gas mileage? Results are mixed. As the foreign cars' weight increases, they do more poorly in relation to domestic cars, but they do better at higher gear ratios. One way to compare the results is to predict what mileage foreign cars would have *if they were manufactured domestically*. The regression provides all the information necessary for making that calculation; mileage for domestic cars is estimated to be

$$-.012 \, \texttt{weight} + 9.81 \cdot 10^{-7} \, \texttt{weightsq} + 1.160 \, \texttt{gratio} + 44.6$$

We can use that equation to predict the mileage of foreign cars and then compare it with the true outcome. The `_b[]` function simplifies reference to the estimated coefficients. We can type

```
. gen asif=_b[weight]*weight + _b[weightsq]*weightsq +
      _b[gratio]*gratio + _b[_cons]
```

`_b[weight]` refers to the estimated coefficient on `weight`, `_b[weightsq]` to the estimated coefficient on `weightsq`, and so on.

    We might now ask how the actual mileage of a Honda compares with the `asif` prediction:

```
. list make asif mpg if index(make,"Honda")

            make         asif       mpg
61. Honda Accord      26.52597        25
62. Honda Civic       30.62202        28
```

Notice the way we constructed our `if` clause to select Hondas. `index()` is the string function that returns the location in the first string where the second string is found or, if the second string does not occur in the first, zero. Thus, any recorded `make` that contains the string "Honda" anywhere in it would be listed; see [U] **16.3.5 String functions**.

We find that both Honda models yield slightly lower gas mileage than the `asif` domestic-car-based prediction. (Please note that we do not endorse this model as a complete model of the determinants of mileage, nor do we single out the Honda for any special scorn. In fact, one should note that the observed values are within the root mean square error of the average prediction.)

We might wish to compare the overall average `mpg` and the `asif` prediction over all foreign cars in the data:

```
. summarize mpg asif if foreign
    Variable |     Obs       Mean    Std. Dev.       Min        Max
-------------+--------------------------------------------------------
         mpg |      22   24.77273    6.611187        14         41
        asif |      22   26.67124    3.142912   19.70466   30.62202
```

We find that, on average, foreign cars yield slightly lower mileage than our `asif` prediction. This might lead us to ask if any foreign cars do better than the `asif` prediction:

```
. list make mpg asif if foreign & mpg>asif
           make               mpg        asif
55. BMW 320i                   25    24.31697
57. Datsun 210                 35    28.96818
63. Mazda GLC                  30    29.32015
66. Subaru                     35    28.85993
68. Toyota Corolla             31    27.01144
71. VW Diesel                  41    28.90355
```

We find six such automobiles.

❑

23.9   Performing hypothesis tests on the coefficients

23.9.1   Linear tests

After estimation, `test` is used to perform tests of linear hypotheses based on the variance–covariance matrix of the estimators (Wald tests).

▷ Example

(`test` has numerous syntaxes and features, so do not use this example as an excuse for not reading [R] **test**.) Using the automobile data, you estimate the following regression:

```
. generate weightsq=weight^2
. regress mpg weight weightsq foreign
      Source |       SS       df       MS              Number of obs =      74
-------------+------------------------------           F(  3,    70) =   52.25
       Model |  1689.15372    3   563.05124            Prob > F      =  0.0000
    Residual |   754.30574   70  10.7757963            R-squared     =  0.6913
-------------+------------------------------           Adj R-squared =  0.6781
       Total |  2443.45946   73  33.4720474            Root MSE      =  3.2827

         mpg |     Coef.   Std. Err.       t    P>|t|     [95% Conf. Interval]
-------------+----------------------------------------------------------------
      weight |  -.0165729   .0039692    -4.175   0.000    -.0244892   -.0086567
    weightsq |   1.59e-06   6.25e-07     2.546   0.013     3.45e-07    2.84e-06
     foreign |    -2.2035   1.059246    -2.080   0.041      -4.3161   -.0909003
       _cons |   56.53884   6.197383     9.123   0.000     44.17855    68.89913
```

You can use the `test` command to calculate the joint significance of `weight` and `weightsq`:

```
. test weight weightsq

 ( 1)  weight = 0.0
 ( 2)  weightsq = 0.0

       F(  2,    70) =    60.83
            Prob > F =     0.0000
```

You are not limited to testing whether coefficients are zero. You can test whether the coefficient on `foreign` is −2 by typing

```
. test foreign = -2

 ( 1)  foreign = -2.0

       F(  1,    70) =     0.04
            Prob > F =     0.8482
```

You can even test more complicated hypotheses since `test` has the ability to perform basic algebra. Here is an absurd hypothesis:

```
. test 2*(weight+weightsq)=-3*(foreign-(weight-weightsq))

 ( 1) - weight + 5.0 weightsq + 3.0 foreign = 0.0

       F(  1,    70) =     4.31
            Prob > F =     0.0416
```

`test` simplified the algebra of our hypothesis and then presented the test results. We discover that the hypothesis may be absurd, but we cannot reject it. You can also use `test`'s `accumulate` option to combine this test with another test:

```
. test foreign+weight=0, accum

 ( 1) - weight + 5.0 weightsq + 3.0 foreign = 0.0
 ( 2)  weight + foreign = 0.0

       F(  2,    70) =     9.12
            Prob > F =     0.0003
```

There are limitations. `test` can test only linear hypotheses. If you attempt to test a nonlinear hypothesis, `test` will tell you that it is not possible:

```
. test weight/foreign=0
not possible with test
r(131);
```

Testing nonlinear hypotheses is discussed in [U] **23.9.4 Nonlinear Wald tests** below.

◁

23.9.2   test can be used after any estimation command

`test` bases its results on the estimated variance–covariance matrix of the estimators (i.e., performs a Wald test) and so can be used after any estimation command. In the case of maximum likelihood estimation, you will have to decide whether you want to perform tests based on the information matrix instead of constraining the equation, reestimating it, and then calculating the likelihood-ratio test (see [U] **23.9.3 Likelihood-ratio tests**). Since `test` bases its results on the information matrix, its results have exactly the same standing as the asymptotic Z statistic presented in the coefficient table.

▷ Example

Let's examine the repair records of the cars in our automobile data as rated by *Consumers Reports*:

```
. tab rep78 for

  Repair |
  Record |        Car type
    1978 |  Domestic   Foreign |     Total
---------+----------------------+----------
       1 |         2         0 |         2
       2 |         8         0 |         8
       3 |        27         3 |        30
       4 |         9         9 |        18
       5 |         2         9 |        11
---------+----------------------+----------
   Total |        48        21 |        69
```

The values are coded 1–5, corresponding to well below average to well above average. We will fit this variable using a maximum-likelihood ordered logit model (the `nolog` option suppresses the iteration log, saving us some paper):

```
. ologit rep78 price for weight weightsq displ, nolog
Ordered logit estimates                     Number of obs   =         69
                                            LR chi2(5)      =      33.12
                                            Prob > chi2     =     0.0000
Log likelihood = -77.133082                 Pseudo R2       =     0.1767

------------------------------------------------------------------------
   rep78 |     Coef.   Std. Err.      z    P>|z|    [95% Conf. Interval]
---------+--------------------------------------------------------------
   price |   -.000034   .0001188   -0.286   0.775    -.0002669    .000199
 foreign |   2.685648   .9320398    2.881   0.004     .8588833   4.512412
  weight |  -.0037447   .0025609   -1.462   0.144    -.0087639   .0012745
weightsq |   7.87e-07   4.50e-07    1.750   0.080    -9.43e-08   1.67e-06
   displ |  -.0108919   .0076805   -1.418   0.156    -.0259455   .0041617
---------+--------------------------------------------------------------
   _cut1 |  -9.417196   4.298201         (Ancillary parameters)
   _cut2 |  -7.581864    4.23409
   _cut3 |   -4.82209   4.147679
   _cut4 |   -2.79344   4.156219
------------------------------------------------------------------------
```

We now wonder whether all of our variables other than `foreign` are jointly significant. We test the hypothesis just as we would after linear regression:

```
. test weight weightsq displ price
 ( 1)   weight = 0.0
 ( 2)   weightsq = 0.0
 ( 3)   displ = 0.0
 ( 4)   price = 0.0

       chi2(  4) =      3.63
     Prob > chi2 =    0.4590
```

It is worthwhile comparing this with the results performed by a likelihood-ratio test; see [U] **23.9.3 Likelihood-ratio tests**. In this case, results differ little.

◁

23.9.3   Likelihood-ratio tests

After maximum likelihood estimation, you can obtain likelihood-ratio tests. This is done by estimating the unconstrained model, typing lrtest, saving(0) to save the model, estimating the constrained model, and then typing lrtest by itself. There are other possibilities, so see [R] **lrtest** for the full details.

▷ Example

In [U] **23.9.2 test can be used after any estimation command** above, we estimated an ordered-logit on rep78 and then tested the significance of all the explanatory variables except foreign.

To obtain the likelihood-ratio test, sometime after estimating the full model, we type lrtest, saving(0).

```
. ologit rep78 price for weight weightsq displ
(output omitted)
. lrtest, saving(0)
```

This command saves the current model as the base with which we shall compare. It is not important that we save the model immediately after estimation.

Next we estimate the constrained model and, after that, typing lrtest by itself compares the current model with the model we saved:

```
. ologit rep78 foreign
Iteration 0:    log likelihood = -93.692061
Iteration 1:    log likelihood = -79.696089
Iteration 2:    log likelihood = -79.044933
Iteration 3:    log likelihood = -79.029267
Iteration 4:    log likelihood = -79.029243

Ordered logit estimates                    Number of obs   =         69
                                           LR chi2(1)      =      29.33
                                           Prob > chi2     =     0.0000
Log likelihood = -79.029243                Pseudo R2       =     0.1565

------------------------------------------------------------------------------
     rep78 |     Coef.   Std. Err.      z     P>|z|     [95% Conf. Interval]
---------+--------------------------------------------------------------------
   foreign |   2.98155   .6203637    4.806   0.000      1.76566    4.197441
---------+--------------------------------------------------------------------
     _cut1 |  -3.158382   .7224269            (Ancillary parameters)
     _cut2 |  -1.362642   .3557343
     _cut3 |   1.232161   .3431227
     _cut4 |   3.246209   .5556646
------------------------------------------------------------------------------

. lrtest
Ologit:  likelihood-ratio test                 chi2(4)    =       3.79
                                               Prob > chi2 =     0.4348
```

When we tested the same constraint with test (which performed a Wald test), we obtained a χ^2 of 3.63 and significance level of 0.4590.

◁

23.9.4 Nonlinear Wald tests

testnl can be used to test nonlinear hypotheses about the parameters of the most recently estimated model. testnl, like test, bases its results on the variance–covariance matrix of the estimators (i.e., performs a Wald test) and so can be used after any estimation command; see [R] **testnl**.

▷ Example

You estimate the model

```
. regress price mpg weight foreign
(output omitted )
```

and then type

```
. testnl (38*_b[mpg]^2 = _b[foreign]) (_b[mpg]/_b[weight]=4)
 (1)   38*_b[mpg]^2 = _b[foreign]
 (2)   _b[mpg]/_b[weight]=4
             F(2, 70) =         0.02
             Prob > F =         0.9806
```

We performed this test on linear regression estimates, but tests of this type could be performed after any estimation command.

◁

23.10   Obtaining linear combinations of coefficients

lincom computes point estimates, standard errors, t or Z statistics, p-values, and confidence intervals for a linear combination of coefficients after any estimation command except anova. Results can optionally be displayed as odds ratios, incidence rate ratios, or relative risk ratios.

▷ Example

We estimate a linear regression

```
. reg y x1 x2 x3

      Source |       SS       df       MS                Number of obs =     148
-------------+------------------------------             F(  3,   144) =   96.12
       Model |   3259.3561      3  1086.45203            Prob > F      =  0.0000
    Residual |  1627.56282    144  11.3025196            R-squared     =  0.6670
-------------+------------------------------             Adj R-squared =  0.6600
       Total |  4886.91892    147  33.2443464            Root MSE      =  3.3619

------------------------------------------------------------------------------
           y |      Coef.   Std. Err.       t    P>|t|     [95% Conf. Interval]
-------------+----------------------------------------------------------------
          x1 |   1.457113    1.07461     1.356   0.177    -.6669339    3.581161
          x2 |   2.221682    .8610358    2.580   0.011     .5197797    3.923583
          x3 |   -.006139    .0005543  -11.076   0.000    -.0072345   -.0050435
       _cons |   36.10135    4.382693    8.237   0.000     27.43863    44.76407
------------------------------------------------------------------------------
```

Suppose that we want to see the difference of the coefficients of x2 and x1. We type

```
. lincom x2 - x1
( 1) - x1 + x2 = 0.0
```

y	Coef.	Std. Err.	t	P>|t|	[95% Conf. Interval]
(1)	.7645682	.9950282	0.768	0.444	-1.20218 2.731316

◁

lincom is very handy for computing the odds ratio of one covariate group relative to another.

▷ Example

We estimate a logistic model of low birth weight:

```
. logit low age lwd black other smoke ptd ht ui

Iteration 0:   log likelihood =  -117.336
Iteration 1:   log likelihood =-99.431174
Iteration 2:   log likelihood =-98.785718
Iteration 3:   log likelihood =   -98.778
Iteration 4:   log likelihood =-98.777998
```

Logit estimates		Number of obs	=	189
		LR chi2(8)	=	37.12
		Prob > chi2	=	0.0000
Log likelihood = -98.777998		Pseudo R2	=	0.1582

low	Coef.	Std. Err.	z	P>|z|	[95% Conf. Interval]
age	-.0464796	.0373888	-1.243	0.214	-.1197603 .0268011
lwd	.8420615	.4055338	2.076	0.038	.0472299 1.636893
black	1.073456	.5150752	2.084	0.037	.0639273 2.082985
other	.815367	.4452979	1.831	0.067	-.0574008 1.688135
smoke	.8071996	.404446	1.996	0.046	.0145001 1.599899
ptd	1.281678	.4621157	2.774	0.006	.3759478 2.187408
ht	1.435227	.6482699	2.214	0.027	.1646415 2.705813
ui	.6576256	.4666192	1.409	0.159	-.2569313 1.572182
_cons	-1.216781	.9556797	-1.273	0.203	-3.089878 .656317

If we want to get the odds ratio for black smokers relative to white nonsmokers (the reference group), we type

```
. lincom black + smoke, or
( 1)  black + smoke = 0.0
```

low	Odds Ratio	Std. Err.	z	P>|z|	[95% Conf. Interval]
(1)	6.557805	4.744692	2.599	0.009	1.588176 27.07811

lincom computed $\exp(\beta_{\text{black}} + \beta_{\text{smoke}}) = 6.56$.

◁

23.11   Obtaining robust variance estimates

Estimates of variance refers to estimated standard errors or, more completely, the estimated variance–covariance matrix of the estimators of which the standard errors are a subset, being the square root of the diagonal elements. We will call this matrix simply the variance. All estimation commands produce an estimate of variance and, using that, produce confidence intervals and significance tests.

In addition to the conventional estimator of variance, there is another. This estimator has been called by various names because it has been derived independently in different ways by different authors. Two popular names associated with the calculation are Huber and White, but it is also known as the sandwich estimator of variance (because of how the calculation formula physically appears) and the robust estimator of variance (because of claims made about it). In addition, this estimator also has an independent and long tradition in the survey literature.

The conventional estimator of variance is derived by starting with a model. To fix ideas, let's assume it is the regression model

$$y_i = \mathbf{x}_i \boldsymbol{\beta} + \epsilon_i, \qquad \epsilon_i \sim N(0, \sigma^2)$$

although it is not important for the discussion that we are using regression. Under the model-based approach, we assume the model is true and thereby derive an estimator for $\boldsymbol{\beta}$ and its variance.

The estimator of the standard error of $\widehat{\boldsymbol{\beta}}$ we develop is based on the assumption that the model is true in every detail. To wit, the only reason y_i is not exactly equal to $\mathbf{x}_i \boldsymbol{\beta}$ (so that we would only need to solve an equation to obtain precisely that value of $\boldsymbol{\beta}$) is because the observed y_i has noise ϵ_i added to it, that noise is Gaussian, and it has constant variance. It is that noise that leads to the uncertainty about $\boldsymbol{\beta}$ and it is from the characteristics of that noise that we are able to calculate a sampling distribution for $\widehat{\boldsymbol{\beta}}$.

The key thought here is that the standard error of $\widehat{\boldsymbol{\beta}}$ arises because of ϵ and is valid only because the model is absolutely, without question, true; we just do not happen to know the particular values of $\boldsymbol{\beta}$ and σ^2 that make the model true. The implication is that, in an infinite-sized sample, the estimator $\widehat{\boldsymbol{\beta}}$ for $\boldsymbol{\beta}$ would converge to the true value of $\boldsymbol{\beta}$ and its variance would go to 0.

Now, here is another interpretation of the estimation problem: We are going to estimate the model

$$y_i = \mathbf{x}_i \mathbf{b} + e_i$$

and, to obtain estimates of $\mathbf{b}$, we are going to use the calculation formula

$$\widehat{\mathbf{b}} = (\mathbf{X}'\mathbf{X})^{-1}\mathbf{X}'\mathbf{y}$$

Please note, we have made no claims that the model is true nor any claims about e_i or its distribution. We shifted our notation from $\boldsymbol{\beta}$ and ϵ_i to $\mathbf{b}$ and e_i to emphasize this. All we have stated are the physical actions we intend to carry out on the data. Interestingly, it is possible to calculate a standard error for $\widehat{\mathbf{b}}$ in this case! At least, it is possible if you will agree with us on what the standard error measures.

We are going to define the standard error as measuring the standard error of the calculated $\widehat{\mathbf{b}}$ were we to repeat the data collection followed by estimation over and over again.

Note well: this is a different concept of the standard error from the conventional, model-based ideas, but it is not unrelated. Both measure uncertainty about $\mathbf{b}$ (or $\boldsymbol{\beta}$). The model-based derivation states from where the variation arises and so is able to make grander statements about the applicability of the measured standard error. The weaker argument makes fewer assumptions and so produces a standard error suitable for one purpose.

There is a subtle difference in interpretation of these identically calculated point estimates. $\widehat{\beta}$ is the estimate of β under the assumption that the model is true. $\widehat{b}$ is the estimate of b, which is merely what the estimator would converge to if we collected more and more data.

Is the estimate of b unbiased? If you mean, does $b = \beta$, that depends on whether the model is true. $\widehat{b}$ is, however, an unbiased estimate of b which, admittedly, is not saying much.

What if x and e are correlated; don't we have a problem in that case? Answer: You may have an interpretation problem—b may not measure what you want to measure, namely β—but we measure $\widehat{b}$ to be such and such and expect, were the experiment and estimation repeated, you would observe results in the range we have reported.

And hence the name, the robust estimate of variance, and its associated authors are Huber (1967) and White (1980, 1982) (who developed it independently), although many others have extended its development including Gail, Tan, and Piantadosi (1988), Kent (1982), Royall (1986), and Lin and Wei (1989). In the survey literature, this same estimator has been developed; see, for example, Kish and Frankel (1974), Fuller (1975), and Binder (1983).

Many of Stata's estimation commands can produce this alternative estimate of variance and, if they can, they have a `robust` option. Without `robust` you get one measure of variance:

```
. regress mpg weight foreign
    Source |       SS       df       MS                  Number of obs =      74
-----------+------------------------------               F(  2,    71) =   69.75
     Model |  1619.2877        2  809.643849             Prob > F      =  0.0000
  Residual |  824.171761       71  11.608053             R-squared     =  0.6627
-----------+------------------------------               Adj R-squared =  0.6532
     Total |  2443.45946       73  33.4720474            Root MSE      =  3.4071

------------------------------------------------------------------------------
       mpg |      Coef.   Std. Err.       t    P>|t|     [95% Conf. Interval]
-----------+------------------------------------------------------------------
    weight |  -.0065879   .0006371    -10.340   0.000    -.0078583   -.0053175
   foreign |  -1.650029   1.075994     -1.533   0.130      -3.7955    .4954421
     _cons |    41.6797   2.165547     19.247   0.000     37.36172   45.99768
------------------------------------------------------------------------------
```

With `robust`, you get another:

```
. regress mpg weight foreign, robust
Regression with robust standard errors                   Number of obs =      74
                                                         F(  2,    71) =   73.81
                                                         Prob > F      =  0.0000
                                                         R-squared     =  0.6627
                                                         Root MSE      =  3.4071

------------------------------------------------------------------------------
           |              Robust
       mpg |      Coef.   Std. Err.       t    P>|t|     [95% Conf. Interval]
-----------+------------------------------------------------------------------
    weight |  -.0065879   .0005462    -12.061   0.000     -.007677   -.0054988
   foreign |  -1.650029   1.132566     -1.457   0.150    -3.908301    .6082423
     _cons |    41.6797   1.797553     23.187   0.000     38.09548   45.26392
------------------------------------------------------------------------------
```

Either way, the point estimates are the same. (See [R] **regress** for an example where specifying `robust` produces strikingly different standard errors.)

The robust estimator of variance has one feature that the conventional estimator does not have: the ability to relax the assumption of independence of the observations. That is, it can produce "correct" standard errors (in the measurement sense) even if the observations are correlated, if you specify the `cluster()` option.

In the case of the automobile data, it is difficult to believe that the models of the various manufacturers are truly independent. Manufacturers, after all, use common technology, engines, and drive trains across their model lines. The VW Dasher in the above regression has a measured residual of −2.80. Having been told that, do you really believe the residual for the VW Rabbit is as likely to be above 0 as below? (The residual is −2.32.) Similarly, the measured residual for the Chevrolet Malibu is 1.27. Does that provide no information about the expected value of the residual of the Chevrolet Monte Carlo (which turns out to be 1.53)?

One wants to be careful about picking examples out of data; we have not told you about the Datsun 210 and 510 (residuals +8.28 and −1.01) or the Cadillac Eldorado and Seville (residuals −1.99 and +7.58), but you should, at least, be questioning the assumption of independence. It may be believable that the measured mpg given the weight of one manufacturer's vehicles is independent of other manufacturers' vehicles, but it is at least questionable whether a manufacturer's vehicles are independent of one another.

In commands with the robust option, another option—cluster()—relaxes the independence assumption and requires only that the observations be independent across the clusters:

```
. regress mpg weight foreign, robust cluster(manuf)

Regression with robust standard errors        Number of obs =      74
                                               F(  2,    22) =   90.93
                                               Prob > F      =  0.0000
                                               R-squared     =  0.6627
Number of clusters (manuf) = 23                Root MSE      =  3.4071

------------------------------------------------------------------------------
             |             Robust
        mpg |      Coef.   Std. Err.       t    P>|t|     [95% Conf. Interval]
---------+--------------------------------------------------------------------
     weight |  -.0065879   .0005339   -12.339   0.000    -.0076952   -.0054806
    foreign |  -1.650029   1.039033    -1.588   0.127    -3.804852    .5047939
      _cons |    41.6797   1.844559    22.596   0.000     37.85432    45.50508
------------------------------------------------------------------------------
```

It turns out that, in this data, whether we specify cluster() makes little difference. The VW and Chevrolet examples we quoted above were not representative; had they been, the confidence intervals would have widened. (In the above, manuf is a variable that takes on values such as "Chev.", "VW", etc., recording the manufacturer of the vehicle. We created this variable from variable make, which contains values such as "Chev. Malibu", "VW Rabbit", etc., by extracting the first word.)

As a demonstration of how well clustering can work, in [R] **regress** we estimate a random-effects model with regress, robust cluster() and then compare the results with ordinary least squares and the GLS random-effects estimator. Here we will simply summarize the results.

We start with a dataset on 4,782 women aged 16 to 46. Subjects appear an average of 7.14 times in this data, so there are a total of 34,139 observations. The model we estimate is log wage on age, age-squared, and grade of schooling completed and the focus of the example is the estimated coefficient on schooling. We obtain the following results:

Estimator	point estimate	confidence interval
(inappropriate) least squares	.081	[.079, .083]
robust, cluster	.081	[.077, .085]
GLS random effects	.080	[.076, .083]

We wish you to start by noticing how well robust with the cluster() option does as compared with the GLS random-effects model. We then run a Hausman specification test, obtaining $\chi^2(2) = 62$, which casts grave doubt on the assumptions justifying the use of the GLS estimator and hence the GLS results. At this point, we will simply quote our comments

Meanwhile, our robust-regression results still stand as long as we are careful about the interpretation. The correct interpretation is that, were the data collection repeated (on women sampled the same way as in the original sample), and were we to reestimate the model, 95% of the time we would expect the estimated coefficient on grade to be in the range $[.077, .085]$.

Even with robust regression, you must be careful about going beyond that statement. In this case the Hausman test is probably picking up something that differs within and between persons and so would cast doubt on our robust-regression model in terms of interpreting $[.077, .085]$ to contain the rate of return to additional schooling, economy wide, for all women, without exception.

The formula for the robust estimator of variance is

$$\widehat{\mathcal{V}} = \widehat{\mathbf{V}}\left(\sum_{j=1}^{N} \mathbf{u}'_j \mathbf{u}_j\right)\widehat{\mathbf{V}}$$

where $\widehat{\mathbf{V}} = (-\partial^2 \ln L/\partial\beta^2)^{-1}$ (the conventional estimator of variance) and $\mathbf{u}_j$ (a row vector) is the contribution from the jth observation to the scores $\partial \ln L/\partial\beta$.

In the above, observations are assumed to be independent. Assume, for a moment, that the observations denoted by j are not independent but that they can be divided into M groups G_1, G_2, $\ldots$, G_M that are independent. Then the robust estimator of variance is

$$\widehat{\mathcal{V}} = \widehat{\mathbf{V}}\left(\sum_{k=1}^{M} \mathbf{u}_k^{(G)'} \mathbf{u}_k^{(G)}\right)\widehat{\mathbf{V}}$$

where $\mathbf{u}_k^{(G)}$ is the contribution of the kth group to the scores $\partial \ln L/\partial\beta$. That is, application of the robust variance formula merely involves using a different decomposition of $\partial \ln L/\partial\beta$, namely $\mathbf{u}_k^{(G)}$, $k = 1, \ldots, M$ rather than $\mathbf{u}_j$, $j = 1, \ldots, N$. Moreover, if the log-likelihood function is additive in the observations denoted by j,

$$\ln L = \sum_{j=1}^{N} \ln L_j$$

then $\mathbf{u}_j = \partial \ln L_j/\partial\beta$ and so

$$\mathbf{u}_k^{(G)} = \sum_{j \in G_k} \mathbf{u}_j$$

In other words, the group scores that enter the calculation are simply the sums of the individual scores within group. That is what the cluster() option does. (This point was first made in writing by Rogers (1993) although he considered the point an obvious generalization of Huber (1967) and the calculation—implemented by Rogers—had appeared in Stata a year earlier.)

❑ Technical Note

What is written above is asymptotically correct but we have ignored a finite-sample adjustment to $\widehat{\mathcal{V}}$. For maximum likelihood estimators, when you specify robust but not cluster(), a better estimate of variance is $\widehat{\mathcal{V}}^* = [N/(N-1)]\widehat{\mathcal{V}}$. When you also specify the cluster() option, this becomes $\widehat{\mathcal{V}}^* = [M/(M-1)]\widehat{\mathcal{V}}$.

In the case of linear regression, the finite sample adjustment is $N/(N - k)$ without cluster() — where k is the number of regressors — and $[M/(M - 1)][(N - 1)/(N - k)]$ with cluster(). In addition, two data-dependent modifications to the calculation for $\widehat{\mathcal{V}}^*$ suggested by MacKinnon and White (1985) are also provided by regress; see [R] **regress**.

❑

23.12 Obtaining scores

Many of the estimation commands that provide the robust option also provide the score(*newvar*) option. score() returns an important ingredient into the robust variance calculation that is sometimes useful in its own right. As explained in [U] **23.11 Obtaining robust variance estimates** above, ignoring the finite-sample corrections, the robust estimate of variance is

$$\widehat{\mathcal{V}} = \widehat{\mathbf{V}} \left(\sum_{j=1}^{N} \mathbf{u}_j' \mathbf{u}_j \right) \widehat{\mathbf{V}}$$

where $\widehat{\mathbf{V}} = (-\partial^2 \ln L/\partial \beta^2)^{-1}$ (the conventional estimator of variance) and $\mathbf{u}_j$ (a row vector) is the contribution from the jth observations to the scores $\partial \ln L/\partial \beta$. Let us consider likelihood functions that are additive in the observations,

$$\ln L = \sum_{j=1}^{N} \ln L_j$$

so that $\mathbf{u}_j = \partial \ln L_j/\partial \beta$. In general, function L_j is a function of $\mathbf{x}_j$ and β, $L_j(\beta; \mathbf{x}_j)$. For many likelihood functions, however, it is only the linear form $\mathbf{x}_j \beta$ that enters the function. In those cases,

$$\frac{\partial \ln L_j(\mathbf{x}_j \beta)}{\partial \beta} = \frac{\partial \ln L_j(\mathbf{x}_j \beta)}{\partial (\mathbf{x}_j \beta)} \frac{\partial (\mathbf{x}_j \beta)}{\partial \beta} = \frac{\partial \ln L_j(\mathbf{x}_j \beta)}{\partial (\mathbf{x}_j \beta)} \mathbf{x}_j$$

Writing $u_j = \partial \ln L_j(\mathbf{x}_j \beta)/\partial (\mathbf{x}_j \beta)$, this becomes simply $u_j \mathbf{x}_j$. Thus, the formula for the robust estimate of variance can be rewritten

$$\widehat{\mathcal{V}} = \widehat{\mathbf{V}} \left(\sum_{j=1}^{N} u_j^2 \mathbf{x}_j' \mathbf{x}_j \right) \widehat{\mathbf{V}}$$

We refer to u_j as the score (in the singular) and it is u_j that is returned when you specify option score(). u_j is like a residual in that

1. $\sum_j u_j = 0$,

2. correlation of u_j and $\mathbf{x}_j$, calculated over $j = 1, \ldots, N$, is 0.

In fact, in the case of linear regression, u_j is the residual, normalized,

$$\frac{\partial \ln L_j}{\partial (\mathbf{x}_j \beta)} = \frac{\partial}{\partial (\mathbf{x}_j \beta)} \ln f((y_j - \mathbf{x}_j \beta)/\sigma)$$

$$= (y_j - \mathbf{x}_j \beta)/\sigma^2$$

where $f()$ is the normal density. As a result, `regress` is one of the few commands that, while providing a `robust` option, does not provide a corresponding `score()` option. It is easy enough to obtain the residuals using `predict`; see [U] **23.7 Obtaining predicted values** above.

▷ Example

Command `probit` does provide both `robust` and `score()` options. The scores play an important role in calculating the robust estimate of variance, but you can specify `score` whether or not you specify `robust`:

```
. probit foreign mpg weight, score(u)
Iteration 0:   log likelihood =  -45.03321
Iteration 1:   log likelihood = -29.244141
  (output omitted )
Iteration 5:   log likelihood = -26.844189
```

```
Probit estimates                                 Number of obs   =         74
                                                 LR chi2(2)      =      36.38
                                                 Prob > chi2     =     0.0000
Log likelihood = -26.844189                      Pseudo R2       =     0.4039
```

foreign	Coef.	Std. Err.	z	P>|z|	[95% Conf. Interval]
mpg	-.1039503	.0515689	-2.016	0.044	-.2050235 -.0028772
weight	-.0023355	.0005661	-4.126	0.000	-.003445 -.0012261
_cons	8.275464	2.554142	3.240	0.001	3.269438 13.28149

```
. summarize u
```

Variable	Obs	Mean	Std. Dev.	Min	Max
u	74	-1.93e-16	.5988325	-1.655439	1.660787

```
. correlate u mpg weight
(obs=74)
```

	u	mpg	weight
u	1.0000		
mpg	0.0000	1.0000	
weight	0.0000	-0.8072	1.0000

```
. list make foreign mpg weight u if abs(u)>1.65
```

	make	foreign	mpg	weight	u
24.	Ford Fiesta	Domestic	28	1,800	-1.6554395
64.	Peugeot 604	Foreign	14	3,420	1.6607871

The light, high-mileage Ford Fiesta is surprisingly domestic while the heavy, low-mileage Peugeot 604 is surprisingly foreign.   ◁

❑ Technical Note

For some estimation commands one score is not enough. Consider a likelihood which can be written as $L_j(\mathbf{x}_j\boldsymbol{\beta}_1, \mathbf{z}_j\boldsymbol{\beta}_2)$. Then $\partial\ln L_j/\partial\boldsymbol{\beta}$ can be written $(\partial\ln L_j/\partial\boldsymbol{\beta}_1, \partial\ln L_j/\partial\boldsymbol{\beta}_2)$. Each of the components can in turn be written as $[\partial\ln L_j/\partial(\beta_1\mathbf{x})]\mathbf{x} = u_1\mathbf{x}$ and $[\partial\ln L_j/\partial(\beta_2\mathbf{z})]\mathbf{z} = u_2\mathbf{z}$. There are then two scores, u_1 and u_2 and, in general, there could be more.

Stata's `weibull` command is an example of this: it estimates β and a shape parameter σ, the latter of which can be thought of as a degenerate linear form $\sigma\mathbf{z}$ with $\mathbf{z} = \mathbf{1}$. `weibull`'s `score()` option requires that you specify two variable names; the first will be defined containing u_1—the score associated with β—and the second will be defined containing u_2—the score associated with σ.   ❑

❑ Technical Note

Using Stata's matrix commands—see [R] **matrix**—we can make the robust variance calculation for ourselves and then compare it with that made by Stata.

```
. quietly probit foreign mpg weight, score(u)
. matrix accum S =  mpg weight [iweight=u^2*74/73]
(obs=26.53642547)
. matrix rV = e(V)*S*e(V)
. matrix list rV
symmetric rV[3,3]
                 mpg        weight        _cons
   mpg     .00352299
weight     .00002216     2.434e-07
 _cons   -.14090346    -.00117031    6.4474172
. quietly probit foreign mpg weight, robust
. matrix list e(V)
symmetric e(V)[3,3]
                 mpg        weight        _cons
   mpg     .00352299
weight     .00002216     2.434e-07
 _cons   -.14090346    -.00117031    6.4474172
```

Results are the same.

There is an important lesson here for programmers. Given the scores, conventional variance estimates can be easily transformed to robust estimates. If one were writing a new estimation command, it would not be difficult to include a `robust` option.

It is, in fact, easy if we ignore clustering. With clustering, it is more work since the calculation involves forming sums within clusters. For programmers interested in implementing robust variance calculations, Stata provides an `_robust` command to ease the task. This is documented in [R] **_robust**.

To use `_robust`, you first produce conventional results (a vector of coefficients and covariance matrix) along with a variable containing the scores u_j (or variables if the likelihood function has more than one stub). You then call `_robust` and it will transform your conventional variance estimate into the robust estimate. `_robust` will handle the work associated with clustering, the details of the finite-sample adjustment, and it will even label your output so that the word Robust appears above standard error when results are displayed.

❑

23.13   Weighted estimation

In [U] **14.1.6 weight** we introduced the syntax for weights. Stata provides four kinds of weights: `fweight`s, or frequency weights; `pweight`s, or sampling weights; `aweight`s, or analytic weights; and `iweight`s, or importance weights. The syntax for using each is the same. Type

```
. regress y x1 x2
```

and you obtain unweighted estimates; type

```
. regress y x1 x2 [pweight=pop]
```

and you obtain (in this example) `pweight`ed estimation.

Below we explain in detail how each kind of weight is used in estimation.

23.13.1 Frequency weights

Frequency weights—fweights—are integers and nothing more than replication counts. The weight is statistically uninteresting but, from a data processing perspective, it is of great importance. Consider the following data

y	x1	x2
22	1	0
22	1	0
22	1	1
23	0	1
23	0	1
23	0	1

and the estimation command

    . regress y x1 x2

Exactly equivalent is the following, more compressed data

y	x1	x2	pop
22	1	0	2
22	1	1	1
23	0	1	3

and the corresponding estimation command

    . regress y x1 x2 [fweight=pop]

When you specify frequency weights, you are treating each observation as one or more real observations.

❑ Technical Note

One will occasionally run across a command that does not allow weights at all, especially among user-written commands. expand (see [R] **expand**) can be used with such commands to obtain frequency-weighted results. The expand command duplicates observations so that the data becomes self-weighting. For example, we want to run the command usercmd, which does something or other, and we would very much like to type usercmd y x1 x2 [fw=pop]. Unfortunately, usercmd does not allow weights. Instead, we type

    . expand pop
    . usercmd y x1 x2

and so obtain our result. Moreover, there is an important principle here: The results of running any command with frequency weights should be the exactly same as running the command on the unweighted, expanded data. Unweighted, duplicated data and frequency-weighted data are merely two ways of recording identical information.

❑

23.13.2 Analytic weights

Analytic weights—analytic is a term made up by us—statistically arise in one particular problem: linear regression on data that are themselves observed means. That is, think of the model

$$y_i = \mathbf{x}_i \boldsymbol{\beta} + \epsilon_i, \qquad \epsilon_i \sim N(0, \sigma^2)$$

and now think about estimating this model on data $(\bar{y}_j, \bar{\mathbf{x}}_j)$ that are themselves observed averages. For instance, a piece of the underlying data for $(y_i, \mathbf{x}_i)$ might be $(3, 1)$, $(4, 2)$, and $(2, 2)$, but you do not know that. Instead, you have a single observation $((3 + 4 + 2)/3, (1 + 2 + 2)/3) = (3, 1.67)$ and know only that the $(3, 1.67)$ arose as the average of 3 underlying observations. All of your data is like that.

regress with aweights is the solution to that problem:

. regress y x [aweight=pop]

There is a history of misusing such weights. A researcher does not have cell-mean data but instead a probability-weighted random sample. Long before Stata existed, some researchers were using aweights to produce estimates from such samples. We will come back to this point in [U] **23.13.3 Sampling weights** below.

Anyway, the statistical problem to which aweights are the solution can be written as

$$y_i = \mathbf{x}_i\boldsymbol{\beta} + \epsilon_i, \qquad \epsilon_i \sim N(0, \sigma^2/w_i)$$

where the w_i are the analytic weights. The details of the solution, it turns out, are to make linear regression calculations using the weights as if they were fweights but to first normalize them to sum to N before doing that.

Most commands that allow aweights handle them in this manner. That is, if you specify aweights, they are

1. normalized to sum to N, and then

2. inserted in the calculation formulas in the same way as fweights.

23.13.3 Sampling weights

Sampling weights—probability weights or pweights—refer to probability-weighted random samples. Actually, what you specify in [pweight=...] is a variable recording the number of subjects in the full population that the sampled observation in your data represents. That is, an observation that had probability $1/3$ of being included in your sample has pweight 3.

We noted above that some researchers have used aweights with this kind of data. If they do, they are probably making a mistake. Consider the regression model

$$y_i = \mathbf{x}_i\boldsymbol{\beta} + \epsilon_i, \qquad \epsilon_i \sim N(0, \sigma^2)$$

Begin by considering the exact nature of the problem of estimating this model on cell-mean data—the problem for which aweights are the solution. That statistical problem is one of heteroscedasticity arising from the grouping. Note that the error term ϵ_i is homoscedastic (meaning it has constant variance σ^2). Pretend the first observation in the data is the mean of three underlying observations. Then,

$$y_1 = \mathbf{x}_1\boldsymbol{\beta} + \epsilon_1, \qquad \epsilon_i \sim N(0, \sigma^2)$$
$$y_2 = \mathbf{x}_2\boldsymbol{\beta} + \epsilon_2, \qquad \epsilon_i \sim N(0, \sigma^2)$$
$$y_3 = \mathbf{x}_3\boldsymbol{\beta} + \epsilon_3, \qquad \epsilon_i \sim N(0, \sigma^2)$$

and taking the mean,

$$(y_1 + y_2 + y_3)/3 = [(\mathbf{x}_1 + \mathbf{x}_2 + \mathbf{x}_3)/3]\boldsymbol{\beta} + (\epsilon_1 + \epsilon_2 + \epsilon_3)/3$$

For another observation in the data—which may be the result of summing of a different number of observations—the variance will be different. Hence, the model for the data is

$$\overline{y}_j = \overline{x}_j \beta + \overline{\epsilon}_j, \qquad \overline{\epsilon}_j \sim N(0, \sigma^2/N_j)$$

This makes intuitive sense. Consider two observations, one recording means over 2 subjects and the other means over 100,000 subjects. You would expect the variance of the residual to be less in the 100,000-subject observation or, said differently, there is more information in the 100,000-subject observation than in the 2-subject observation.

Now, instead pretend you are estimating the same model, $y_i = \mathbf{x}_i \beta + \epsilon_i$, $\epsilon_i \sim N(0, \sigma^2)$, on probability-weighted data. Each observation in your data is a single subject, it is just that the different subjects have differing chances of being included in your sample. Therefore, for each subject in your data, it is true that

$$y_i = \mathbf{x}_i \beta + \epsilon_i, \qquad \epsilon_i \sim N(0, \sigma^2)$$

That is, there is no heteroscedasticity problem. The use of the aweighted estimator cannot be justified on these grounds.

As a matter of fact, based on the argument just given, you do not need to adjust for the weights at all, although the argument does not justify not making an adjustment. If you do not adjust, you are holding tightly to the assumed truth of your model. There are two issues when considering adjustment for sampling weights:

1. the efficiency of the point estimate $\widehat{\beta}$ of β;

2. the reported standard errors (and, more generally, variance matrix of $\widehat{\beta}$).

Efficiency argues in favor of adjustment and that, by the way, is why many researchers have used aweights with pweighted data. The adjustment implied by pweights to the point estimates is the same as the adjustment implied by aweights.

It is with regard to the second issue that the use of aweights produces incorrect results because it interprets larger weights as designating more accurately measured points. In the case of pweights, however, the point is no more accurately measured—it is still just one observation with a single residual ϵ_j and variance σ^2. In [U] **23.11 Obtaining robust variance estimates** above, we introduced another estimator of variance that measures the variation that would be observed were the data collection followed by the estimation repeated. Those same formulas provide the solution to pweights and they have the added advantage that they are not conditioned on the model being true. If one has any hopes of measuring the variation that would be observed were the data collection followed by estimation repeated, one must include the probability of the observations being sampled in the calculation.

In Stata, when you type

    . regress y x1 x2 [pw=pop]

results are the same as if you typed

    . regress y x1 x2 [pw=pop], robust

That is, specifying pweights implies the **robust** option and hence the robust variance calculation (but weighted). In this example we use **regress** simply for illustration. The same is true of **probit** and all of Stata's estimation commands. Estimation commands that do not have a **robust** option (there are a few) do not allow pweights.

pweights are adequate for handling random samples where the probability of being sampled varies. pweights may be all you need. If, however, the observations were not sampled independently but were sampled in groups—called clusters in the jargon—you should specify the estimator's **cluster()** option as well:

```
. regress y x1 x2 [pw=pop], cluster(block)
```

There are two ways of thinking about this:

1. The robust estimator answers the question of the variation that would be observed were the data collection followed by the estimation repeated and, if that question is to be answered, the estimator must account for the clustered nature of how observations are selected. If observations 1 and 2 are in the same cluster, then one cannot select observation 1 without selecting observation 2 (and, by extension, one cannot select observations like 1 without selecting observations like 2).

2. If you prefer, you can think about potential correlations. Observations in the same cluster may not really be independent—that is an empirical question to be answered by the data. For instance, if the clusters are neighborhoods, it would not be surprising that the individual neighbors are similar in their income, their tastes, and their attitudes, and even more similar than two randomly drawn persons from the area at large with similar characteristics such as age and sex.

Either way of thinking leads to the same (robust) estimator of variance.

Sampling weights usually arise from complex sampling designs, and these designs often involve not only unequal probability sampling and cluster sampling, but also stratified sampling. There is a family of commands in Stata designed to work with the features of complex survey data, and those are the commands that begin with svy. To estimate a linear regression model with stratification, for example, one would use the svyreg command.

Non-svy commands that allow pweights and clustering give essentially identical results to the svy commands. So if the sampling design is simple enough that it can be accommodated by the non-svy command, that is a fine way to perform the analysis. The svy commands differ in that they have additional bells and whistles, and they do all the little details correctly for bona fide survey data. See [U] **30 Overview of survey estimation** for a brief discussion of some of the issues involved in the analysis of survey data, and a list of all the differences between the svy and non-svy commands.

Not all model estimation commands in Stata allow pweights. In many of these cases, this is because they are computationally or statistically difficult to implement. Nevertheless, in the future, more commands will be rewritten to allow pweights and clustering, and the svy family will be expanded.

23.13.4 Importance weights

Stata's iweights—importance weights—are the emergency exit. These weights are for those who want to take control and create special effects. For example, programmers have used regress with iweights to compute iteratively reweighted least-squares solutions for various problems.

iweights are treated much like aweights except that they are not normalized. To wit, Stata's iweight rule is

1. the weights are not normalized;

2. they are generally inserted into calculation formulas in the same way as fweights. There are exceptions; see the *Methods and Formulas* for the particular command.

iweights are used mostly by programmers who are often on the way to implementing one of the other kinds of weights.

23.14   A list of post-estimation commands

The following commands can be used after estimation:

[R] **adjust**	Tables of adjusted means and proportions
[R] **level**	Set default significance level
[R] **lincom**	Obtain linear combinations of coefficients
[R] **linktest**	Specification link test for single-equation models
[R] **lrtest**	Likelihood-ratio test after model estimation
[R] **predict**	Obtain predictions, residuals, etc., after estimation
[R] **test**	Test linear hypotheses after model estimation
[R] **testnl**	Test nonlinear hypotheses after model estimation
[R] **vce**	Display covariance matrix of the estimators

Also see [U] **16.5 Accessing coefficients and standard errors** for accessing coefficients and standard errors.

23.15   References

Binder, D. A. 1983. On the variances of asymptotically normal estimators from complex surveys. *International Statistical Review* 51: 279–292.

Fuller, W. A. 1975. Regression analysis for sample survey. *Sankhyā, Series C* 37: 117-132.

Gail, M. H., W. Y. Tan, and S. Piantadosi. 1988. Tests for no treatment effect in randomized clinical trials. *Biometrika* 75: 57–64.

Huber, P. J. 1967. The behavior of maximum likelihood estimates under non-standard conditions. In *Proceedings of the Fifth Berkeley Symposium on Mathematical Statistics and Probability*. Berkeley, CA: University of California Press, 1, 221–233.

Kent, J. T. 1982. Robust properties of likelihood ratio tests. *Biometrika* 69: 19–27.

Kish, L. and M. R. Frankel. 1974. Inference from complex samples. *Journal of the Royal Statistical Society* B 36: 1–37.

Lin, D. Y. and L. J. Wei. 1989. The robust inference for the Cox proportional hazards model. *Journal of the American Statistical Association* 84: 1074–1078.

MacKinnon, J. G. and H. White. 1985. Some heteroskedasticity consistent covariance matrix estimators with improved finite sample properties. *Journal of Econometrics* 29: 305–325.

Rogers, W. H. 1993. sg17: Regression standard errors in clustered samples. *Stata Technical Bulletin* 13: 19–23. Reprinted in *Stata Technical Bulletin Reprints*, vol. 3, 88–94.

Royall, R. M. 1986. Model robust confidence intervals using maximum likelihood estimators. *International Statistical Review* 54: 221–226.

White, H. 1980. A heteroskedasticity-consistent covariance matrix estimator and a direct test for heteroskedasticity. *Econometrica* 48: 817–830.

——. 1982. Maximum likelihood estimation of misspecified models. *Econometrica* 50: 1–25.

Advice on Stata

Chapters

24 Commands to input data

Contents

24.1   Six ways to input data

The six ways to input data into Stata are

  [R] **edit** and [R] **input**   to enter data from the keyboard
  [R] **insheet**   to read tab- or comma-separated data
  [R] **infile (free format)**   to read unformatted data
  [R] **infile (fixed format)** or [R] **infix (fixed format)** to read formatted data
  [U] **24.4 Transfer programs**   to transfer data

Since dataset formats differ, you should familiarize yourself with each method.

Note that [R] **infile (fixed format)** and [R] **infix (fixed format)** are alternatives. These are two different commands that do the same thing. Read about both and then use whichever appeals to you.

After you have read this chapter, also see [R] **infile** for additional examples of the different commands to input data.

24.2   Eight rules for determining which input method to use

Below are eight rules which, when applied sequentially, will direct you to the appropriate method for entering your data. Following the eight rules is a description of each command as well as a reference to the corresponding entry in the *Reference Manual*. The rules are

1. If you have a small amount of data and simply wish to type it directly into Stata at the keyboard, see [R] **input**—there are many examples and you should have little difficulty. Stata for Windows and Stata for Macintosh users should also see [R] **edit**.

2. If your data is in binary format or the "internal" format of some software package, you can

    a. Translate it into ASCII (also known as character) format using the other software. For instance, Lotus 1-2-3 datasets can be used in Lotus, the data marked, and then written to a `.prn` file, which is what Lotus calls an ASCII file. (The key is that the file is printable.) Then see [R] **insheet**.

    b. There are also software packages available that will automatically convert non-Stata format data files into Stata format files. See [U] **24.4 Transfer programs** for information.

    c. If the data is in a spreadsheet, Stata for Windows or Stata for Macintosh users may also be able to copy-and-paste it into Stata's data editor. See [R] **edit** for details.

3. If the data has one observation per line and is tab- or comma-separated, see [R] **insheet**. This is the easiest way to read data.

4. If the data is formatted and that formatting information is required in order to interpret the data, see [R] **infile (fixed format)** or [R] **infix (fixed format)** (your choice).

5. If there are no string variables, see [R] **infile (free format)**.

6. If all the string variables in the data are enclosed in (single or double) quotes, see [R] **infile (free format)**.

7. If the undelimited string variables have no blanks, see [R] **infile (free format)**.

8. If you make it to here, see [R] **infile (fixed format)** or [R] **infix (fixed format)** (your choice).

Let us now back up and start again.

24.2.1　If you wish to enter data interactively: Rule 1

Rule 1 simply says that if you have a small amount of data, you can type it directly into Stata; see [R] **input**. Otherwise, we assume your data is stored on disk.

24.2.2　If the data is in binary format: Rule 2

Stata can read ASCII datasets, which is technical jargon for datasets composed of characters — datasets that can be typed on your screen or printed on your printer. The alternative, binary datasets, cannot be read by Stata. Binary datasets are quite popular and almost every software package has its own binary format. Stata `.dta` datasets are an example, although this is a binary format Stata can read. The Lotus `.wks` format is another binary format and one that Stata cannot read.

Thus, rule 2: If your data is in binary format or the "internal" format of another software package, you must either translate it into ASCII or use some other program to convert it.

Detecting whether data is stored in binary format can be tricky. For instance, many Windows users wish to read data that has been entered into a word processor — let's assume WordPerfect. Unwittingly, they have stored the data as a WordPerfect document. The data looks like ASCII to them: When they look at it in WordPerfect, they see readable characters. The data seems to even pass the printing test in that WordPerfect can print it. Nevertheless, the data is not ASCII; it is stored in an internal WordPerfect format and the data cannot really pass the printing test since only WordPerfect can print it. To read this data, they must use it in WordPerfect and then store it as a DOS Text file, DOS Text being the term WordPerfect decided to use to mean ASCII.

So how are you to know whether your data is binary? Here's a simple test: Regardless of the operating system you use, enter Stata and type `type` followed by the name of the file:

```
. type myfile.raw
output will appear
```

You do not have to print the entire file, press *Break* when you have seen enough.

Do you see things that look like hieroglyphics? If so, the data is binary. See [U] **24.4 Transfer programs** below.

If it looks like data, however, it is (probably) ASCII.

Let us assume you have an ASCII dataset that you wish to read. The data's format will determine the command you need to use. The different formats are discussed in the following sections.

24.2.3 If the data is simple: Rule 3

The easiest way to read data is with `insheet`; see [R] **insheet**.

`insheet` is smart: it looks at the data, determines what it contains, and then reads it. That is, `insheet` is smart given certain restrictions, such as that the data has one observation per line and that the values are tab- or comma-separated. `insheet` can read this

```
———————————————————————————————— top of data1.raw ————————
M,Joe Smith,288,14
M,K Marx,238,12
F,Farber,211,7
———————————————————————————————— end of data1.raw ————————
```

or this (which has variable names on the first line)

```
———————————————————————————————— top of data2.raw ————————
sex, name, dept, division
M,Joe Smith,288,14
M,K Marx,238,12
F,Farber,211,7
———————————————————————————————— end of data2.raw ————————
```

or this (which has one tab character separating the values)

```
———————————————————————————————— top of data3.raw ————————
M        Joe Smith        288    14
M        K Marx   238     12
F        Farber   211     7
———————————————————————————————— end of data3.raw ————————
```

(which looks odd because of how tabs work; `data3.raw` could similarly have a variable header), but `insheet` cannot read

```
———————————————————————————————— top of data4.raw ————————
M        Joe Smith        288    14
M        K Marx           238    12
F        Farber           211    7
———————————————————————————————— end of data4.raw ————————
```

which has spaces rather than tabs!

There is a way to tell `data3.raw` from `data4.raw`: Ask Stata to type the data and show the tabs:

```
. type data3.raw, showtabs
M<T>Joe Smith<T>288<T>14
M<T>K Marx<T>238<T>12
F<T>Farber<T>211<T>7

. type data4.raw, showtabs
M        Joe Smith        288    14
M        K Marx           238    12
F        Farber           211    7
```

24.2.4   If the data is formatted and the formatting is significant: Rule 4

Rule 4 says that if the data is formatted and that formatting information is required in order to interpret the data, see [R] **infile (fixed format)** or [R] **infix (fixed format)**, which being a matter of preference.

Using `infix` or `infile` with a data dictionary is something new users want to avoid if at all possible.

The purpose of this rule is only to take you to the most complicated of all cases if there is no alternative. Otherwise, let's wait and see if it is necessary. Do not misinterpret the rule and say, "Ah, my data is formatted, at last a solution."

Just because data is formatted does not mean you have to exploit the formatting information. The following data is formatted,

```
 ─────────────────────────────────── top of data5.raw ───────────
    1    27.39     12
    2     1.00      4
    3   100.10    100
 ─────────────────────────────────── end of data5.raw ───────────
```

in that the numbers line up in neat columns, but you do not need to know the information to read it. Alternatively, consider the same data run together:

```
 ─────────────────────────────────── top of data6.raw ───────────
    1 27.39 12
    2  1.00  4
    3100.10100
 ─────────────────────────────────── end of data6.raw ───────────
```

This data is formatted, too, and we must know the formatting information in order to make sense of "3100.10100". We must know that variable 2 starts in column 4 and is 6 characters long to extract the 100.10. It is datasets like `data6.raw` that we are looking for at this stage—datasets that can only be made sense of if we know the starting and ending columns of data elements. In order to read data such as `data6.raw`, we must use either `infix` or `infile` with a data dictionary.

It should be obvious why reading unformatted data is easier. If the formatting information is required to interpret the data, then you must communicate that information to Stata, which means you will have to type it. This is the hardest kind of data to read, but Stata can do it. See [R] **infile (fixed format)** or [R] **infix (fixed format)**.

Looking back at `data4.raw`,

```
 ─────────────────────────────────── top of data4.raw ───────────
    M      Joe Smith      288     14
    M      K Marx         238     12
    F      Farber         211      7
 ─────────────────────────────────── end of data4.raw ───────────
```

you may be uncertain whether you have to read it with a data dictionary. If you are uncertain, do not jump yet.

Finally, here is an obvious example of unformatted data:

```
 ─────────────────────────────────── top of data7.raw ───────────
    1 27.39          12
    2 1 4
    3 100.1 100
 ─────────────────────────────────── end of data7.raw ───────────
```

In this case, blanks separate one data element from the next and, in one case, lots of blanks, although there is no special meaning attached to more than one blank.

In the following sections, we will look at data that is unformatted or formatted in a way that does not require a data dictionary.

24.2.5   If there are no string variables: Rule 5

Rule 5 says that if there are no string variables, see [R] **infile (free format)**.

Although the dataset `data7.raw` is unformatted, it still could be read using `infile` without a dictionary. This is not the case with `data4.raw`, because this dataset contains undelimited string variables with embedded blanks.

❑ Technical Note

Some Stata users prefer to read data with a data dictionary even when we suggest differently, as above. They like the convenience of the data dictionary—one can sit in front of an editor and carefully compose the list of variables and attach variable labels rather than having to type the variable list (correctly) on the Stata command line. What they should understand, however, is that one can create a do-file containing the `infile` statement, and thus have all the advantages of a data dictionary without some of the (extremely technical) disadvantages of data dictionaries.

Nevertheless, we do tend to agree with such users—we, too, prefer data dictionaries. Our recommendations, however, are designed to work in all cases. If the data is unformatted and contains no string variables, it can always be read without a data dictionary, whereas only in some cases can it be read with a data dictionary.

The distinction is that `infile` without a data dictionary performs stream I/O whereas with a data dictionary it performs record I/O. The difference is intentional—it guarantees that you will be able to read your data into Stata somehow. Some datasets require stream I/O, others require record I/O, and still others can be read either way. Recommendations 1–5 identify datasets that either require stream I/O or can be read either way.

❑

We are now left with datasets which contain at least one string variable:

24.2.6   If all the string variables are enclosed in quotes: Rule 6

Rule 6 reads: if all the string variables in the data are enclosed in (single or double) quotes, see [R] **infile (free format)**.

See [U] **26 Commands for dealing with strings** for a formal definition of strings, but as a quick guide, a string variable is a variable that takes on values such as "bob", "joe", etc., as opposed to numeric variables which take on values like 1, 27.5, and –17.393. Undelimited strings—strings not enclosed in quotes—can be difficult to read.

Here is an example including delimited string variables:

```
──────────────────────────────────────── top of data8.raw ──────────
    "M"  "Joe Smith" 288 14
    "M"  "K Marx" 238 12
    "F"  "Farber" 211 7
──────────────────────────────────────── end of data8.raw ──────────
```

or

-- top of data8.raw, alternative format ---------------

```
"M" "Joe Smith" 288  14
"M" "K Marx"    238  12
"F" "Farber"    211   7
```
-- end of data8.raw, alternative format ---------------

Both of these are merely variations on `data4.raw` except that the strings are enclosed in quotes. In this case, `infile` without a dictionary can be used to read the data.

Here is another version of `data4.raw`, but this time without delimiters or even formatting:

-- top of data9.raw ----------------

```
M Joe Smith 288 14
M K Marx 238 12
F Farber 211 7
```
-- end of data9.raw ----------------

What makes this data difficult? Blanks sometimes separate values and sometimes are nothing more than a blank within a string. For instance, you cannot tell whether Farber has first initial F with missing sex or is instead female with a missing first initial.

Fortunately, such data rarely happens. Either the strings are delimited, as we showed in `data8.raw`, or the data is in columns, as in `data4.raw`.

24.2.7   If the undelimited strings have no blanks: Rule 7

There is a case in which uncolumnized, undelimited strings cause no confusion—when they contain no blanks. For instance, if our data contained only last names:

-- top of data10.raw ----------------

```
Smith 288 14
Marx 238 12
Farber 211 7
```
-- end of data10.raw ----------------

Stata could read it without a data dictionary. Caution: the last names must contain no blanks—no Van Owen's or von Beethoven's.

Thus, rule 7: If the undelimited string variables have no blanks, see [R] **infile (free format)**.

Which leaves us with our final rule:

24.2.8   If you make it to here: Rule 8

If you make it to here, see [R] **infile (fixed format)** or [R] **infix (fixed format)** (your choice). Remember `data4.raw`?

-- top of data4.raw ----------------

```
M       Joe Smith    288    14
M       K Marx       238    12
F       Farber       211     7
```
-- end of data4.raw ----------------

It must be read using either `infile` with a dictionary or `infix`.

24.3 If you run out of memory

You can increase the amount of memory allocated to Stata; see [U] **7 Setting the size of memory**.

You can also try to conserve memory.

When you read the data, did you specify variable types? Stata can store integers more compactly than floats, and small integers more compactly than large integers; see [U] **15 Data**.

If that is not sufficient, then you will have to resort to reading the data in pieces. Both `infile` and `infix` allow specifying an `in` *range* modifier and, in this case, the range is interpreted as the observation range to read. Thus, `infile ... in 1/100` would read observations 1 through 100 of your data and stop.

`infile ... in 101/200` would read observations 101 through 200. The end of the range may be specified as larger than the actual number of observations in the data. If the data contained only 150 observations, `infile ... in 101/200` would read observations 101 through 150.

Another way of reading the data in pieces is to specify the `if` *exp* modifier. Say your data contained an equal number of males and females, coded as the variable `sex` (which you will read) being 0 or 1 respectively. You could type `infile ... if sex==0` to read the males. `infile` will read an observation, ask itself if `sex` is zero, and if not, throw the observation away. Obviously, you could read just the females by typing `infile ... if sex==1`.

If the dataset is really big, perhaps you only need a random sample of the data—it was never your intention to analyze the entire dataset. Since `infile` and `infix` allow `if` *exp*, you could type `infile ... if uniform()<.1`. `uniform()` is the uniformly distributed random number generator; see [U] **16.3.2 Statistical functions**. This method would read an approximate 10% sample of the data. If you are serious about random samples, do not forget to set the seed before using `uniform()`; see [R] **generate**.

The final approach is to read all the observations but only a subset of the variables. When reading data without a data dictionary, you can specify `_skip` for variables, indicating that the variable is to be skipped over. When reading with a data dictionary or using `infix`, you can specify the actual columns to read, skipping any columns you wish to ignore.

24.4 Transfer programs

To import data from, say, Lotus 1-2-3, either you can write the data out as a Lotus `.prn` file and then read it in according to the rules above, or you can purchase a program to translate the dataset from Lotus's format to Stata's format.

One of the programs is Stat/Transfer, which is available for Windows 98/95/NT and Windows 3.1. It reads and writes data in a variety of formats, including Microsoft Access, dBASE, Epi Info, Excel, GAUSS, LIMDEP, Lotus 1-2-3, MATLAB, Paradox, Quattro Pro, S-Plus, SAS, SPSS, SYSTAT, and, of course, Stata.

Stat/Transfer, available from...	and manufactured by...
Stata Corporation	Circle Systems
702 University Drive East	1001 Fourth Avenue Plaza, Suite 3200
College Station, Texas 77840	Seattle, Washington 98154
Telephone: 409-696-4600	Telephone: 206-682-3783
Fax: 409-696-4601	Fax: 206-328-4788
Email: *stata@stata.com*	*stsales@circlesys.com*

There are other transfer programs available, too. Our web site, *http://www.stata.com*, lists programs available from other sources.

Access and Excel are trademarks of Microsoft Corporation. dBASE, Paradox, and Quattro Pro are trademarks of Borland International, Inc. Epi Info is a trademark of The Centers for Disease Control and Prevention. GAUSS is a trademark of Aptech Systems, Inc. LIMDEP is a trademark of Econometric Software, Inc. Lotus and 1-2-3 are trademarks of Lotus Development Corporation. MATLAB is a trademark of The Math Works, Inc. SAS is a trademark of the SAS Institute Inc. S-Plus is a trademark of MathSoft, Inc. SPSS and SYSTAT are trademarks of SPSS Inc.

24.5 References

Swagel, P. 1994. os14: A program to format raw data files. *Stata Technical Bulletin* 20: 10–12. Reprinted in *Stata Technical Bulletin Reprints*, vol. 4, pp. 80–82.

25 Commands for combining data

Pretend you have two datasets and wish to combine them. Below, we will draw a dataset as a box where, in the box, the variables go across and the observations go down.

See [R] **append** if you want to combine datasets vertically:

append adds observations to the existing variables. That is an oversimplification because **append** does not require that the datasets have exactly the same variables, but **append** is appropriate, for instance, when you have data on hospital patients and then receive data on more patients.

See [R] **merge** if you want to combine datasets horizontally:

merge adds variables to the existing observations. That is an oversimplification because **merge** does not require that the datasets have exactly the same observations, but **merge** is appropriate, for instance, when you have data on survey respondents and then receive data on part 2 of the questionnaire.

See [R] **joinby** when you want to combine datasets horizontally but form all pairwise combinations within group:

joinby is like **merge**, but forms all combinations of the observations where it makes sense. **joinby** would be appropriate, for instance, where A contained data on parents and B their children. **joinby** *familyid* would form a dataset of each parent joined with each of his or her children.

26 Commands for dealing with strings

Contents

Please read [U] **15 Data** before reading this entry.

26.1   Description

The word *string* is shorthand for a string of characters. "Male" and "Female"; "yes" and "no"; "R. Smith" and "P. Jones" are examples of strings. The alternative to strings is numbers—0, 1, 2, 5.7, and so on. Variables containing strings—called string variables—occur in data for a variety of reasons. Four of these reasons are listed below.

A variable might contain strings because it is an *identifying* variable. Employee names in a payroll file, patient names in a hospital file, city names in a city data file are all examples of this. This is a proper use of string variables.

A variable might contain strings because it records categorical information. "Male" and "Female", "yes and "no" are examples of such use, but this is not an appropriate use of string variables. It is not appropriate because the same information could be coded numerically and, if it were, (1) it would take less memory to store the data and (2) the data would be more useful. We will explain how to convert categorical strings to categorical numbers below.

In addition, a variable might contain strings because of a mistake. The variable contains things like 1, 5, 8.2, but due to an error in reading the data, it was mistakenly put into a string variable. We will explain how to fix such mistakes.

Finally, a variable might contain strings because the data simply could not be coerced into being stored numerically. "15 Jan 1992", "1/15/92", and "1A73" are examples of such use. We will explain how to deal with such complexities.

26.2   Categorical string variables

A variable might contain strings because it records categorical information.

Suppose that you have read in a dataset that contains a variable called **sex**, recorded as "Male" and "Female", yet when you attempt to run an ANOVA, the following message is displayed:

```
. anova bp sex
no observations
```

There are no observations because **anova**, along with most of Stata's "analytic" commands, cannot deal with string variables. They want to see numbers and when they do not, they treat the variable as if it contained numeric missing values. Despite this limitation, it is possible to obtain tables:

```
. encode sex, gen(gender)
. anova bp gender
```

```
                   Number of obs =      74    R-squared     =  0.1548
                   Root MSE      = 5.35582    Adj R-squared =  0.1430
          Source |  Partial SS    df       MS             F    Prob > F
      -----------+----------------------------------------------------
           Model |  378.153515     1  378.153515        13.18    0.0005
                 |
          gender |  378.153515     1  378.153515        13.18    0.0005
                 |
        Residual |  2065.30594    72  28.6848048
      -----------+----------------------------------------------------
           Total |  2443.45946    73  33.4720474
```

The magic here is to convert the string variable sex into a numeric variable called gender with an associated value label, a trick accomplished by encode; see [U] **15.6.3 Value labels** and [R] **encode**.

26.3   Mistaken string variables

A variable might contain strings because of a mistake.

Suppose that you have numeric data in a variable called x, but due to a mistake, x was made a string variable when you read the data. When you list the data, it looks fine:

```
. list x

            x
  1.        2
  2.      2.5
  3.       17
etc.
```

Yet, when you attempt to obtain summary statistics on x:

```
. summarize x
Variable |    Obs      Mean    Std. Dev.     Min       Max
---------+-----------------------------------------------------
       x |      0
```

If this happens to you, type describe to confirm that x is stored as a string:

```
. describe
Contains data
  obs:          10
  vars:          3
  size:        160 (100.0% of memory free)
-------------------------------------------------------------------------------
  1. x          str4    %9s
  2. y          float   %9.0g
  3. z          float   %9.0g
-------------------------------------------------------------------------------
Sorted by:
```

x is stored as an str4.

The problem is that summarize does not know how to calculate the mean of string variables—how to calculate the mean of "Joe" plus "Bill" plus "Roger"—even when the string variable contains what could be numbers. By using the real() function described in [U] **16.3.5 String functions**, the variable mistakenly stored as a str4 can be converted to a numeric variable:

```
. gen newx = real(x)
```

```
. summarize newx
    Variable |      Obs        Mean    Std. Dev.         Min         Max
-------------+--------------------------------------------------------
        newx |      214    17.88451    11.14695    .2317163    37.70975
```

It would now be a good idea to

```
. drop x
```

```
. rename newx x
```

26.4   Complex strings

A variable might contain strings because the data simply could not be coerced into being stored numerically.

A complex string is a string that contains more than one piece of information. The most common example of complex strings is dates: "15 Jan 1992" contains three pieces of information—a day, a month, and a year. If your complex strings are dates, see [U] **27 Commands for dealing with dates**.

Although Stata has functions for dealing with dates, you will have to deal with other complex strings yourself. For instance, assume that you have data including part numbers:

```
. list partno

             partno
    1.       5A2713
    2.       2B1311
    3.       8D2712
   etc.
```

The first digit of the part number is a division number and the character that follows identifies the plant at which the part was manufactured. The next three digits represent the major part number and the last digit is a modifier indicating the color. This complex variable can be decomposed using the `substr()` and `real()` functions described in [U] **16.3.5 String functions**:

```
. gen byte div = real(substr(partno,1,1))
```

```
. gen str1 plant = substr(partno,2,1)
```

```
. gen int part = real(substr(partno,3,3))
```

```
. gen byte color = real(substr(partno,6,1))
```

We use the `substr()` function to extract pieces of the string and the `real()` function, when appropriate, to translate the piece into a number.

27 Commands for dealing with dates

27.1   Overview

You can record dates however you want but there is one format that Stata understands, called elapsed dates or **%d** dates. A **%d** date is the number of days from January 1, 1960. In this format,

0	corresponds to	01jan1960			
1	corresponds to	02jan1960	−1	corresponds to	31dec1959
2	corresponds to	03jan1960	−2	corresponds to	30dec1959
⋮			⋮		
31	corresponds to	01feb1960	−31	corresponds to	01dec1959
⋮			⋮		
15,000	corresponds to	25jan2001	−15,000	corresponds to	01dec1918
⋮			⋮		
2,936,549	corresponds to	31dec9999	−679,350	corresponds to	01jan0100

This format can be used with dates 01jan0100–31dec9999, although caution should be exercised in dealing with dates before Friday, 15oct1582, because that is when the Gregorian calendar went into effect.

Stata provides functions to convert dates into **%d** dates, formats to print **%d** dates in understandable forms, and other functions to manipulate **%d** dates.

Use of **%d** dates is described in [U] **27.2 Dates** below.

In addition to %d dates, Stata has five other date formats, called %t dates, that it understands:

Format	Description	Coding		
%td	daily (same as %d)	−1 = 31dec1959,	0 = 01jan1960,	1 = 02jan1960
%tw	weekly	−1 = 1959w52,	0 = 1960w1,	1 = 1960w2
%tm	monthly	−1 = 1959m12,	0 = 1960m1,	1 = 1960m2
%tq	quarterly	−1 = 1959q4,	0 = 1960q1,	1 = 1960q2
%th	half-yearly	−1 = 1959h2,	0 = 1960h1,	1 = 1960h2
%ty	yearly	1959 = 1959,	1960 = 1960,	1961 = 1961

Use of %t dates is described in [U] **27.3 Time-series dates** below.

27.2   Dates

In this section we discuss %d dates, also called elapsed dates.

27.2.1   Inputting dates

The trick to inputting dates in Stata is to forget they are dates: input them as strings and then later convert them into Stata's elapsed dates. You might have

```
──────────────────────────────── top of bdays.raw ────────────
   Bill   21 Jan 1952   22
   May    11 Jul 1948   18
   Sam    12 Nov 1960   25
   Kay     9 Aug 1975   16
─────────────────────────────────── end of bdays.raw ─────────
```

and if you did, you could read this data by typing

```
. infix str name 1-5  str bday 7-17  x 20-21 using bdays
(4 observations read)
```

If you now listed the data, it would look fine,

```
. list
            name          bday         x
   1.       Bill   21 Jan 1952        22
   2.        May   11 Jul 1948        18
   3.        Sam   12 Nov 1960        25
   4.        Kay    9 Aug 1975        16
```

but you would find there is not much you could do with bday because it is just a string variable. Turning it into a date Stata understands is easy,

```
. gen birthday = date(bday,"dmy")
. list
            name          bday         x   birthday
   1.       Bill   21 Jan 1952        22      -2902
   2.        May   11 Jul 1948        18      -4191
   3.        Sam   12 Nov 1960        25        316
   4.        Kay    9 Aug 1975        16       5699
```

and then making the numeric birthday variable look like a date is equally easy:

```
. format birthday %d
. list
```

	name	bday	x	birthday
1.	Bill	21 Jan 1952	22	21jan1952
2.	May	11 Jul 1948	18	11jul1948
3.	Sam	12 Nov 1960	25	12nov1960
4.	Kay	9 Aug 1975	16	09aug1975

The convenient thing about the variable `birthday` is that it is numeric, which means you can make calculations on it. How old will each of these people be on January 1, 2000? It is easy to add such a variable:

```
. gen age2000 = (mdy(1,1,2000)-birthday)/365.25
. list
```

	name	bday	x	birthday	age2000
1.	Bill	21 Jan 1952	22	21jan1952	47.94524
2.	May	11 Jul 1948	18	11jul1948	51.47433
3.	Sam	12 Nov 1960	25	12nov1960	39.13484
4.	Kay	9 Aug 1975	16	09aug1975	24.39699

27.2.2 Conversion into %d elapsed dates

Two functions are provided—`mdy()` and `date()`—for converting variables into elapsed dates.

27.2.2.1 The mdy() function

`mdy()` takes three numeric arguments—a month, day, and year—and returns the corresponding elapsed date. For instance:

```
. list
```

	month	day	year
1.	7	11	1948
2.	1	21	1952
3.	11	2	1994
4.	8	12	93

```
. gen edate = mdy(month,day,year)
(1 missing value generated)
. list
```

	month	day	year	edate
1.	7	11	1948	-4191
2.	1	21	1952	-2902
3.	11	2	1994	12724
4.	8	12	93	.

Note that in the last observation, `mdy()` produced missing. It did this because the year was 93 and `mdy()` does not assume 93 means 1993.

27.2.2.2 The date() function

The second way to convert to elapsed dates is with the `date()` function. `date()` takes two string arguments. There is a variation on this—two strings arguments and a numeric argument—but let's postpone that. The first argument is the date to be converted. The second argument tells `date()` the order of the month, day, and year in the first argument. Typing `date(strvar,"mdy")` means that *strvar* contains the month, day, and year in that order. Typing `date(strvar,"dmy")` means *strvar* contains the day, month, and year. Knowing the order, `date()` allows *strvar* to be in almost any format. For instance:

```
. list
                  mydate
    1.           7-11-1948
    2.            1/21/52
    3.           11.2.1994
    4.        Aug 12, 1993
    5.        Sept 11,2002
    6.   November 13, 2005

. gen edate = date(mydate, "mdy")
(1 missing value generated)

. list
                  mydate       edate
    1.           7-11-1948      -4191
    2.            1/21/52          .
    3.           11.2.1994      12724
    4.        Aug 12, 1993      12277
    5.        Sept 11,2002      15594
    6.   November 13, 2005      16753
```

Or, if you prefer:

```
. list
                  mydate
    1.           11-7-1948
    2.            21/1/52
    3.           2.11.1994
    4.          12 Aug 1993
    5.           11Sept2002
    6.      13 November 2005

. gen edate = date(mydate, "dmy")
(1 missing value generated)

. list
                  mydate       edate
    1.           11-7-1948      -4191
    2.            21/1/52          .
    3.           2.11.1994      12724
    4.          12 Aug 1993     12277
    5.           11Sept2002     15594
    6.      13 November 2005     16753
```

date() can deal with virtually any date format: all it needs to know is the order of the month, day, and year, and that you indicate by the second argument using the letters m, d, and y. Second argument "mdy" means month–day–year order, "dmy" means day–month–year order, and so on.

Note however that, like mdy(), date() refused to translate two-digit years: 1/21/52 and 21/1/52 both translated to missing. Unlike mdy(), date() would be willing to assume that 52 means 1952 or 2052 if you will tell it which. There are two ways you can do this.

The first way involves specifying a default century and you do that using date()'s second argument. Specify "md19y" or "dm20y" and date() will assume that two-digit years should be interpreted as being prefixed by 19 or 20; four-digit years will still be correctly interpreted no matter which default you specify.

The second way involves specifying date()'s third argument. Specify date(...,...,2040), and date() will assume that two-digit years should be interpreted as the maximum year not greater than 2040. 52 would be interpreted as 1952 but 39 would be interpreted as 2039. 40 would be interpreted as 2040. You can specify whatever third argument works best for your data.

If you do neither, then two-digit years cannot be translated and that is why we saw the missing values in the examples above; `date()` could not translate 1/21/52 (21/1/52 when we varied the order). We could have translated 1/21/52 had we typed

```
. gen edate = date(mydate, "md19y")
```

or had we typed

```
. gen edate = date(mydate, "mdy",2040)
```

Either method would translate 1/21/52 as 01jan1952 but they would differ on how they would treat dates with two-digit years 00, 01, ..., 40. The first method would treat them as 1900, 1901, ..., 1940 whereas the second would treat them as 2000, 2001, ..., 2040.

To summarize: To get dates into Stata, either create three numeric variables containing the month, day, and year and use `mdy()` to convert them, or create a string variable containing the date in whatever format and use `date()` to translate it. If you are reading date data into Stata, by far the easiest is to read the data into a string and then use `date()`.

❏ Technical Note

How `date()` *works.* The date to be converted has three pieces of information—the month, day, and year—and the second argument specifies the order, of which there are six possibilities, `"dmy"`, `"mdy"`, `"ymd"`, `"ydm"`, `"dym"`, and `"myd"`, although the last three rarely occur. Knowing the order, `date()` examines the contents of the first argument and looks for transitions, meaning any separating character such as blanks, commas, dashes, and slashes, or changes from numeric to alpha or alpha to numeric. This allows dividing the source into its three components for translation. If the source divides into other than three components, or if it divides into three but they do not make sense, `date()` returns a missing value.

If you have two-digit years, `date()` will return missing unless you specify a default century on the second argument—e.g., `"md19y"`—or you specify a third argument.

`date()` can translate virtually any format except formats where all three elements run together and the months are indicated numerically, such as 012152 or 520121. It is the lack of blanks or other separating characters that confuses `date()`; `date` could translate 01 21 52 or 52 01 21. `date()` could also translate 21Jan52—note the absence of blanks—because the string Jan makes clear the separation.

Let us assume you have a date of the form 520121—the order is year, month, and day—stored in a numeric variable called `ymd`. Here is how you might translate it:

```
. gen year = int(ymd/10000)
. gen month = int((ymd-year*10000)/100)
. gen day = ymd - year*10000  - month*100
. gen edate = mdy(month, day, 1900 + year)
```

❏

27.2.3 Displaying dates

`%d` elapsed dates are convenient for computers and sometimes even for humans—you can, for instance, subtract them to obtain the number of days between dates. Nevertheless, they are unreadable. For instance, here are some birth dates:

```
. list
```

```
        bdate
1.      -2902
2.      -4191
3.        316
4.       5699
```

All you need do to make such dates readable is assign Stata's **%d** format to the variable:

```
. format bdate %d

. list
        bdate
1. 21jan1952
2. 23jun1971
3. 12nov1960
4. 09aug1975
```

If you now saved the data, the date would forever more be displayed in this format.

You may find the format 21jan1952 unappealing and, if so, you can modify it. The **%d** format is equivalent to **%dD1CY**, meaning first display the day (**D**), then the month abbreviation in lowercase (**1**), then the century (**C**), and finally the year without the century (**Y**). This default was selected because, by using an abbreviated month, it makes clear the order of the day and month and, by omitting the blanks, it is short.

Here is a variation on the format:

```
. format bdate %dD_m_CY

. list
        bdate
1. 21 Jan 1952
2. 11 Jul 1948
3. 12 Nov 1960
4. 09 Aug 1975
```

And here are two more variations:

```
. format bdate %dN/D/Y

. list
        bdate
1. 01/21/52
2. 07/11/48
3. 11/12/60
4. 08/09/75

. format bdate %dM_d,_CY

. list
           bdate
1.   January 21, 1952
2.      July 11, 1948
3.  November 12, 1960
4.     August 9, 1975
```

You can specify simply **%d** or you can follow the **%d** with up to 11 characters that tell Stata what to display. Here is what the characters mean:

C	Display the century of the year with a leading 0; year 500 is 05, year 1994 is 19, year 2002 is 20.
c	Display the century of the year without a leading 0; year 500 is 5, year 1994 is 19.
Y	Display the year within century with a leading 0; 1908 is 08, 1994 is 94, 2002 is 02.
y	Display the year within century without a leading 0; 1908 is 8, 1994 is 94, 2002 is 2.
M	Display the month spelled out; January is January, February is February, . . . , December is December.
m	Display the 3-letter abbreviation of month; January is Jan, February is Feb, . . . , December is Dec.
L	Same as M except the month is presented in all lowercase; January is january, February is february, . . . , December is december.
l	Same as m except the 3-letter abbreviation is in all lowercase; January is jan, February is feb, . . . , December is dec.
N	Display the numeric month with a leading 0; January is 01, February is 02, . . . , December is 12.
n	Display the numeric month without a leading 0; January is 1, February is 2, . . . , December is 12.
D	Display the day with a leading 0; 1 is 01, 2 is 02, . . . , 31 is 31.
d	Display the day without a leading 0; 1 is 1, 2 is 2, . . . , 31 is 31.
J	display day-within-year with leading 0s (001 to 366)
j	display day-within-year without leading 0s (1 to 366)
h	display half of year (1 to 2)
q	display quarter of year (1, 2, 3, or 4)
W	display week of year with leading 0 (01 to 52)
w	display week of year without leading 0 (1 to 52)
_	Display a blank.
.	Display a period.
,	Display a comma.
:	Display a colon.
-	Display a dash.
/	Display a slash.
'	Display a close single quote.
!c	display character c (code !! to display exclamation point)

When using the detail characters, you need not specify all the components of the date. In our birth dates, if we wanted to see just the month and year, we might

```
. format bdate %dm,_CY

. list

        bdate
1.  Jan, 1952
2.  Jul, 1948
3.  Nov, 1960
4.  Aug, 1975
```

27.2.4   Other date functions

How you display the date does not matter; the variable itself always contains the number of days from January 1, 1960. Given a date variable, the following functions extract information from it:

year(*date*)	returns four-digit year; e.g., 1980, 2002
month(*date*)	returns month; 1, 2, . . . , 12
day(*date*)	returns day within month; 1, 2, . . . , 31
halfyear(*date*)	returns the half of year; 1 or 2
quarter(*date*)	returns quarter of year; 1, 2, 3, or 4
week(*date*)	returns week of year; 1, 2, . . . , 52
dow(*date*)	returns day of week; 0, 1, . . . , 6; 0 = Sunday
doy(*date*)	returns day of year; 1, 2, . . . , 366

For example,

```
. gen m = month(bdate)
. gen d = day(bdate)
. gen y = year(bdate)
. gen week_d = dow(bdate)
. list
```

	bdate	m	d	y	week_d
1.	21jan52	1	21	1952	1
2.	11jul48	7	11	1948	0
3.	12nov60	11	12	1960	6
4.	09aug75	8	9	1975	6

dow() returns 0–6, 0 meaning Sunday, 1 Monday, . . . , 6 Saturday. Thus, the person born on January 21, 1952 was born on a Monday.

27.2.5   Specifying particular dates (date literals)

If you work with dates, you will want to type dates in expressions. For instance, in a previous example when we needed to calculate the age of persons as of 1jan2000, we typed

```
. gen age2000 = (mdy(1,1,2000)-birthday)/365.25
```

We used mdy() to obtain 1jan2000 as an elapsed date. Alternatively, we could have used Stata's d(*constant*) function

```
. gen age2000 = (d(1jan2000)-birthday)/365.25
```

d() is not a usual function in that you cannot type any expression inside the parentheses; instead you must type a day followed by a month followed by a four-digit year. Do that, however, and d() returns the corresponding %d date value.

You may type the day, month, and year however you wish but you must specify them in the order day, month, and year. Stata will understand d(1/1/2000) or d(1-1-2000) or d(1.1.2000) or d(1 jan 2000) or d(1 January 2000), etc., but it would not understand d(jan.1,2000):

```
. gen age2000 = (d(jan.1,2000)-birthday)/365.25
d(jan.1,2000) invalid
r(198);
```

In addition, if you type the date in a style where two numbers appear one after the other, you must put some form of punctuation other than space between the two numbers: You would think Stata would understand d(1 1 2000), but it does not because spaces disappear in expressions:

```
. gen age2000  = (d(1 1 2000)-birthday)/365.25
d(112000) invalid
r(198);
```

When you spell out the month, you can include spaces or not:

```
. gen age2000 = (d(1 jan 2000)-birthday)/365.25
```

Finally, d() is allowed only in expressions. There may be occasions when you need to specify the numeric value of a date in the option of some command, such as

```
. graph value time, xline(d(15apr1998))
xline() invalid
r(198);
```

Unfortunately, Stata does not understand xline(d(15apr1998)). In such cases, you can use display to obtain the numeric equivalent:

```
. display d(15apr1998)
13984
```

```
. graph value time, xline(13984)
(output omitted)
```

27.3 Time-series dates

In addition to %d date formats, Stata has five other date formats called %t or time-series dates. These dates work like %d in that 1jan1960 is mapped to 0, but the meaning of 1 varies:

Format	Description	Coding		
%td	daily (same as %d)	−1 = 31dec1959,	0 = 01jan1960,	1 = 02jan1960
%tw	weekly	−1 = 1959w52,	0 = 1960w1,	1 = 1960w2
%tm	monthly	−1 = 1959m12,	0 = 1960m1,	1 = 1960m2
%tq	quarterly	−1 = 1959q4,	0 = 1960q1,	1 = 1960q2
%th	half-yearly	−1 = 1959h2,	0 = 1960h1,	1 = 1960h2
%ty	yearly	1959 = 1959,	1960 = 1960,	1961 = 1961

To best understand these formats, think of what happens when variable *d* contains a date and you add 1 to it. If *d* is in %td format, you obtain the next day. If *d* is in %tw format, you obtain the next week. If *d* is in %tm format, you obtain the next month. If *d* is in %tq format, you obtain the next quarter. If *d* is in %th format, you obtain the next half-year. If *d* is in %ty format, you obtain the next year.

Or think of it like this: subtract two dates and you obtain the number of days, weeks, months, quarters, half-years, or years between them.

%td daily format.
   This format is equivalent to %d. In any context, whether a variable is %td or %d makes no difference. Examples of dates in this format are 12jun1998, 22dec2002, etc.

%tw weekly format.
   The year is divided into 52 weeks: week 1 is the first 7 days of year, week 2 the second 7 days, and so on. Since years have just over 52 weeks in them, the 52nd week is defined as having 8 or 9 days. Examples of dates in this format are 1998w24, 2002w52, etc.

%tm monthly format.
   The year is divided into the 12 calendar months. Examples of dates in this format are 1998m6, 2002m12, etc.

%tq quarterly format.
   The year is divided into 4 quarters based on months; quarter 1 is January through March; quarter 2 is April through June; quarter 3 is July through September; quarter 4 is October through December. Examples of dates in this format are 1998q2, 2002q4, etc.

%th half-yearly format.
   The year is divided into 2 halves based on months; half 1 is January through June and half 2 is July through December. Examples of dates in this format are 1998h1 and 2002h2.

%ty yearly format.
   The year is not divided at all. Examples of dates in this format are 1998 and 2002.

27.3.1 Inputting time variables

   Our advice for inputting %d variables, summarized in [U] **27.2.1 Inputting dates** and [U] **27.2.2 Conversion into %d elapsed dates**, was

   1. use the mdy() function if you have three integers recording the month, day, and year; or

   2. input the date as a string and then use the date() function to convert it.

We offer the same advice for %d variables; just the names of the functions change:

format	integer conversion	string conversion
%td	mdy(*month*,*day*,*year*)	date(*string*, "md[*##*]y" or "dm[*##*]Y" ..., [*topyear*])
%tw	yw(*year*, *week*)	weekly(*string*, "w[*##*]y" or "[*##*]yw", [*topyear*])
%tm	ym(*year*, *month*)	monthly(*string*, "m[*##*]y" or "[*##*]ym", [*topyear*])
%tq	yq(*year*, *quarter*)	quarterly(*string*, "q[*##*]y" or "[*##*]yq", [*topyear*])
%th	yh(*year*, *halfyear*)	halfyearly(*string*, "q[*##*]y" or "[*##*]yq", [*topyear*])
%ty	*year*	yearly(*string*, "[*##*]y", [*topyear*])

   For instance, just as mdy(5,30,1998) = 14,029 (30may1998),
      yw(1998,22) = 1,997 (1998w22)
      ym(1998,5) = 460 (1998m5)
      yq(1998,2) = 153 (1998q2)
      yh(1998,1) = 76 (1998h1)
      1998 = 1998

So if one had a dataset containing numeric variables **year** and **quarter**, one could translate them into a Stata %tq variable by typing

```
. gen date = yq(year,quarter)
(1 missing value generated)

. format date %tq

. list

          year    quarter      date
  1.      1998          1     1998q1
  2.      1999          5          .
  3.      2005          3     2005q3
```

Note that the mistaken year 1999 quarter 5 translated to a missing value. Had our years all been in the 20th century and recorded in two digits (e.g., 95, 98, etc.), we would have typed 'generate date = yq(1900+year,quarter)'.

   The string-conversion functions work just like the date() function; the second argument specifies the order in which the components of the date are expected to occur in the string; the y of the second argument may be prefixed with 19 or 20 as one way of handling two-digit years; a third argument specifying the maximum year may be specified as another way of handling two-digit years; if you do not prefix y and do not specify a third argument, years in the string must be four digits.

For example, `monthly(s,"my")` could translate s containing "jan 1999" or "January, 1999" or "jan1999" or "1/1999" or "1-1999", but would return missing for "jan99". "jan99" could be decoded by specifying `monthly(s,"m19y")` (in which case it would be interpreted as 1jan1999) or by specifying, say, `monthly(s,"my",2040)` (in which case it would also be interpreted as 1jan1999 because 1999 is the maximum year not greater than 2040).

Below we have string variable `sdate` containing quarterly dates and translate it to a Stata date:

```
. use a different dataset
. gen date = quarterly(sdate,"yq",2040)
(1 missing value generated)
. format date %tq
. list
                sdate       date
   1.          1995q2     1995q2
   2.          1996 3     1996q3
   3.          1996 5          .
   4. 1997 quarter 1     1997q1
   5.            98q.4    1998q4
   6.          2001-3     2001q3
   7.          2002q2     2002q2
```

27.3.2   Specifying particular dates (date literals)

Just as you can use the `d(`*constant*`)` function to type a date in an expression, such as

```
. gen age2000 = (d(1jan2000)-birthday)/365.25
```

Stata provides one-letter functions for typing weekly, monthly, quarterly, half-yearly, and yearly dates:

format	function	argument	examples
%td	d()	type day, month, year	d(15feb1998), d(15-5-2002)
%tw	w()	type year, week	w(1998w7), w(2002-25)
%tm	m()	type year, month	m(1998m2), m(2002-5)
%tq	q()	type year, quarter	q(1998q1), q(2002-2)
%th	h()	type year, half	h(1998h1), h(2002-1)
%ty	y()	type year	y(1998), y(2002)

For instance, if variable `qtr` contained a `%tq` date, you could type

```
. list if qtr>=q(1998q1)
(output omitted)
```

The `y()` function is included largely for completeness. For the `%ty` format, the year maps to the year, so 1960 means 1960 and there is little reason to type `y(1960)`. Programmers, however, sometimes find `y()` useful in terms of syntax checking. `y()`—just as all the single-letter functions—produces an error when given an invalid date which, in this case, means year < 100 or year > 9999.

27.3.3   Time-series formats

Just as with the `%d` format, the `%td`, `%tw`, `%tm`, `%tq`, `%th`, and `%ty` formats may be modified so that you can display the date in the form you wish. This is done using the same letter codes used with the `%d` format; see [U] **27.2.3 Displaying dates**. The default formats for each of the types is

format	means
%td	%tdD1CY
%tw	%twCY!ww
%tm	%tmCY!mn
%tq	%tqCY!qq
%th	%thCY!hh
%ty	%tyCY

Think of the %t format as

   %t⟨*character stating how data encoded*⟩⟨*optional characters saying how displayed*⟩

If you had variable qtr containing %tq dates and you wanted the dates displayed as, for instance, 1980 Q.1, you could type

   . format qtr %tqCY_!Q.q

27.3.4   Translating between time units

A time unit, such as %tq quarterly, can be translated to any other time unit, such as %th half-yearly. Stata provides functions to translate any time unit to and from %td daily units. The trick is to combine these functions:

Input	Output %td daily	%tw weekly	%tm monthly	%tq quarterly	%th half-yearly	%ty yearly
%td		wofd(d)	mofd(d)	qofd(d)	hofd(d)	yofd(d)
%tw	dofw(w)		mofd(dofw(w))	qofd(dofw(w))	hofd(dofw(w))	yofd(dofw(w))
%tm	dofm(m)	wofd(dofm(m))		qofd(dofm(m))	hofd(dofm(m))	yofd(dofm(m))
%tq	dofq(q)	wofd(dofq(q))	mofd(dofq(q))		hofd(dofq(q))	yofd(dofq(q))
%th	dofh(h)	wofd(dofh(h))	mofd(dofh(h))	qofd(dofh(h))		yofd(dofh(h))
%ty	dofy(y)	wofd(dofy(y))	mofd(dofy(y))	qofd(dofy(y))	hofd(dofy(y))	

The functions that translate *from* %td dates—wofd(d), mofd(d), qofd(d), hofd(d), and yofd(d)—return the date containing d. Thus, qofd() of 6apr1998 is 1998q2.

The functions that translate *to* %td dates—dofw(), dofm(), dofq(), dofh(), and dofy()—return the %td date of the beginning of the period. Thus, dofq() of 1998q2 is 01apr1998.

27.3.5   Extracting components of time

With %d dates, functions month(), day(), and year() extract the month, day, and year. In fact, if you look back at [U] **27.2.4 Other date functions**, you will find that there are functions to extract the week (week()), quarter (quarter()), and so on.

To extract values from a %t date, combine these extraction functions with the dof*() functions. For instance, if variable qtr contains a %tq date and you want new variable q to contain the quarter, type

   . gen q = quarter(dofq(qtr))

If you wanted to create a variable equal to 1 for the first two quarters of each year and 0 otherwise, you could type

   . gen first = quarter(dofq(qtr))<=2

or

   . gen first = halfyear(dofq(qtr))==1

27.3.6   Creating time variables

If you have data for which you know the first observation is for the first quarter of 1990, the second for the second quarter of 1990, and so on, but the data contains no variable recording that fact, you can generate one by typing

```
. generate time = d(1990q1)+_n-1
. format time %tq
```

Remember that _n is Stata's built-in observation counter variable; _n = 1 in the first observation, 2 in the second, and so on. The single-letter functions make it easy to type a date.

27.3.7   Setting the time variable

If you find the %t format useful, it is likely you are performing time-series analysis and will find Stata's other time-series features, such as time-series varlists, useful as well. If so, you need to set the time variable to turn those other features on:

```
. tsset time
```

The time variable you set does not need to be a %t variable, but results will be more readable if it is. See [R] **tsset**.

27.3.8   Selecting periods of time

Once you have tsset a time variable, two new functions become available to you: tin() and twithin(). These functions are useful for selecting contiguous subsets of data:

```
. regress y x l.x if tin(01jan1998,31dec2000)
```

```
. list if tin(01jan1998,)
```

$tin(t_0,t_1)$ and $twithin(t_0,t_1)$ select observations in the range t_0 to t_1 and differ only in whether observations for which $t = t_0$ and $t = t_1$ are included (tin() includes them and twithin() excludes them).

The time variable t is assumed to be the time variable you previously tsset.

The arguments you specify should be typed in the same way you would type them using the single-letter d(), w(), m(), q(), h(), or y() functions; see [U] **27.3.2 Specifying particular dates (date literals)**. That means what you type depends on the time variable's %t format.

If the time variable t does not have a %t variable, then you type numbers:

```
. list if tin(5,20)
```

means the same as

```
. list if t>=5 & t<=20
```

where t is the name of the time variable you previously tsset.

If the time variable t has a %td or %d format, then you can type things like

```
. list if tin(5jun1995,20jun1995)
```

and that means the same as

```
. list if t>=d(5jun1995) & t<=d(20jun1995)
```

If the time variable t has a `%tq` format, then you can type things like

. `list if tin(1998q1,1998q4)`

which means the same as

. `list if` t`>=q(1998q1) &` t`<=d(1998q4)`

`tin()` and `twithin()` work by accessing the time variable previously `tsset` and then examining its display format to determine how the arguments you type should appear.

In typing `tin()` and `twithin()`, you may omit either or both arguments, but you must type the comma unless you omit both arguments. The following are all valid:

. `list if tin(1998q1,)`   (meaning 1998q1 and thereafter)
. `list if tin(,1998q4)`   (meaning up to and including 1998q4)
. `list if tin(,)`   (meaning all the data)
. `list if tin()`   (also meaning all the data)

There is no reason to type `if tin(,)` or `if tin()`, but you can.

27.3.9   The %tg format

In addition to the `%t` formats documented above, there is one more: `%tg`. The **g** stands for generic. The `%tg` format is like not putting a `%t` format on your variable at all—it is equivalent to `%9.0g`. It is included for those who have a time variable, who wish to emphasize that it is a time variable, but the variable is in units that Stata does not understand.

When you have such a variable, it does not matter whether you place a `%tg` format on it.

28 Commands for dealing with categorical variables

28.1   Continuous, categorical, and indicator variables

Categorical and indicator variables occur so often in statistical data that it is worth emphasizing certain elements of Stata that might escape your attention. Although to Stata a variable is a variable, it is helpful to distinguish among three conceptual types—*continuous*, *categorical*, and *indicator* variables.

A *continuous variable* measures something. Such a variable might measure, for instance, a person's age, height, or weight; a city's population or land area; or a company's revenues or costs.

A *categorical variable* identifies a group to which the thing belongs. For example, one can categorize persons according to their race or ethnicity, cities according to their geographic location, or companies according to their industry. Often, but not always, categorical variables are stored as strings.

An *indicator variable* denotes whether something is true. For example, is a person a veteran, does a city have a mass transit system, or is a company profitable?

Indicator variables are a special case of categorical variables. Consider a variable that records a person's sex. Examined one way, it is a categorical variable. A categorical variable identifies the group to which a thing belongs, and in this case the thing is a person and the basis for categorization is anatomy. Looked at another way, however, it is an indicator variable. It indicates whether a person is, say, female.

In fact, we can use the same logic on any categorical variable that divides the data into two groups. It is a categorical variable since it identifies whether an observation is a member of this or that group; it is an indicator variable since it denotes the truth value of the statement "the observation is in this group".

All indicator variables are categorical variables, but the opposite is not true. A categorical variable might divide the data into more than two groups. For clarity, let's reserve the term *categorical variable* for variables that divide the data into more than two groups, and use the term *indicator variable* for categorical variables that divide the data into exactly two groups.

Stata has the capability to convert continuous variables to categorical and indicator variables and categorical variables to indicator variables.

28.1.1   Converting continuous to indicator variables

Stata treats logical expressions as taking on the values *true* or *false*, which it identifies with the numbers 1 and 0; see [U] **16 Functions and expressions**. For instance, if you have a continuous variable measuring a person's age and you wish to create an indicator variable denoting persons aged 21 and over, you could type

        . generate age21p = age>=21

The variable **age21p** takes on the value 1 for persons aged 21 and over and 0 for persons under 21.

Since **age21p** can take on only 0 or 1, it would be more economical to store the variable as a **byte**. Thus, it would be better to type

        . generate byte age21p=age>=21

This solution has a problem. The value of **age21** is set to 1 for all persons whose **age** is missing because Stata defines missing to be larger than all other numbers. In our data we have no such missing ages, but it still would have been safer to type

        . generate byte age21p = age>=21 if age~=.

That way, persons whose age is missing would also have a missing **age21p**.

❏ Technical Note

Put aside missing values and consider the following alternative to **gen age21p = age>=21** that may have occurred to you:

        . gen age21p = 1 if age>=21

That does not produce the desired result. This statement makes **age21p** 1 (*true*) for all persons aged 21 and above but makes **age21p** missing for everyone else.

If you followed this second approach, you would have to combine it with

        . replace age21p = 0 if age<21

to make the result identical to that produced by the single statement **gen age21p = age>=21**.

❏

28.1.2   Converting continuous to categorical variables

Suppose you wish to categorize persons into four groups based on their age. You want a variable to denote whether a person is 21 or under, between 22 and 38, between 39 and 64, and 65 and above. Although most people would label these categories 1, 2, 3, and 4, there is really no reason to restrict ourselves to such a meaningless numbering scheme. Let's call this new variable **agecat** and make it so that it takes on the topmost value for each group. Thus, persons in the first group will be identified with an **agecat** of 21, persons in the second with 38, persons in the third with 64, and persons in the last (drawing a number out of the air) with 75. Here is a way that will work, but it is not the best method:

        . generate byte agecat=21 if age<=21
        (176 missing values generated)
        . replace agecat=38 if age>21 & age<=38
        (148 real changes made)

```
. replace agecat=64 if age>38 & age<=64
(24 real changes made)
. replace agecat=75 if age>64 & age~=.
(4 real changes made)
```

We mechanically created the categorical variable according to the definition using the `generate` and `replace` commands. The only thing that deserves comment is the opening `generate`. We (wisely) told Stata to `generate` the new variable `agecat` as a `byte`, thus conserving memory.

We can create the same result with one command using the `recode()` function:

```
. generate byte agecat=recode(age,21,38,64,75)
```

`recode()` takes three or more arguments. It examines the first argument (in this case, `age`) against the remaining arguments in the list. It returns the first element in the list that is greater than or equal to the first argument, or, failing that, the last argument in the list. Thus, for each observation, `recode()` asked if `age` was less than or equal to 21. If so, the value is 21. If not, is it less than or equal to 38? If so, the value is 38. If not, is it less than or equal to 64? If so, the value is 64. If not, then the value is 75.

One typically makes tables of categorical variables, so we will `tabulate` the result:

```
. tabulate agecat
    agecat |      Freq.     Percent        Cum.
-----------+-----------------------------------
        21 |         28       13.73       13.73
        38 |        148       72.55       86.27
        64 |         24       11.76       98.04
        75 |          4        1.96      100.00
-----------+-----------------------------------
     Total |        204      100.00
```

There is another way to convert continuous variables into categorical variables, and it is even more automated: `autocode()`. `autocode()` works something like `recode()`, except that all you tell the function is the range and the total number of cells you desire that range broken into:

```
. generate agecat=autocode(age,4,18,65)
. tabulate agecat
    agecat |      Freq.     Percent        Cum.
-----------+-----------------------------------
     29.75 |         96       47.06       47.06
      41.5 |         92       45.10       92.16
     53.25 |          8        3.92       96.08
        65 |          8        3.92      100.00
-----------+-----------------------------------
     Total |        204      100.00
```

In one instruction we told Stata to break `age` into four evenly spaced categories from 18 to 65. When we `tabulate agecat`, we see the result. In particular, we see that the break points of the four categories are 29.75, 41.5, 53.25, and 65. The first category contains everyone aged 29.75 years or less; the second category contains persons over 29.75 who are 41.5 years old or less; the third category contains persons over 41.5 who are 53.25 years old or less; the last category contains all persons over 53.25.

❑ Technical Note

  We chose the range 18 to 65 arbitrarily. Although you cannot tell from the table above, there are persons in this data who are under 18 and persons over 65. Those persons are counted in the first and last cells, but we have not divided the age range in the data evenly. We could split the full age range into four categories by obtaining the overall minimum and maximum ages (by typing `summarize`) and substituting the overall minimum and maximum for the 18 and 65 in the `autocode()` function:

```
. summarize age
Variable |    Obs        Mean   Std. Dev.       Min        Max
---------+-----------------------------------------------------
     age |    204    29.64706    9.805645         2         66
. generate agecat2=autocode(age,4,2,66)
```

  Alternatively, we could `sort` the data into ascending order of `age`, and tell Stata to construct four categories over the range `age[1]` (the minimum) to `age[_N]` (the maximum):

```
. sort age
. generate agecat2=autocode(age,4,age[1],age[_N])
. tabulate agecat2
    agecat2 |      Freq.     Percent        Cum.
------------+-----------------------------------
         18 |         20        9.80        9.80
         34 |        148       72.55       82.35
         50 |         28       13.73       96.08
         66 |          8        3.92      100.00
------------+-----------------------------------
      Total |        204      100.00
```

❑

28.1.3   Converting categorical to indicator variables

  The easiest way to convert categorical to indicator variables is to use the `xi` command, which will construct indicator variables on the fly. Here we use `xi` with the `logistic` command; `grp` is a variable taking on values 1, 2, 3, and 4:

```
. xi: logistic outcome i.grp age
i.grp                     Igrp_1-4     (naturally coded; Igrp_1 omitted)
Logit estimates                             Number of obs   =        100
                                            LR chi2(4)      =      24.12
                                            Prob > chi2     =     0.0001
Log likelihood = -47.236618                 Pseudo R2       =     0.2034

------------------------------------------------------------------------
 outcome | Odds Ratio   Std. Err.       z    P>|z|    [95% Conf. Interval]
---------+--------------------------------------------------------------
  Igrp_2 |   .1479688    .1142328    -2.475   0.013    .0325865    .6718969
  Igrp_3 |   .3533434    .2395335    -1.535   0.125    .0935756   1.334232
  Igrp_4 |   .0662626    .0607619    -2.960   0.003    .0109832    .3997684
     age |   1.388413    .2415687     1.886   0.059    .9872315   1.952623
------------------------------------------------------------------------
```

See [R] **xi**. `xi` is an important command; do not ignore it.

  There will be circumstances, however, where we will want to convert categorical to indicator variables permanently, so let's consider how to do that.

We should ask ourselves how this variable is stored. Is it a set of numbers, with different numbers reflecting the different categories, or is it a string? Things will be easier if it is numeric, so if it is not, use `encode` to convert it; see [U] **26.2 Categorical string variables**. Making categorical variables numeric is not really necessary, but it is a good thing to do if only because numeric variables can be stored more compactly than string variables. More importantly, all of Stata's statistical commands know how to deal with numeric variables; some do not know what to make out of a string.

So let's suppose you have a categorical variable that divides your data into four groups. To make matters concrete, we will assume that an observation in your data is a state and that the categorical variable denotes the geographical region for each state. Each state is in one of the four Census regions known as the Northeast, North Central, South, and West.

Typing one command will create four new variables, the first indicating whether it is true that the state is in the North Central, the second whether it is true the state is in the Northeast, and so on. Such variables are sometimes called *dummy* variables, and you can use them in regressions to control for the effects of, for instance, geographic region.

Here is the data before we type this miraculous command:

```
. describe

Contains data from state.dta
  obs:            50                          State data
  vars:            6                          1 May 1998 13:54
  size:         1,800 (99.7% of memory free)
-------------------------------------------------------------------------------
  1. state      str8    %9s
  2. medage     float   %9.0g                 Median Age
  3. region     str10   %10s                  Census Region
  4. mrgrate    long    %12.0g                Marriages per 100,000
  5. dvcrate    long    %12.0g                Divorces per 100,000
  6. reg        int     %8.0g        reg      Census Region
-------------------------------------------------------------------------------
Sorted by:

. label list reg
reg:
           1 N. Centr
           2 N. East
           3 South
           4 West
```

`reg` is the categorical variable and, in our example, it is numeric, although that is not important for what we are about to do. The regions are numbered 1 to 4, and a value label, also named `reg`, maps those numbers into the words `N. Centr`, `N. East`, `South`, and `West`.

We can make the four indicator variables from this categorical variable by typing

```
. tabulate reg, generate(reg)

   Census  |
   Region  |      Freq.     Percent        Cum.
-----------+-----------------------------------
 N. Centr  |         12       24.00       24.00
  N. East  |          9       18.00       42.00
    South  |         16       32.00       74.00
     West  |         13       26.00      100.00
-----------+-----------------------------------
    Total  |         50      100.00
```

```
. describe

Contains data from state.dta
   obs:              50                          State data
  vars:              10                          1 May 1998 13:54
  size:          2,000  (99.7% of memory free)
-------------------------------------------------------------------------------
    1. state         str8     %9s
    2. medage        float    %9.0g              Median Age
    3. region        str10    %10s               Census Region
    4. mrgrate       long     %12.0g             Marriage rate
    5. dvcrate       long     %12.0g             Divorce rate
    6. reg           int      %8.0g     reg      Census Region
    7. reg1          byte     %8.0g              reg==N. Centr
    8. reg2          byte     %8.0g              reg==N. East
    9. reg3          byte     %8.0g              reg==South
   10. reg4          byte     %8.0g              reg==West
-------------------------------------------------------------------------------
Sorted by:
     Note:  dataset has changed since last save
```

Typing `tabulate reg, generate(reg)` produced a table of the number of states in each region (which is, after all, what `tabulate` does) *and*, because we specified the `generate()` option, it silently created four new variables—one for each line of the table.

Describing the data, we see that there are four new variables called `reg1`, `reg2`, `reg3`, and `reg4`. They are called this because we said `generate(reg)`. If we had said `tabulate reg, gen(junk)`, they would have been called `junk1`, `junk2`, `junk3`, and `junk4`.

Each of the new variables is stored as a `byte` and each has been automatically labeled for us. The variable `reg1` takes on the value 1 if the state is in the North Central and 0 otherwise. (It also takes on the value missing if `reg` is missing for the observation, which never occurs in our data.)

Just to be clear about the relationship of `reg1` to `reg`, here is a tabulation:

```
. tabulate reg reg1
Census    | reg==N. Centr
Region    |         0         1 |     Total
----------+----------------------+----------
 N. Centr |         0        12 |        12
 N. East  |         9         0 |         9
    South |        16         0 |        16
     West |        13         0 |        13
----------+----------------------+----------
    Total |        38        12 |        50
```

If `reg1` is 1, the region is North Central and vice versa.

28.2   Using indicator variables in estimation

Indicator variables allow you to control for the effects of a variable in a regression. Using the indicator variables we generated in the previous example, we can control for region in the following regression

$$y_j = \beta_0 + \beta_1 age_j + \beta_2 \delta_{2j} + \beta_3 \delta_{3j} + \beta_4 \delta_{4j} + \epsilon_j$$

where y_j represents the marriage rate in state j, age_j represents the median age of the state's population, and δ_{ij} is 1 if state j is in region i and 0 otherwise. We also eliminate the state of Nevada from our regression.

```
. regress mrgrate medage reg1 reg3 reg4 if state~="NEVADA"

      Source |       SS       df       MS              Number of obs =      49
-------------+------------------------------           F(  4,    44) =    8.71
       Model |  .000232847     4   .000058212          Prob > F      =  0.0000
    Residual |  .000294193    44   6.6862e-06          R-squared     =  0.4418
-------------+------------------------------           Adj R-squared =  0.3910
       Total |  .000527039    48    .00001098          Root MSE      =  .00259

-----------------------------------------------------------------------------
     mrgrate |    Coef.   Std. Err.      t     P>|t|    [95% Conf. Interval]
-------------+---------------------------------------------------------------
      medage | -.0008815   .0002723    -3.237   0.002   -.0014303   -.0003326
        reg1 |  .0003526   .0012315     0.286   0.776   -.0021293    .0028345
        reg3 |   .002699   .0011637     2.319   0.025    .0003537    .0050442
        reg4 |  .0022186   .0014201     1.562   0.125   -.0006434    .0050807
       _cons |   .039585   .0085496     4.630   0.000    .0223544    .0568155
-----------------------------------------------------------------------------
```

We see from the results above that the marriage rate, after controlling for age, is significantly higher in region 3, the South.

28.2.1 Testing significance of indicator variables

After seeing these results, you might wonder if region, taken as a whole, significantly contributes toward the explanatory power of the regression. We can find out using the `test` command:

```
. test reg1=0
 ( 1)  reg1 = 0.0
        F(  1,    44) =    0.08
             Prob > F =    0.7760
. test reg3=0, accumulate
 ( 1)  reg1 = 0.0
 ( 2)  reg3 = 0.0
        F(  2,    44) =    4.07
             Prob > F =    0.0239
. test reg4=0, accumulate
 ( 1)  reg1 = 0.0
 ( 2)  reg3 = 0.0
 ( 3)  reg4 = 0.0
        F(  3,    44) =    2.86
             Prob > F =    0.0478
```

We typed three commands. The first, `test reg1=0`, tested the coefficient on the variable `reg1` against 0. The resulting F test showed the same significance level as the corresponding t test presented in the regression output.

We next typed `test reg3=0, accumulate`. This command tests whether `reg3` is zero and accumulates that test with any previous tests. Thus, we are now testing the hypothesis that `reg1` and `reg3` are jointly zero. The F statistic is 4.07, and the result is significant at the 2.4% level; thus, at any significance level above 2.4%, it appears that `reg1` and `reg3` are not both zero.

We finally typed `test reg4=0, accumulate`. As before, this command tests the newly introduced constraint that `reg4` is zero and accumulates that with the previous tests. We are now testing whether `reg1`, `reg3`, and `reg4` are jointly zero. The F statistic is 2.86, and its significance level is roughly 4.8%; thus, at the 5% level, we can reject the hypothesis that, taken together, region has no effect on the marriage rate after controlling for age.

Stata's `test` command has a shorthand for tests of two or more variables simultaneously equal to zero. Type `test` followed by the names of the variables:

```
. test reg1 reg3 reg4
 ( 1)  reg1 = 0.0
 ( 2)  reg3 = 0.0
 ( 3)  reg4 = 0.0
       F(  3,     44) =     2.86
             Prob > F =     0.0478
```

❑ Technical Note

Sometimes tests of this kind are embedded in an ANOVA or ANCOVA model. In the language of ANOVA, we are doing a one-way layout after controlling for the effect of age. Hence, we did not have to estimate the model with `regress`: we could have used Stata's `anova` command. We would have typed `anova mrgrate medage reg, continuous(medage)` to obtain the ANOVA table and the desired test directly. We could have seen the underlying regression after estimation of the model by then typing `regress` without arguments. Typing any estimation command without arguments is taken as a request to reshow the last estimation results. Since all of Stata's estimation commands are tightly coupled, after estimation of an ANOVA or ANCOVA model, you can ask `regress` to show you the underlying regression coefficients.

❑

28.2.2   Importance of omitting one of the indicators

Some people prefer to estimate models with dummy variables in the context of ANOVA, and others prefer the regression context. The choice is yours.

If you opt for the regression method, however, remember to leave one of the indicator variables out of the regression so that the coefficients have the interpretation of changes from a base group. If you fail to follow that advice, your regression will still be correct, but it is important that you understand what it is you are estimating and testing.

In the example above, for instance, we omitted `reg2`, the dummy for the Northeast. Let's rerun that regression and include the `reg2` dummy:

```
. regress mrgrate medage reg1-reg4 if state~="NV"

    Source |       SS       df       MS                  Number of obs =      49
-----------+------------------------------              F(  4,     44) =    8.71
     Model | .000232847        4  .000058212            Prob > F      =  0.0000
  Residual | .000294193       44  6.6862e-06            R-squared     =  0.4418
-----------+------------------------------              Adj R-squared =  0.3910
     Total | .000527039       48   .00001098            Root MSE      =  .00259

-------------------------------------------------------------------------------
   mrgrate |      Coef.   Std. Err.       t    P>|t|     [95% Conf. Interval]
-----------+-------------------------------------------------------------------
    medage | -.0008815   .0002723     -3.237   0.002    -.0014303   -.0003326
      reg1 |  .0399376   .0080754      4.946   0.000     .0236628    .0562124
      reg2 |  .039585    .0085496      4.630   0.000     .0223544    .0568155
      reg3 |  .0422839   .0080922      5.225   0.000     .0259752    .0585926
      reg4 |  .0418036   .0076958      5.432   0.000     .0262938    .0573135
     _cons | (dropped)
-------------------------------------------------------------------------------
```

If you compare the top half of the regression output with that of the previous example, you will find that they are identical. You will also find that the estimated coefficient, standard error, and related statistics for the variable `medage` are identical as well. Yet the estimates for each of the `reg` variables are different.

You will also note that the models have been estimated on different variables. In the first case, dummies for regions 1, 3, and 4 were included along with a constant. In the last case, dummies for regions 1, 2, 3, and 4 were included, and the constant was mysteriously dropped by `regress`. There are the same number of variables in each regression, but their identities have changed. In one case, there is a constant; in the other, a dummy for region 2.

Let's first explain why the constant was dropped; it was dropped because it was unnecessary. You can think of this model as making four different predictions for a given median age, one for each region. Each prediction is given by $-.0008815$ multiplied by the median age plus a constant for the region. The constant for the region is the estimated coefficient for the region from the table above. We can write this mathematically as

$$-.0008815age + c_i$$

For instance, for region 2, $c_2 = .039585$.

An overall constant is unnecessary in the sense that it is arbitrary and zero is a good choice if one is choosing an arbitrary number. Let's choose another arbitrary number—say, 1. Then each prediction would be given by $-.0008815$ multiplied by the median age plus a (different) constant for the region plus 1:

$$-.0008815age + c_i' + 1$$

The estimated constants for each of the regions would then have to change, and in fact they would have to change by exactly 1, so that

$$c_i' + 1 = c_i$$

The constant was dropped because including all four dummy variables in the regression made it unnecessary. In our first model we turned the problem around. We included a constant but left out the dummy for region 2.

Let's spend a moment proving that the results are identical. The model we just estimated indicates that the marriage rate in region 2 is given by

$$-.0008815age + .039585$$

Now turn back to the previous model. The marriage rate in region 2 is equal to the same thing! The constant in the original regression is identical to the coefficient on `reg2` in the second regression.

Let's look at region 1. The model we just estimated indicates the marriage rate is given by

$$-.0008815age + .0399376$$

In the first model it is

$$-.0008815age + .0003526 + .039585 = -.0008815age + .0399376$$

which is again the same thing! If you perform the calculations for region 3 and region 4, you will discover two more equivalences.

The models are identical in the sense that they make the same predictions. They differ, however, in other ways. In the first model, the coefficients on `reg1`, `reg3`, and `reg4` measure the difference between that region and region 2. In the second model, the coefficients measure the region's level directly.

Notice that the t statistic on the coefficient for `reg2` in the second model is equal to the t statistic on the constant in the first—namely, 4.630. So they should be. Both test the same hypothesis—that the marriage rate is zero in Northeastern states with a median age of zero. (Yes, there are no such states, and yes, the hypothesis sounds silly. Perhaps you prefer the hypothesis stated as "the constant for the Northeast is zero".)

The comparison of the t statistics for region 1, however, does not yield such equivalent results. In the first model, the t statistic is 0.286; in the second model, the statistic is 4.946. They are different because they test different hypotheses. In the first case, the statistic tests whether the constant for region 1 is the same as that for region 2; in the second case, it tests the hypothesis that the constant for region 1 is zero.

The differences in meaning of these statistical tests carry over to any `test` commands you type. In the technical note above, we tested that `reg1`, `reg3`, and `reg4` were all simultaneously zero. We obtained an F statistic of 2.86. If we were to type the same statements after estimating our second model, we would obtain a whopping F statistic of 13.54! What we are testing, however, is different. If we wanted to perform the same test—namely, that all the regions have the same intercept—we would type `test reg1=reg2` followed by `test reg3=reg2, accumulate` followed by `test reg4=reg2, accumulate`. That test gives identical results.

29 Overview of model estimation in Stata

Contents

29.1   Introduction

By estimation commands we mean commands that estimate models such as linear regression, probit, and the like. Stata has many such commands, so many that it is easy to overlook a few. Some of these commands differ greatly from each other, others are gentle variations on a theme, and still others are outright equivalent.

If you have not yet read [U] **23 Estimation and post-estimation commands**, do so soon. Estimation commands share features which we shall not deal with here. We especially direct your attention to [U] **23.11 Obtaining robust variance estimates**, which discusses an alternative calculation for the estimated variance matrix (and hence standard errors) that many of Stata's estimation commands provide, and we direct your attention to [U] **23.9 Performing hypothesis tests on the coefficients**.

Here, however, we will put aside all of that—and all issues of syntax—and deal solely with matching commands to their statistical concepts. We will also put aside cross-referencing when it is obvious. We will not say "the `regress` command—see [R] **regress**—allows ..." nor will we even say "the `tobit` command—see [R] **tobit**—is related ...". To find the details on a particular command, look up its name in the index.

29.2   Linear regression with simple error structures

Let us begin by considering models of the form

$$y_j = \mathbf{x}_j \boldsymbol{\beta} + \epsilon_j$$

for a continuous y variable. In this category we restrict ourselves to estimation when σ_ϵ^2 is constant across observations j. The model is called the linear regression model and the estimator is often called the (ordinary) least squares estimator.

regress is Stata's linear-regression command. (regress will produce the robust estimate of variance as well as the conventional estimate and regress has a collection of commands that can be run after it to explore the nature of the fit.)

In addition, the following commands will estimate linear regressions as does regress, but offer special features:

1. ivreg will estimate instrumental variables models.

2. areg estimates models $y_j = \mathbf{x}_j\boldsymbol{\beta} + \mathbf{d}_j\boldsymbol{\gamma} + \epsilon_j$ where $\mathbf{d}_j$ is a mutually exclusive and exhaustive dummy variable set. Through numerical trickery, areg obtains estimates of $\boldsymbol{\beta}$ (and associated statistics) without ever forming $\mathbf{d}_j$, meaning it also does not report the estimated $\boldsymbol{\gamma}$. If your interest is in estimating fixed-effects models, Stata has a better command—xtreg—discussed in [U] **29.13.1 Linear regression with panel data** below. Most users who find areg appealing are probably seeking xtreg because it provides more useful summary and test statistics. areg literally duplicates the output regress would produce were you to generate all the dummy variables. This means, for instance, that the reported R^2 includes the effect of $\boldsymbol{\gamma}$.

3. boxcox estimates linear-regression models on dependent variables subjected to the Box–Cox transform. The model estimated is $y_j^{(\lambda)} = \mathbf{x}_j\boldsymbol{\beta} + \epsilon_j$, where λ is estimated from the data and $y_j^{(\lambda)}$ is $(y^\lambda - 1)/\lambda$.

4. tobit allows estimation of linear regression models when y_i has been subject to left censoring, right censoring, or both. For instance, say y_i is not observed if $y_i < 1000$, but for those observations, it is known that $y_i < 1000$. tobit estimates such models.

5. cnreg (censored-normal regression) is a generalization of tobit. The lower and upper censoring points, rather than being constants, are allowed to vary observation-by-observation. Any model tobit can estimate, cnreg can estimate.

6. intreg (interval regression) is a generalization of cnreg. In addition to allowing open-ended intervals, intreg allows closed intervals, too. Rather than observing y_j, it is assumed that y_{0j} and y_{1j} are observed, where $y_{0j} \le y_j \le y_{1j}$. Survey data might report that a subject's monthly income was in the range $1,500 to $2,500. intreg allows such data to be used to estimate a regression model. intreg allows $y_{0j} = y_{1j}$ and so can reproduce results reported by regress. intreg allows y_{0j} to be $-\infty$ and y_{1j} to be $+\infty$ and so can reproduce results reported by cnreg and tobit.

7. cnsreg allows placing linear constraints on the coefficients.

8. eivreg adjusts estimates for errors in variables.

9. nl provides the nonlinear least-squares estimator of $y_j = f(\mathbf{x}_j, \boldsymbol{\beta}) + \epsilon_j$.

10. rreg estimates robust regression models, a term not to be confused with regression with robust standard errors. Robust standard errors are discussed in [U] **23.11 Obtaining robust variance estimates**. Robust regression concerns point estimates more than standard errors and it implements a data-dependent method for downweighting outliers.

11. `qreg` produces quantile-regression estimates, a variation which is not linear regression at all but which is an estimator of $y_j = \mathbf{x}_j\boldsymbol{\beta} + \epsilon_j$. In the basic form of this model, sometimes called median regression, $\mathbf{x}_j\boldsymbol{\beta}$ measures not the predicted mean of y_j conditional on $\mathbf{x}_j$, but its median. As such, `qreg` is of most interest when ϵ_j does not have constant variance. `qreg` allows you to specify the quantile, so you can produce linear estimates for the predicted 1st, 2nd, ..., 99th percentile.

Another command, `bsqreg`, is identical to `qreg` but presents bootstrapped standard errors.

The `sqreg` command estimates multiple quantiles simultaneously; standard errors are obtained via the bootstrap.

The `iqreg` command estimates the difference between two quantiles; standard errors are obtained via the bootstrap.

12. `vwls` (variance-weighted least squares) produces estimates of $y_j = \mathbf{x}_j\boldsymbol{\beta} + \epsilon_j$ where the variance of ϵ_j is calculated from group data or is known *a priori*. As such, `vwls` is of most interest to categorical-data analysts and physical scientists.

29.3   ANOVA and ANCOVA

ANOVA and ANCOVA are certainly related to linear regression, but we classify them separately. The related Stata commands are `anova`, `oneway`, and `loneway`. Stata does not currently provide a MANOVA estimator.

`anova` estimates ANOVA and ANCOVA models, one-way and up—including two-way factorial, three-way factorial, etc.—and it estimates nested and mixed-design models and repeated-measures models. It is probably what you are looking for.

`oneway` estimates one-way ANOVA models. It is quicker at producing estimates than `anova`, although `anova` is so fast this probably does not matter. The important difference is that `oneway` can report multiple-comparison tests.

`loneway` is an alternative to `oneway`. The results are numerically the same, but `loneway` can deal with more levels (limited only by dataset size; `oneway` is limited to 376 levels and `anova` to 798, but for `anova` to reach 798 requires a lot of memory) and `loneway` reports some additional statistics including the intraclass correlation.

29.4   General linear model

The general linear model is

$$g(E(y_j)) = \mathbf{x}_j\boldsymbol{\beta}, \qquad y_j \sim F$$

where $g()$ is called the link function and F is a member of the exponential family, both of which you specify prior to estimation. `glm` estimates this model.

The GLM framework encompasses a surprising array of models known by other names, including linear regression, Poisson regression, exponential regression, and others. Stata provides dedicated estimation commands for many of these. Stata has, for instance, `regress` for linear regression, `poisson` for Poisson regression, and `ereg` and `stereg` for exponential regression, and that is not all of the overlap.

For each family F there is a corresponding link function $g()$, called the canonical link, for which GLM estimation produces results identical to maximum likelihood estimation. You can, however, match families and link functions as you wish and, when you match a family to a link function other than the canonical link, you obtain a different but valid estimator of that model. The estimator you obtain is asymptotically equivalent to the maximum likelihood estimator which, in small samples, will produce slightly different results.

For example, the canonical link for the binomial family is logit. `glm` with that combination produces results identical to the maximum-likelihood `logit` (and `logistic`) command. The binomial family with the probit link produces the probit model, but probit is not the canonical link in this case. Hence, `glm` produces results that in small samples differ slightly from those produced by Stata's maximum-likelihood `probit` command.

Many researchers feel that the maximum likelihood estimates are preferable to GLM estimates (when they are not identical), but they would have a difficult time justifying that feeling. Maximum likelihood probit is an estimator with (solely) asymptotic properties; GLM with the binomial family and probit link is an estimator with (solely) asymptotic properties, and in finite samples, the estimators differ a little.

Still, we recommend that you use Stata's maximum likelihood estimators whenever possible. GLM—the theory—and `glm`—the command—are all-encompassing in their generality and that means they rarely use quite the right jargon or provide things in quite the way you wish they would. The narrower commands, such as `logit`, `probit`, `poisson`, etc., focus on the issue at hand and are invariably more convenient.

`glm` is useful when you want to match a family to a link function that is not provided elsewhere.

29.5   Binary outcome qualitative dependent variable models

There are lots of ways to write these models; one way is

$$\Pr(y_j \neq 0) = F(\mathbf{x}_j \boldsymbol{\beta})$$

where F is some cumulative distribution. Two popular choices for $F()$ are the normal and logistic and the models are called the probit and logit (or logistic regression) models. A third is the complementary log–log function; maximum likelihood estimates are obtained by Stata's `cloglog` command.

The two parent commands for the maximum likelihood estimator of probit and logit are `probit` and `logit`, although `logit` has a sibling `logistic` that provides the same estimates but displays results in a slightly different way.

Do not read anything into the names logit and logistic although, even with that warning, we know you will. Logit and logistic have two completely interchanged definitions in two scientific camps. In the medical sciences, logit means the minimum χ^2 estimator and logistic means maximum likelihood. In the social sciences, it is the other way around. From our experience, it appears that neither reads the other's literature, since both talk (and write books) asserting that logit means one thing and logistic the other. Our solution is to provide both `logit` and `logistic` that do the same thing so that each camp can latch on to the maximum likelihood command under the name each expects.

There are two slight differences between `logit` and `logistic`. `logit` reports estimates in the coefficient metric whereas `logistic` reports exponentiated coefficients—odds ratios. This is in accordance with the expectations of each camp and it makes no substantive difference. The other difference is that `logistic` has a family of post-`logistic` commands that you can run to explore the nature of the fit. Actually, that is not exactly true because all the commands for use after `logistic` can be used after `logit` and a note is even made of that fact in the `logit` documentation.

If you have not already selected one of `logit` or `logistic` as your favorite, we recommend you try `logistic`. Logistic regression (logit) models are more easily interpreted in the odds-ratio metric.

In addition to `logit` and `logistic`, Stata provides `glogit` and `blogit` commands.

`blogit` is the maximum likelihood estimator (same as `logit` or `logistic`), but applied on data organized in a different way. Rather than individual observations, your data is organized so that each observation records the number of successes and the number of failures observed.

`glogit` is the weighted-regression, grouped-data estimator.

Related to logit, the skewed logit estimator scobit adds a power to the logit link function and is estimated by Stata's `scobit` command.

Turning to probit, you have two choices: `probit` and `dprobit`. Both are maximum likelihood and it makes no substantive difference which you use. They differ only in how they report results. `probit` reports coefficients. `dprobit` reports changes in probabilities. Many researchers find changes in probabilities easier to interpret.

As in the logit case, Stata also provides `bprobit` and `gprobit`. `bprobit` is maximum likelihood—equivalent to `probit` or `dprobit`—but works with data organized in the different way outlined above. `gprobit` is the weighted-regression, grouped-data estimator.

Continuing with probit, `hetprob` estimates heteroscedastic probit models. In these models, the variance of the error term is parameterized.

In addition, Stata's `biprobit` command will estimate bivariate probit models, meaning two correlated outcomes. `biprobit` will also estimate partial-observability models in which only the outcomes $(0,0)$ and $(1,1)$ are observed.

29.6   Conditional logistic regression

`clogit` is Stata's conditional logistic regression estimator. In this model, observations are assumed to be partitioned into groups and a predetermined number of events occur in each group. The model measures the risk of the event according to the observation's characteristics x_j. The model is used in matched case–control studies (`clogit` allows $1:1$, $1:k$, and $m:k$ matching) and is also used in natural experiments whenever observations can be grouped into pools in which a fixed number of events occur.

29.7   Multiple outcome qualitative dependent variable models

For more than two outcomes, Stata provides ordered logit, ordered probit, multinomial logistic regression, and McFadden's choice model.

`oprobit` and `ologit` provide maximum-likelihood ordered probit and logit. These are generalizations of probit and logit models known as the proportional odds model and are used when the outcomes have a natural ordering from low to high. The idea is there is an unmeasured $z_j = x_j\beta$ and the probability that the kth outcome is observed is $\Pr(c_{k-1} < z_j < c_k)$, where $c_0 = -\infty$, $c_k = +\infty$, and $c_1, \ldots, c_{k-1}$ along with β are estimated from the data.

`mlogit` estimates maximum-likelihood multinomial logistic models, also known as polytomous logistic regression. It is intended for use when the outcomes have no natural ordering and all that is known are the characteristics of the outcome chosen (and, perhaps, the chooser).

`clogit` estimates McFadden's choice model, also known as conditional logistic regression. In the context denoted by the name McFadden's choice model, the model is used when the outcomes have no natural ordering, just as multinomial logistic regression, but the characteristics of the outcomes chosen and not chosen are known (along with, perhaps, the characteristics of the chooser).

In the context denoted by the name conditional logistic regression—which we have already mentioned above—subjects are members of pools and one or more are chosen, typically to be infected by some disease or to have some other unfortunate event befall them. Thus, the characteristics of the chosen and not chosen are known and the issue of the characteristics of the chooser never arises. Said either way, it is the same model.

29.8   Simple count dependent variable models

These models concern dependent variables that count the number of occurrences of an event. In this category, we include Poisson and negative-binomial regression. For the Poisson model,

$$E(\text{count}) = E_j \exp(\mathbf{x}_j \boldsymbol{\beta})$$

where E_j is the exposure time. `poisson` estimates this model.

Negative-binomial regression refers to estimating with data that is a mixture of Poisson counts. One derivation of the negative-binomial model is that individual units follow a Poisson regression model but there is an omitted variable that follows a gamma distribution with variance α. Negative-binomial regression estimates $\boldsymbol{\beta}$ and α. `nbreg` estimates such models. A variation on this, unique to Stata, allows you to model α. `gnbreg` estimates those models.

Zero inflation refers to count models in which the number of 0 counts is more than would be expected in the regular model and that is due to there being a probit or logit process that must first generate a positive outcome before the counting process can begin.

Stata's `zip` command estimates zero-inflated Poisson models.

Stata's `zinb` command estimates zero-inflated negative-binomial models.

29.9   Linear regression with heteroscedastic errors

We now consider estimating the model $y_j = \mathbf{x}_j \boldsymbol{\beta} + \epsilon_j$ where the variance of ϵ_j is nonconstant.

First, `regress` can estimate such models if you specify the `robust` option. What we call robust is also known as the White correction for heteroscedasticity.

For scientists who have data where the variance of ϵ_j is known *a priori*, `vwls` is the command. `vwls` produces estimates of the model given each observation's variance which is recorded in a variable in the data.

Finally, as mentioned above, `qreg` estimates quantile regression and it is in the presence of heteroscedasticity that this is most of interest. Median regression (one of `qreg`'s capabilities) is an estimator of $y_j = \mathbf{x}_j \boldsymbol{\beta} + \epsilon_j$ when ϵ_j is heteroscedastic. Even more usefully, one can estimate models of other quantiles and so model the heteroscedasticity. Also see the `sqreg` and `iqreg` commands; `sqreg` estimates multiple quantiles simultaneously. `iqreg` estimates differences in quantiles.

29.10   Linear regression with systems of equations (correlated errors)

If by correlated errors, you mean that observations are grouped, that within group, the observations might be correlated but, across groups, they are uncorrelated, realize that `regress` with the `robust` and `cluster()` options can produce "correct" estimates, which is to say, inefficient estimates with correct standard errors and lots of robustness; see [U] **23.11 Obtaining robust variance estimates**. Obviously, if you know the correlation structure (and are not mistaken), you can do better, so `xtreg` and `xtgls` are also of interest in this case; we discuss them in [U] **29.13.1 Linear regression with panel data** below.

Turning to simultaneous multiple-equation models, Stata can produce three-stage least squares (3SLS) and two-stage least squares (2SLS) estimates using the `reg3` and `ivreg` commands. Two-stage models can be estimated by either `reg3` or `ivreg`. Three-stage models require use of `reg3`. The `reg3` command can produce constrained and unconstrained estimates.

In the case where we have correlated errors across equations but no endogenous right-hand side variables,

$$y_{1j} = \mathbf{x}_{1j}\boldsymbol{\beta} + \epsilon_{1j}$$
$$y_{2j} = \mathbf{x}_{2j}\boldsymbol{\beta} + \epsilon_{2j}$$
$$\vdots$$
$$y_{mj} = \mathbf{x}_{mj}\boldsymbol{\beta} + \epsilon_{mj}$$

where ϵ_k. and ϵ_l. are correlated with correlation ρ_{kl}, a quantity to be estimated from the data. This is called Zellner's seemingly unrelated least squares and `sureg` estimates such models. In the case where $\mathbf{x}_{1j} = \mathbf{x}_{2j} = \cdots = \mathbf{x}_{mj}$, the model is known as multivariate regression and the corresponding command is `mvreg`.

Estimation in the presence of autocorrelated errors are discussed in [U] **29.12 Models with time-series data**.

29.11   Models with endogenous sample selection

What has become known as the Heckman model refers to estimating linear regression in the presence of sample selection: $y_j = \mathbf{x}_j\boldsymbol{\beta} + \epsilon_j$ is not observed unless some event occurs which itself has probability $p_j = F(\mathbf{z}_j\boldsymbol{\gamma} + \nu_j)$ where ϵ and ν might be correlated and $\mathbf{z}_j$ and $\mathbf{x}_j$ may contain variables in common.

`heckman` estimates such models by either maximum likelihood or Heckman's original two-step procedure.

This model has recently been generalized to replacing the linear regression equation with another probit equation and that model is estimated by `heckprob`.

29.12   Models with time-series data

ARIMA refers to models with autoregressive integrated moving average processes and Stata's `arima` command estimates models with ARIMA disturbances via the Kalman filter and maximum likelihood. These models may be estimated with or without confounding covariates.

Stata's `prais` command estimates regression with AR(1) disturbances using the Prais–Winsten or Cochrane–Orcutt transformation. Both two-step and iterative solutions are available and a version of the Hildreth–Lu search procedure is also available. The Prais–Winsten estimates of the model are an improvement over the Cochrane–Orcutt estimates in that the first observation is preserved in the estimation. This is particular important with trended data in small samples.

`prais` automatically produces the Durbin–Watson d-statistic and this can also be obtained after `regress` using `dwstat`.

`newey` produces linear regression estimates with the Newey–West variance estimates that are robust to heteroscedasticity and autocorrelation of specified order.

Stata provides estimators for regression models with autoregressive conditional heteroscedastic (ARCH) disturbances:

$$y_t = \mathbf{x}_t \boldsymbol{\beta} + \mu_t$$

where μ_t is distributed $N(0, \sigma_t^2)$ and σ_t^2 is given by some function of the lagged disturbances.

Stata's `arch`, `aparch`, and `egarch` commands provide different parameterizations of the conditional heteroscedasticity. All three of these commands also allow ARMA disturbances and/or multiplicative heteroscedasticity.

29.13   Panel-data models

29.13.1   Linear regression with panel data

This section could just as well be called linear regression with complex error structures. The letters `xt` are the prefix for the commands in this class.

`xtreg` estimates models of the form

$$y_{it} = \mathbf{x}_{it} \boldsymbol{\beta} + \nu_i + \epsilon_{it}$$

`xtreg` can produce the fixed (between)-effects estimator, the within-effects estimator, or the random-effects (matrix weighted average of between and within) estimator. In addition, it can produce the full maximum-likelihood random-effects estimator.

`xtgls` produces generalized least squares estimates of models of the form

$$y_{it} = \mathbf{x}_{it} \boldsymbol{\beta} + \epsilon_{it}$$

where you may specify the variance structure of ϵ_{it}. If you specify the ϵ_{it} is independent for all i and t, `xtgls` produces the same results as `regress` up to a small-sample degrees-of-freedom correction applied by `regress` but not by `xtgls`.

You may choose among three variance structures concerning i and three concerning t, producing a total of nine different models. Assumptions concerning i deal with heteroscedasticity and cross-sectional correlation. Assumptions concerning t deal with autocorrelation and, more specifically, AR(1) serial correlation.

In the jargon of GLS, the random-effects model estimated by `xtreg` has exchangeable correlation within i—`xtgls` does not model this particular correlation structure. `xtgee`, however, does.

`xtgee` will estimate population-averaged models and it will optionally provide robust estimates of variance. Moreover, `xtgee` will allow other correlation structures. One that is of particular interest to those with lots of data goes by the name unstructured. The within-panel correlations are simply estimated in an unconstrained way. In [U] **29.13.3 General linear model with panel data**, we have more to say about this estimator since it is not restricted to just linear regression models.

29.13.2   Censored linear regression with panel data

xttobit estimates random-effects tobit models and generalizes that to observation-specific censoring.

xtintreg estimates random-effects interval regression and generalizes that to observation-specific censoring. Interval regression, in addition to allowing open-ended intervals, allows closed intervals, too.

29.13.3   General linear model with panel data

In [U] **29.4 General linear model** above we discussed the model

$$g(E(y_j)) = \mathbf{x}_j \boldsymbol{\beta}, \qquad y_j \sim F$$

where $g()$ is the link function and F is a member of the exponential family, both of which you specify prior to estimation. This model can be further generalized to work with cross-sectional time-series data, so let us rewrite it

$$g(E(y_{it})) = \mathbf{x}_{it} \boldsymbol{\beta}, \qquad y_{it} \sim F \text{ with parameters } \theta_{it}$$

We refer to this as the GEE method for panel data models, where GEE stands for General Estimating Equations. xtgee estimates this model and allows specifying the correlation structure of the errors.

If you specify errors are independent within i, xtgee is equivalent to glm. An advantage of xtgee is that it can provide the robust estimates of variance which, currently, glm cannot (an oversight on our part that will be corrected some day). Thus, since glm can reproduce (to name a few), the estimates produced by regress, logit, and poisson, so can xtgee.

If you specify errors are exchangeable within i, xtgee estimates equal-correlation models. This means that with the identity link and Gaussian family, xtgee can reproduce the models estimated by xtreg. The only difference is that xtgee can provide standard errors that are robust to the correlations not being exchangeable.

xtgee provides other correlation structures, including multiplicative, AR(m); stationary(m); nonstationary(m); unstructured; and fixed (meaning user-specified). Unstructured should be of particular interest to those with large datasets even if you ultimately plan to impose a structure such as exchangeability (equal correlation). If relaxing the equal-correlation assumption in a large dataset causes your results to change importantly, there is an issue before you worthy of some thought.

xtgee provides 175 models from which to choose.

29.13.4   Qualitative dependent variable models with panel data

xtprobit estimates random-effects probit regression via maximum likelihood. It will also estimate population-averaged models via GEE. This last is nothing more than xtgee with the binomial family, probit link, and exchangeable error structure.

xtlogit estimates random-effects logistic regression models via maximum likelihood. It will also estimate conditional fixed-effects models via maximum likelihood. Finally, as with xtprobit, it will estimate population-averaged models via GEE.

xtclog estimates random-effects complementary log–log regression via maximum likelihood. It will also estimate population-averaged models via GEE.

clogit is also of interest since it provides the conditional fixed-effects logistic estimator.

29.13.5 Count dependent variable models with panel data

xtpois estimates two different random-effects Poisson regression models via maximum likelihood. The two distributions for the random effect are gamma and normal. It will also estimate conditional fixed-effects models. It will also estimate population-averaged models via GEE. This last is nothing more than xtgee with the Poisson family, log link, and exchangeable error structure.

xtnbreg estimates random-effects negative-binomial regression models via maximum likelihood (the distribution of the random effect is assumed to be beta). It will also estimate conditional fixed-effects models. It will also estimate population-averaged models via GEE.

29.13.6 Random coefficient models with panel data

xtrchh will estimate the Hildreth–Houck random-coefficients model. In this model, rather than just the intercept being constant within group and varying across groups, all the coefficients vary across groups.

29.14 Survival-time (failure-time) models

Commands are provided to estimate Cox proportional hazards models as well as several parametric survival models including exponential, Weibull, Gompertz, log-normal, log-logistic, and generalized log gamma. The commands all allow for right censoring, left truncation, gaps in histories, and time-varying regressors. The commands are appropriate for use with single- or multiple-failure per subject data. Conventional and robust standard errors are available, with and without clustering.

The commands are stcox and cox for the Cox model and streg and a host of commands (all documented in [R] **weibull**) for the parametric models. Whether you use stcox or cox, or streg or the other commands, makes no difference.

The commands without the st prefix are straightforward estimation commands: you specify the time-of-failure variable, the outcome variable, etc. The st variants require less typing because you first use another command, stset, to declare the key survival-time variables. If you are in the mode of performing survival analysis, the st commands are more convenient. If you are doing something else and simply need to estimate one of these models along the way, the non-st variants will be more convenient.

29.15 Commands for estimation with survey data

Many of Stata's estimation commands allow sampling weights and, if they do, provide a cluster() option as well; see [U] **23.13 Weighted estimation**. That still leaves the issue of stratification and a parallel set of commands beginning with the letters svy are provided. The list currently includes

1. svyreg for linear regression,

2. svyivreg for instrumental-variables regression,

3. svyintrg for censored and interval regression,

4. svylogit for logistic regression,

5. svyprobt for probit,

6. svymlog for multinomial logistic regression,

7. svyolog for ordered logistic regression,

8. svyoprob for ordered probit, and

9. svypois for Poisson regression,

See [U] **30 Overview of survey estimation**.

29.16   Multivariate analysis

In this category we include canonical correlation, principal components, and factor analysis. See [U] **29.10 Linear regression with systems of equations (correlated errors)** above for multivariate regression.

Canonical correlation attempts to describe the relationship between two sets of variables. Given $\mathbf{x}_j = (x_{1j}, x_{2j}, \ldots, x_{Kj})$ and $\mathbf{y}_j = (y_{1j}, y_{2j}, \ldots, y_{Lj})$, the goal is to find linear combinations

$$\widehat{x}_{1j} = \widehat{b}_{11} x_{1j} + \widehat{b}_{12} x_{2j} + \cdots + \widehat{b}_{1K} x_{Kj}$$
$$\widehat{y}_{1j} = \widehat{g}_{11} y_{1j} + \widehat{g}_{12} y_{2j} + \cdots + \widehat{g}_{1L} y_{Lj}$$

such that the correlation of $\widehat{x}_1$ and $\widehat{y}_1$ is maximized. That is called the first canonical correlation. The second canonical correlation is defined similarly with the added proviso that $\widehat{x}_2$ and $\widehat{y}_2$ are orthogonal to $\widehat{x}_1$ and $\widehat{y}_1$, and the third canonical correlation with the proviso that $\widehat{x}_3$ and $\widehat{y}_3$ are orthogonal to $\widehat{x}_1$, $\widehat{y}_1$, $\widehat{x}_2$, and $\widehat{y}_2$, and so on. canon estimates canonical correlations and their corresponding loadings.

Principal components concerns finding

$$\widehat{x}_{1j} = \widehat{b}_{11} x_{1j} + \widehat{b}_{12} x_{2j} + \cdots + \widehat{b}_{1K} x_{Kj}$$
$$\widehat{x}_{2j} = \widehat{b}_{21} x_{1j} + \widehat{b}_{22} x_{2j} + \cdots + \widehat{b}_{2K} x_{Kj}$$
$$\vdots$$
$$\widehat{x}_{Kj} = \widehat{b}_{K1} x_{1j} + \widehat{b}_{K2} x_{2j} + \cdots + \widehat{b}_{KK} x_{Kj}$$

such that $\widehat{x}_1$ has maximum variance, $\widehat{x}_2$ has maximum variance subject to being orthogonal to $\widehat{x}_1$, $\widehat{x}_3$ has maximum variance subject to being orthogonal to $\widehat{x}_1$ and $\widehat{x}_2$, and so on. factor with the pc option extracts principal components and reports eigenvalues and loadings.

Factor analysis is concerned with finding a small number of common factors $\widehat{\mathbf{z}}_k$, $k = 1, \ldots, q$, that linearly reconstruct the original variables $\mathbf{y}_i$, $i = 1, \ldots, L$.

$$\widehat{y}_{1j} = \widehat{z}_{1j} \widehat{b}_{11} + \widehat{z}_{2j} \widehat{b}_{12} + \cdots + \widehat{z}_{qj} \widehat{b}_{1q} + e_{1j}$$
$$\widehat{y}_{2j} = \widehat{z}_{1j} \widehat{b}_{21} + \widehat{z}_{2j} \widehat{b}_{22} + \cdots + \widehat{z}_{qj} \widehat{b}_{2q} + e_{2j}$$
$$\vdots$$
$$\widehat{y}_{Lj} = \widehat{z}_{1j} \widehat{b}_{L1} + \widehat{z}_{2j} \widehat{b}_{L2} + \cdots + \widehat{z}_{qj} \widehat{b}_{Lq} + e_{Lj}$$

Note that everything on the right hand-side is estimated so the model has an infinite number of solutions. Various constraints are introduced along with a definition of "reconstruct" to make the model determinate. Reconstruction, for instance, is typically defined in terms of prediction of the covariance of the original variables. factor estimates such models and provides principal factors, principal component factors, iterated principal components, and maximum likelihood solutions.

29.17   Not elsewhere classified

Two commands are not really estimation commands but estimation-command modifiers: sw and fracpoly.

sw, typed in front of an estimation command as a separate word, provides stepwise estimation. You can use the sw prefix with some, but not all estimation commands. In [R] **sw** is a table of which estimation commands are currently supported, but do not take it too literally. It was accurate as of the day Stata 6.0 was released but, if you install the official STB updates (see [U] **20.8 How do I install STB updates?**), sw may now work with other commands, too. If you want to use sw with some estimation command, our advice is to try it. Either it will work or you will get the message that the estimation command is not supported by sw.

fracpoly is a command to assist performing specification searches.

29.18   Have we forgotten anything?

We have discussed all the estimation commands included in Stata 6.0 the day it was released. By now, there may be more; see [U] **20.8 How do I install STB updates?**. To obtain an up-to-date list, type search estimation.

And of course, you can always write your own; see [R] **ml**.

30 Overview of survey estimation

Contents

30.1   Introduction

There is a family of commands in Stata designed especially for data from sample surveys. These commands all begin with `svy`. The `svy` commands are

`svyset`	[R] **svyset**	Set variables for survey data
`svydes`	[R] **svydes**	Describe strata and PSUs of survey data
`svyreg`	[R] **svy estimators**	Linear regression for survey data
`svyivreg`	[R] **svy estimators**	Instrumental variables regression for survey data
`svyintrg`	[R] **svy estimators**	Censored and interval regression for survey data
`svylogit`	[R] **svy estimators**	Logistic regression for survey data
`svyprobt`	[R] **svy estimators**	Probit models for survey data
`svymlog`	[R] **svy estimators**	Multinomial logistic regression for survey data
`svyolog`	[R] **svy estimators**	Ordered logistic regression for survey data
`svyoprob`	[R] **svy estimators**	Ordered probit models for survey data
`svypois`	[R] **svy estimators**	Poisson regression for survey data
`svymean`	[R] **svymean**	Estimation of population and subpopulation means
`svyprop`	[R] **svymean**	Estimation of population and subpopulation proportions
`svyratio`	[R] **svymean**	Estimation of population and subpopulation ratios
`svytotal`	[R] **svymean**	Estimation of population and subpopulation totals
`svytab`	[R] **svytab**	Two-way tables for survey data
`svylc`	[R] **svylc**	Estimate linear combinations of parameters (e.g., differences of means, regression coefficients)
`svytest`	[R] **svytest**	Hypotheses tests for survey data

Before using any of the svy commands, first take a quick look at [R] **svyset**. The svyset command allows one to specify a variable containing the sampling weights, strata and PSU identifier variables, and finite-population correction variable (if any). Once set, the svy commands will automatically use these design specifications until they are cleared or changed.

After the design variables are set, the svy estimation commands can be used in a manner essentially identical to that of the corresponding nonsurvey command. The svyreg command is the parallel of regress; svyivreg corresponds to ivreg; svyintrg corresponds to intreg (and can estimate basic tobit models as well as fancier generalizations); svylogit corresponds to logit; svyprobt to probit; svymlog to mlogit; svyolog to ologit; svyoprob to oprobit; and svypois to poisson.

The [R] **svymean** entry describes the commands for the estimation of means, totals, ratios, and proportions: svymean, svytotal, svyratio, and svyprop.

The svytab command can be used to produce tests of independence for two-way contingency tables with survey data. svytab will also produce estimates of proportions with standard errors and confidence intervals.

All of the svy commands that compute standard errors use the "linearization" variance estimator—so called because it is based on a first-order Taylor series linear approximation. See the *Methods and Formulas* sections of [R] **svymean** and [R] **svy estimators** for full details.

The _robust command is a programmer's command that can be used to compute the linearization variance estimator after user-programmed estimators, such as those implemented using ml. Even if you are not interested in programming your own estimator, the [R] **_robust** entry may be worth reading since it contains an elementary introduction to the linearization variance estimator.

There are two special svy post-estimation commands: svytest (a version of test for survey data) and svylc (a version of lincom for survey data). svytest will compute p-values for multidimensional hypothesis tests after any svy estimation command (namely, the commands described in [R] **svy estimators** and [R] **svymean**).

The svylc command has many capabilities. It can be used after svymean, by() to do the equivalent of a two-sample t test for survey data. It can be used after svylogit to produce odds ratios for any linear combination of coefficients; after svymlog, it can calculate relative risk ratios; and after svypois, incidence rate ratios.

The svydes command describes the strata and primary sampling units for a survey dataset, and is useful in tracking down the cause of the error message "stratum with only one PSU detected, r(460)".

Most of the svy commands appeared in the *Stata Technical Bulletin* (STB) before they appeared in a new release of Stata, and this pattern will likely be followed in the future, with new additions to the svy commands appearing in the STB between releases. See [U] **2.4 The Stata Technical Bulletin** for more information on the STB.

30.2   Accounting for the sample design in survey analyses

Why should one use these svy commands rather than, say, the ci command for means, or regress for linear regression? To answer this question, we need to discuss some of the characteristics of survey design and survey data collection because these characteristics affect how we must do our analysis if we want to "get it right".

Survey data generally have three important characteristics:

1. sampling weights—also called probability weights,

2. clustering,

3. stratification.

These factors arise from the design of the data collection procedure. Here's a brief description of how these design features affect the analysis of the data.

1. *Sampling weights.* In sample surveys, observations are selected through a random process, but different observations may have different probabilities of selection. Weights are equal to (or proportional to) the inverse of the probability of being sampled. Various post-sampling adjustments to the weights are sometimes done as well. A weight of w_j for the jth observation means, roughly speaking, that the jth observation represents w_j elements in the population from which the sample was drawn.

   Including sampling weights in the analysis gives estimators that are approximately unbiased for whatever we are attempting to estimate in the full population. If we omit the weights in the analysis, our estimates may be very biased. Weights also affect the standard errors of our estimates.

2. *Clustering.* Observations are not sampled independently in almost all survey designs. Groups (for example, counties, city blocks, or households) may be sampled as a group—what we term a "cluster".

   There may also be further subsampling within the clusters. For example, counties may be sampled, then city blocks within counties, then households within city blocks, and then finally persons within households. The units at the first level of sampling are called "primary sampling units"—in this example, counties are the primary sampling units (PSUs). The PSUs play a special role in the analysis of the data, as we will discuss in the following subsection.

   Because of the sampling design, observations in the same cluster are not independent. If we use estimators that assume independence (e.g., `regress` or `regress, robust`), the standard errors may be too small—the difference can be as much as a factor of 2 or more. Accounting for clustering is necessary for "honest" estimates of standard errors, valid p-values, and confidence intervals whose true coverage is close to 95% (or whatever level you use).

3. *Stratification.* In surveys, different groups of clusters are often sampled separately. These groups are called strata. For example, the 254 counties of a state might be divided into two strata, say, urban counties and rural counties. Then ten counties might be sampled from the urban stratum, and fifteen from the rural stratum.

   Sampling is done independently across strata, with the stratum divisions fixed in advance. Thus, strata are statistically independent and can be analyzed as such. In many cases, this produces smaller (and honestly so) estimates of standard errors.

To put it succinctly: It is important to use sampling weights in order to get the point estimates right. We must consider the clustering and stratification of the survey design to get the standard errors right. If our analysis ignores the clustering in our design, we would likely produce standard errors that are much smaller than they should be. Stratification and weighting can also have a substantial effect on standard errors.

30.2.1   Multistage sample designs

Many surveys involve multistage sampling, i.e., cluster sampling with two or more levels of clustering. We earlier mentioned an example of multistage sampling, with selection stages at the county, city block, and household levels.

The variance estimators in the current svy commands are suitable for use with multistage sample data. Note, however, that the variance estimates are based only on computations at the primary-sampling-unit level, i.e., the first stage of sampling. They do not require information about the secondary (and beyond) sampling units.

These variance estimators make minimal assumptions about the nature of the sample. They allow any amount of correlation within the primary sampling units. Thus, elements within a primary sampling do not have to be independent; that is, there can be secondary clustering. Hence, the variance estimator of the svy commands can be used for multistage designs; they produce variance estimates that generally will be either approximately unbiased or, if biased, they will be biased toward more conservative estimates, i.e., larger standard errors.

There are other variance estimation methods (not yet implemented in Stata) that explicitly account for secondary sampling. However, these methods require more assumptions than the svy variance estimator. If these assumptions are correct, then these other methods may yield more efficient variance estimates. However, the svy estimator, because it makes fewer assumptions, can be more robust than these methods.

30.2.2   Finite population corrections

Some surveys are based on simple random sampling *without* replacement of individual persons. Without-replacement sampling generally increases the precision of our estimators. If the number of persons sampled is large relative to the total number of persons in that stratum, then this increase in precision can be substantial.

For these cases, we can reflect this increase in precision by using a "finite population correction" term in our variance estimator. The svy commands will compute this correction when an fpc variable is set using svyset.

The same applies to PSUs, but only when all the elements of PSUs are, by design, included in the study. For example, if our PSUs were households and if we included every member of the household in our study, then a finite population correction term would be appropriate when the households are sampled using simple random sampling without replacement within each stratum.

If, however, we subsampled within the household, the finite population correction calculation of the svy commands should not be used. Thus, for multistage sampling, the fpc option should not be specified with the svy commands, no matter how the sampling was conducted.

30.2.3   Design effects: deff and deft

In addition to handling stratification and clustering effects, the svy commands have another special feature that other Stata estimation commands do not have: they calculate the design effects deff and deft.

The design effect deff is equal to the design-based variance estimate divided by an estimate of the variance we would have obtained if we had carried out a similar survey using simple random sampling. This ratio is a measure of how the survey design affects variance estimates. The related measure deft is approximately equal to the square root of deff.

See [R] **svymean** for detailed descriptions of deff and deft.

30.3   Example of the effects of weights, clustering, and stratification

Below we present an example of the effects of weights, clustering, and stratification. This is a typical case, but it is still, nevertheless, dangerous to draw general rules from any single example. One could find particular analyses from other surveys that are counterexamples for each of the trends for standard errors that we exhibit here.

We use data from the Second National Health and Nutrition Examination Survey (NHANES II) (McDowell et al. 1981) as our example. This is a national survey, and the dataset has sampling weights, strata, and clustering. In this example, we will consider the estimation of the mean serum zinc level of all adults in the U.S.

First, consider a proper design-based analysis, which accounts for weighting, clustering, and stratification. Before we issue our svy estimation command, we set the weight, strata, and PSU identifier variables:

```
. svyset pweight finalwgt
. svyset strata stratid
. svyset psu psuid
```

We now estimate the mean using the proper design-based analysis:

```
. svymean zinc

Survey mean estimation

pweight: finalwgt                         Number of obs    =       9202
Strata:  stratid                          Number of strata =         31
PSU:     psuid                            Number of PSUs   =         62
                                          Population size  = 1.043e+08

------------------------------------------------------------------------
    Mean |  Estimate   Std. Err.   [95% Conf. Interval]        Deff
---------+--------------------------------------------------------------
    zinc |  87.17107    .493689    86.16418   88.17795      10.33092
------------------------------------------------------------------------
```

If we ignore the survey design and use ci to estimate the mean, we get

```
. ci zinc

Variable |      Obs        Mean    Std. Err.       [95% Conf. Interval]
---------+--------------------------------------------------------------
    zinc |     9202    86.50739    .1509294        86.21153    86.80324
```

Note that the point estimate from the unweighted analysis is smaller by more than one standard error from the proper design-based estimate. In addition, note that design-based analysis produced a standard error that is 3.27 times larger than the standard error produced by our incorrect ci analysis.

30.3.1   Halfway isn't enough: the importance of stratification and clustering

When some people analyze survey data, they say, "I know I have to use my survey weights, but I'll just ignore the stratification and clustering information." If we follow this strategy, we will obtain the proper design-based point estimates, but our standard errors, confidence intervals, and test statistics will usually be wrong.

To illustrate this, suppose we used the svymean procedure with pweights only.

```
. svyset, clear
. svyset pweight finalwgt
```

```
. svymean zinc
Survey mean estimation
pweight:  finalwgt               Number of obs     =      9202
Strata:   <one>                  Number of strata  =         1
PSU:      <observations>         Number of PSUs    =      9202
                                 Population size    = 1.043e+08
-------------------------------------------------------------------------
   Mean |  Estimate   Std. Err.  [95% Conf. Interval]       Deff
--------+----------------------------------------------------------------
   zinc |  87.17107   .1827547   86.81283   87.52931      1.415693
-------------------------------------------------------------------------
```

This gives us the same point estimate as our design-based analysis, but the reported standard error is less than one-half of the design-based standard error.

In addition, we emphasize that it is important to account for stratification, as well as weighting and clustering, in the analysis of survey data. For example, if we only accounted for clustering and weights, and ignored stratification in NHANES II, we would obtain the following analysis.

```
. svyset psu psuid
. svyset
pweight is finalwgt
psu is psuid

. svymean zinc
Survey mean estimation
pweight:  finalwgt               Number of obs     =      9202
Strata:   <one>                  Number of strata  =         1
PSU:      psuid                  Number of PSUs    =        62
                                 Population size    = 1.043e+08
-------------------------------------------------------------------------
   Mean |  Estimate   Std. Err.  [95% Conf. Interval]       Deff
--------+----------------------------------------------------------------
   zinc |  87.17107   .442225    86.28678   88.05535      8.289311
-------------------------------------------------------------------------
```

Here our standard error is still about 10% different from what we obtained in our proper design-based analysis.

❏ Technical Note

Although ci with aweights gives the same point estimates as svymean with pweights, it gives different standard errors.

```
. ci zinc [aweight=finalwgt]
Variable |     Obs       Mean    Std. Err.    [95% Conf. Interval]
---------+-----------------------------------------------------------
   zinc |    9202    87.17107    .1535975     86.86998   87.47215
```

svymean with pweights gave a slightly different standard error of 0.183.

ci with aweights uses a different formula for the standard error. svymean with pweights handles sampling weights properly. Indeed, any command in Stata that allows pweights handles pweights properly in this regard. But, as we illustrated above, it is also of critical importance to include clustering and stratification in the analysis.

❏

30.4   Linear regression and other models

Let's look at a regression. We model zinc based on age, sex, race, and rural or urban residence. We compare a proper design-based analysis with an ordinary regression (which assumes i.i.d. error).

Here is our design-based analysis:

```
. svyreg zinc age age2 weight female black orace rural

Survey linear regression

pweight:  finalwgt                        Number of obs    =        9202
Strata:   stratid                         Number of strata =          31
PSU:      psuid                           Number of PSUs   =          62
                                          Population size  = 1.043e+08
                                          F(  7,     25)   =       62.90
                                          Prob > F         =      0.0000
                                          R-squared        =      0.0699

------------------------------------------------------------------------------
     zinc |     Coef.   Std. Err.      t    P>|t|     [95% Conf. Interval]
----------+-------------------------------------------------------------------
      age |  -.1728042   .0847617    -2.039   0.050    -.3456769    .0000685
     age2 |   .0008989   .0008692     1.034   0.309    -.0008737    .0026716
   weight |   .0538921   .0139968     3.850   0.001     .0253455    .0824387
   female |  -6.132373   .4397907   -13.944   0.000    -7.029332   -5.235414
    black |  -2.873718   1.076066    -2.671   0.012     -5.06837    -.679066
    orace |  -4.105586   1.619298    -2.535   0.016    -7.408166   -.8030067
    rural |  -.5323477   .6168877    -0.863   0.395    -1.790498    .7258029
    _cons |   92.50021   2.226381    41.547   0.000     87.95948    97.04094
------------------------------------------------------------------------------
```

If we had improperly ignored our survey weights, stratification, and clustering (i.e., if we used the usual Stata **regress** command), we would have obtained the following results:

```
. reg zinc age age2 weight female black orace rural

    Source |       SS       df       MS              Number of obs =    9202
-----------+------------------------------           F(  7,  9194) =   79.82
     Model |  110497.994     7  15785.4277           Prob > F      =  0.0000
  Residual |  1818204.00  9194  197.759844           R-squared     =  0.0573
-----------+------------------------------           Adj R-squared =  0.0566
     Total |  1928702.00  9201  209.618737           Root MSE      =  14.063

------------------------------------------------------------------------------
     zinc |     Coef.   Std. Err.      t    P>|t|     [95% Conf. Interval]
----------+-------------------------------------------------------------------
      age |  -.0929365   .0637866    -1.457   0.145    -.2179725    .0320994
     age2 |  -7.25e-06   .0006782    -0.011   0.991    -.0013366    .0013221
   weight |   .0608037   .0105845     5.745   0.000     .0400558    .0815517
   female |  -5.014644   .319133    -15.713   0.000    -5.640215   -4.389072
    black |  -2.304524   .5071736    -4.544   0.000    -3.298697   -1.310351
    orace |   -3.38212   1.060691    -3.189   0.001    -5.461309   -1.302931
    rural |  -.0924905   .3095838    -0.299   0.765    -.6993434    .5143624
    _cons |    89.532    1.476464    60.639   0.000      86.6378    92.42619
------------------------------------------------------------------------------
```

The point estimates differ by 3–100% and the standard errors for the proper designed-based analysis are 30–110% larger. The differences are not as dramatic as we saw with the estimation of the mean, but are still very substantial.

30.5   Hypothesis testing using svytest

After any of the `svy` estimation commands, you should use `svytest`, rather than `test`, for hypothesis testing. The `svytest` command computes an adjusted Wald F test. This adjustment to the Wald test statistic is a simple multiplicative adjustment that accounts for the sample design. See [R] **svytest** for formulas, references, and a description of the use of `svytest`.

Consider our previous example:

```
. svyreg zinc age age2 weight female black orace rural
Survey linear regression
pweight:  finalwgt                    Number of obs    =       9202
Strata:   stratid                     Number of strata =         31
PSU:      psuid                       Number of PSUs   =         62
                                      Population size  = 1.043e+08
                                      F(   7,     25)  =      62.90
                                      Prob > F         =     0.0000
                                      R-squared        =     0.0699
```

zinc	Coef.	Std. Err.	t	P>|t|	[95% Conf.	Interval]
age	-.1728042	.0847617	-2.039	0.050	-.3456769	.0000685
age2	.0008989	.0008692	1.034	0.309	-.0008737	.0026716
weight	.0538921	.0139968	3.850	0.001	.0253455	.0824387
female	-6.132373	.4397907	-13.944	0.000	-7.029332	-5.235414
black	-2.873718	1.076066	-2.671	0.012	-5.06837	-.679066
orace	-4.105586	1.619298	-2.535	0.016	-7.408166	-.8030067
rural	-.5323477	.6168877	-0.863	0.395	-1.790498	.7258029
_cons	92.50021	2.226381	41.547	0.000	87.95948	97.04094

Since we fit an age term (`age`) and an age-squared term (`age2`), it would be reasonable to carry out a joint test for these coefficients:

```
. svytest age age2
Adjusted Wald test
 ( 1)   age = 0.0
 ( 2)   age2 = 0.0
        F(  2,     30) =    27.20
            Prob > F =     0.0000
```

Tested jointly, we see that the age terms have a p-value of less than 0.0001.

`svytest` can be used after any of the model estimation commands described in [R] **svy estimators**. It can also be used after `svymean`, `svytotal`, and `svyratio`. See [R] **svytest** for details.

30.6   Pseudo likelihoods

Many of the `svy` estimators are based on maximum-likelihood estimators; namely, `svylogit`, `svyprobt`, `svymlog`, `svyolog`, `svyoprob`, `svypois`, and `svyintrg`. But it should be noted that the "likelihood" for weighted or clustered designs is not a true likelihood; that is, the "likelihood" is not the distribution of the sample. When there is clustering, individual observations are no longer independent, and the "likelihood" does not reflect this. When there are sampling weights, the "likelihood" does not fully account for the "randomness" of the weighted sampling.

The "likelihood" for the `svy` maximum-likelihood estimators is used only for the computation of the point estimates. It should not be used for variance estimation using standard formulas, and the standard likelihood-ratio test should not be used after the `svy` maximum-likelihood estimators. This is why the `svy` maximum-likelihood commands do not display or save the log likelihood. Instead of likelihood-ratio tests, `svytest` should be used to calculate adjusted Wald tests.

30.7 Estimation of differences and other linear combinations of parameters

In addition to `svytest`, there is another post-estimation `svy` command called `svylc`. `svylc` will display an estimate of a linear combination of parameters, along with its standard error, a confidence interval, and a test that the difference is zero. `svylc` is a `svy` version of the `lincom` command (see [R] **lincom**).

Most commonly, `svylc` is used to compute the differences of two subpopulation means. For example, suppose we wish to estimate the difference of zinc levels in white males versus black males. First, we estimate the subpopulation means:

```
. svymean zinc, by(race) subpop(male)

Survey mean estimation

pweight:  finalwgt                      Number of obs   =       9202
Strata:   stratid                       Number of strata =        31
PSU:      psuid                         Number of PSUs  =         62
Subpop.:  male==1                       Population size = 1.043e+08

------------------------------------------------------------------------------
Mean    Subpop. |   Estimate    Std. Err.   [95% Conf. Interval]        Deff
----------------+-------------------------------------------------------------
zinc            |
         White  |   91.15725     .541625     90.0526     92.2619     4.816818
         Black  |    88.269     1.208336    85.80458    90.73342     2.475099
         Other  |   85.54716    2.608974    80.22612     90.8682     3.147972
------------------------------------------------------------------------------
```

Then we run `svylc`:

```
. svylc [zinc]White - [zinc]Black
 ( 1)  [zinc]White - [zinc]Black = 0.0

------------------------------------------------------------------------------
    Mean |   Estimate    Std. Err.       t     P>|t|      [95% Conf. Interval]
---------+--------------------------------------------------------------------
     (1) |   2.888249    1.103999    2.616    0.014      .6366288    5.139868
------------------------------------------------------------------------------
```

Note that the t statistic and its p-value give a survey analysis equivalent of a two-sample t test.

`svylc` can also be used after any of the model estimation commands described in [R] **svy estimators**. It can, for example, display results as odds ratios after `svylogit`, and can be used to compute odds ratios for one covariate group relative to another. See [R] **svylc** for full details.

30.8 Two-way contingency tables

The `tabulate` command with `iweight`s will produce tabulations for weighted data. It will not, however, compute tests of independence for two-way tables. For this, you should use the `svytab` command. The `svytab` command produces tests of independence appropriate for complex survey data.

Here is an example of its use:

```
. svyset strata stratid
```

```
. svyset psu psuid

. svyset pweight finalwgt

. svytab race diabetes, row
---------------------------------------------------------------------
pweight:  finalwgt                  Number of obs     =      10349
Strata:   stratid                   Number of strata  =         31
PSU:      psuid                     Number of PSUs    =         62
                                    Population size    = 1.171e+08
    ---------------------------------------------------------------

    ----------+--------------------
            |       Diabetes
      Race  |    no    yes  Total
    ----------+--------------------
     White  |  .968   .032      1
     Black  |  .941   .059      1
     Other  | .9797  .0203      1
            |
     Total  | .9658  .0342      1
    ----------+--------------------
    Key:  row proportions

    Pearson:
       Uncorrected   chi2(2)      =    21.3483
       Design-based  F(1.52,47.26) =   15.0056      P = 0.0000
```

Actually, `svytab` has several capabilities that `tabulate` does not have. Not only can it display proportions (or percentages), but it can compute standard errors and confidence intervals for these proportions as well:

```
. svytab race diabetes, row se ci format(%7.4f)
---------------------------------------------------------------------
pweight:  finalwgt                  Number of obs     =      10349
Strata:   stratid                   Number of strata  =         31
PSU:      psuid                     Number of PSUs    =         62
                                    Population size    = 1.171e+08
    ---------------------------------------------------------------

    ----------+-----------------------------------------------------
            |                    Diabetes
      Race  |              no               yes           Total
    ----------+-----------------------------------------------------
     White  |          0.9680            0.0320          1.0000
            |         (0.0020)          (0.0020)
            | [0.9638,0.9718]   [0.0282,0.0362]
            |
     Black  |          0.9410            0.0590          1.0000
            |         (0.0061)          (0.0061)
            | [0.9271,0.9523]   [0.0477,0.0729]
            |
     Other  |          0.9797            0.0203          1.0000
            |         (0.0076)          (0.0076)
            | [0.9566,0.9906]   [0.0094,0.0434]
            |
     Total  |          0.9658            0.0342          1.0000
            |         (0.0018)          (0.0018)
            | [0.9619,0.9693]   [0.0307,0.0381]
    ----------+-----------------------------------------------------
    Key:  row proportions
          (standard errors of row proportions)
          [95% confidence intervals for row proportions]

    Pearson:
```

```
Uncorrected    chi2(2)        =   21.3483
Design-based   F(1.52,47.26)  =   15.0056      P = 0.0000
```

The test of independence that is displayed by default is based on the usual Pearson χ^2 statistic for two-way tables. To account for the survey design, the statistic is turned into an F-statistic with noninteger degrees of freedom using a second-order Rao and Scott (1984) correction.

svytab will actually compute a total of eight statistics for the test of independence. It will compute a Rao-and-Scott corrected Pearson statistic and a Rao-and-Scott corrected likelihood-ratio statistic, using either of two variants of the correction, yielding four of the statistics. It will compute a "Pearson" Wald statistic and a log-linear Wald statistic, both either adjusted or unadjusted, yielding the other four statistics.

This dizzying array of statistics is not intended to dazzle the user. The two Wald statistics in their unadjusted form have been in use for many years, and so we felt compelled to implement them in Stata. In many situations, however, they possess poor statistical properties. The adjusted variants of these Wald statistics have better statistical properties, but based on simulations (Sribney 1998) they do not appear to be as good as the best Rao-and-Scott corrected statistic, which is the statistic displayed by default. Hence, we advise researchers to use the default statistic in all situations, and conversely, we recommend that the other statistics only be used for comparative or pedagogical purposes.

A summary of the properties of these statistics is given in the [R] **svytab** entry, but anyone wishing to see the results of the simulations should look at the article by Sribney (1998) that appeared in the *Stata Technical Bulletin.*

30.9   Differences between the svy commands and other commands

As we mentioned earlier, all of the svy model estimation commands (those commands documented in [R] **svy estimators**) have corresponding non-svy commands: svyreg corresponds to regress, svylogit to logit, etc. The corresponding non-svy commands all allow pweights and have a cluster() option that corresponds to the psu() option of the svy commands. When pweights, cluster(), or the robust option is specified, the non-svy commands use the same robust (linearization) variance estimator that is used by the svy commands.

The point estimates from the svy commands are exactly the same as the weighted point estimates from the non-svy commands. Indeed, this must be the case since the svy commands call the corresponding non-svy commands to compute the point estimates.

Despite their many similarities, the svy commands and corresponding non-svy commands have a number of differences:

1. All of the svy commands handle stratified sampling, but none of the non-svy commands do. Since stratification usually makes standard errors smaller, ignoring stratification is usually conservative. So using the non-svy commands for stratified sampling is not a terrible thing to do. However, in order to get the smallest possible "honest" standard error estimates for stratified sampling, one should use the svy commands.

2. All of the svy commands use t-statistics with $n - L$ degrees of freedom for the testing the significance of coefficients, where n is the total number of sampled PSUs (clusters) and L is the number of strata. Some of the non-svy commands use t-statistics, but most use z-statistics. If the non-svy command uses z-statistics for its standard variance estimator, then it also uses z-statistics with the robust (linearization) variance estimator. Strictly speaking, t-statistics are always appropriate with the robust (linearization) variance estimator; see [R] **_robust** for the theoretical rationale. But using z rather than t-statistics only yields a nontrivial difference when there is a small number of clusters (< 50). If a non-svy command uses t-statistics and the

`cluster()` option is specified, then the degrees of freedom used are the same as that of an `svy` command.

3. `svy` commands produce adjusted Wald tests for the model test, and `svytest` can be used to produce adjusted Wald tests for other hypotheses. Non-`svy` commands can only produce unadjusted Wald tests. The adjustment can be important when the degrees of freedom $n - L$ are small relative to the dimension of the test. (If the dimension is one, then the adjusted and unadjusted Wald tests are identical.) This fact along with point (2) makes it important to use the `svy` command if the number of sampled PSUs (clusters) is small (< 50).

4. `svyreg` differs slightly from `regress` and `svyivreg` differs slightly from `ivreg` in that they use different multipliers for the variance estimator. `regress` and `ivreg` use a multiplier of $((N - 1)/(N - k))(n/(n - 1))$, where N is the number of observations, n is the number of clusters (PSUs), and k is the number of regressors including the constant. `svyreg` and `svyivreg` use a multiplier of merely $n/(n - 1)$. Thus, they produce slightly different standard errors. The $(N - 1)/(N - k)$ is ad hoc and has no rigorous theoretical justification; hence, the purist `svy` commands do not use it. The `svy` commands tacitly assume that $N \gg k$. If $(N - 1)/(N - k)$ is not close to 1, then you may be well advised to use `regress` or `ivreg`, so that some punishment is inflicted on your variance estimates. Note that maximum-likelihood estimators in Stata (e.g., `logit`) do no such adjustment, but rely on the sensibilities of the analyst to ensure that N is reasonably larger than k. Thus, the `svy` maximum-likelihood estimators (e.g., `svylogit`) produce exactly the same standard errors as the corresponding non-`svy` commands (e.g., `logit`)—but p-values are slightly different because of point (2).

5. `svy` commands can produce proper estimates for subpopulations using the `subpop()` option. Use of an `if` restriction with `svy` or non-`svy` commands can yield incorrect standard error estimates for subpopulations. Many times an `if` restriction will yield exactly the same standard error as `subpop()`; most other times the two standard errors will be slightly different; but, in some cases—usually for thinly sampled subpopulations—the standard errors can be appreciably different. Hence, only `svy` commands with the `subpop()` option should be used to obtain estimates for thinly sampled subpopulations. See [R] **svymean** for more information.

6. `svy` commands handle zero sampling weights properly. Non-`svy` commands ignore any observation with a weight of zero. Most times this will yield exactly the same standard errors, but sometimes they will differ. Sampling weights of zero can arise from various post-sampling adjustment procedures. If the sum of weights for one or more PSUs is zero, then `svy` and non-`svy` commands will produce different standard errors; but usually this difference is very small.

7. The `svy` commands that allow `iweight`s will let these weights be negative. Negative sampling weights can arise from various post-sampling adjustment procedures. If you want to use negative sampling weights, then you must use `svy` commands with `iweight`s; no other commands will allow negative sampling weights.

8. Only the `svy` commands will compute finite population corrections (FPC). Finite population corrections are only justified for some special sampling designs. Omitting the FPC is conservative, so failing to specify an FPC cannot be harshly condemned. See [U] **30.2.2 Finite population corrections** earlier in this chapter for details.

9. Only the `svy` commands will compute the design effects deff and deft.

30.10 References

McDowell, A., A. Engel, J. T. Massey, and K. Maurer. 1981. Plan and operation of the Second National Health and Nutrition Examination Survey, 1976–1980. *Vital and Health Statistics* 15(1). National Center for Health Statistics, Hyattsville, MD.

Rao, J. N. K. and A. J. Scott. 1984. On chi-squared tests for multiway contingency tables with cell proportions estimated from survey data. *Annals of Statistics* 12: 46–60.

Sribney, W. M. 1998. svy7: Two-way contingency tables for survey or clustered data. *Stata Technical Bulletin* 45: 33–49.

31 Commands everyone should know

Contents

31.1   Thirty-seven commands

Putting aside the statistical commands that might particularly interest you, here are 37 commands everyone should know:

Getting on-line help
  search, help   [U] **8 Stata's on-line help and search facilities**

Operating system interface
  pwd, cd   [R] **cd**

Using and saving data from disk
  use, save   [R] **save**
  append, merge   [U] **25 Commands for combining data**
  compress   [R] **compress**

Inputting data into Stata   [U] **24 Commands to input data**
  input   [R] **input**
  edit   [R] **edit**
  infile   [R] **infile (free format)**; [R] **infile (fixed format)**
  infix   [R] **infix (fixed format)**
  insheet   [R] **insheet**

Basic data reporting
  describe   [R] **describe**
  codebook   [R] **codebook**
  list   [R] **list**
  browse   [R] **edit**
  count   [R] **count**
  inspect   [R] **inspect**
  table   [R] **table**
  tabulate   [R] **tabulate**

Data manipulation   [U] **16 Functions and expressions**
  generate, replace   [R] **generate**
  egen   [R] **egen**
  rename   [R] **rename**
  drop, keep   [R] **drop**
  sort   [R] **sort**
  encode, decode   [R] **encode**
  order   [R] **order**
  by   [U] **14.5 by varlist: construct**
  reshape   [R] **reshape**

Keeping track of your work
    `log`   [U] **18 Printing and preserving output**
    `notes`   [R] **notes**

Convenience
    `display`   [R] **display**

31.2   The by construct

If you do not understand the **by** *varlist*: construct, _n, and _N, and their interaction, and if you process data where observations are related, you are missing out on something. See

[U] **16.7 Explicit subscripting**
[U] **14.5 by varlist: construct**

For instance, say you have a dataset with multiple observations per person, and you want the average value of each person's blood pressure (**bp**) for the day. You could

```
. egen avgbp = mean(bp), by(person)
```

but you should understand that you could also

```
. sort person
. by person: gen avgbp = sum(bp)/_N
. by person: replace avgbp = bp[_N]
```

Yes, typing three commands is more work than one, but understanding the three-command construct is the key to generating more complicated things that no one ever thought about adding to **egen**.

For instance, say your data also contains **time** recording when each observation was made. If you want to add the total time the person is under observation (last time minus first time) to each observation:

```
. sort person time
. by person: gen ttl = time[_N]-time[1]
```

Or suppose you want to add how long it has been since the person was last observed to each observation:

```
. sort person time
. by person: gen howlong = time - time[_n-1]
```

If instead you wanted how long it would be until the next observation:

```
. sort person time
. by person: gen whennext = time[_n+1] - time
```

by *varlist*:, _n, and _N are often the solution to many difficult calculations.

32 Using the Internet to keep up to date

Contents

This chapter concerns features of Stata for Windows 98/95/NT, Stata for Power Macintosh, and Stata for Unix, which is to say, not Stata for Windows 3.1 and not Stata for 680x0 Macintosh.

32.1 Overview

Stata has the ability to read files over the Internet. Just to prove that to yourself, type the following,

```
. use http://www.stata.com/manual/chapter32, clear
```

You have just reached out and gotten a dataset that we put on our web site. The dataset is not in HTML format, nor does this have anything to do with how browsers work. We just copied the Stata data file `chapter32.dta` onto our server and now people all over the world can use it. If you have a home page, you can do the same thing. It is a very convenient way of sharing datasets with colleagues.

Now type the following:

```
. update from http://www.stata.com
```

We promise nothing bad will happen. `update` will read a short file from www.stata.com that will allow Stata to report whether your copy of Stata is up to date. Is your copy up to date? Now you know. If it is not, we will show you how to update it—it is no more difficult than it was to type `update`.

Now type the following:

```
. net from http://www.stata.com
```

That will go to www.stata.com and tell you what is available from our user-download site. The material there is not official, but it is useful.

To summarize, Stata has the ability to read files over the Internet:

1. You can share datasets, do-files, etc., with colleagues all over the world. This requires no special expertise, but you do need to have a home page.

2. You can update Stata; it is free, easy, and nearly instant.

3. You can find and add new features to Stata; it is also free, easy, and nearly instant.

Finally, you can create a site to distribute new features for Stata.

32.2   Sharing datasets (and other files)

There is just nothing to it: you copy the file as-is (in binary) onto the server and then let your colleagues know the file is there. This works for .dta files, .do files, .ado files, and in fact, all files.

On the receiving end, you can use the file (if it is a .dta dataset) or you can copy it:

```
. use http://www.stata.com/manual/chapter32, clear
. copy http://www.stata.com/manual/chapter32.dta mycopy.dta
```

Stata includes a copy-file command and it works over the Internet just as use does; see [R] **copy**.

❏ Technical Note

If you are concerned about transmission errors, you can create a checksum file before you copy the file onto the server. In placing chapter32.dta on our site, we started with chapter32.dta in our working directory and typed

```
. checksum chapter32.dta, save
```

This created new file chapter32.sum. We then placed both files on our server. We did not have to create this second file but, since we did, when you use the data, Stata will be able to detect transmission errors and warn you if there are problems.

How would Stata know? chapter32.sum is a very short file containing the result of a mathematical calculation made on the contents of chapter32.dta. When your Stata receives chapter32.dta, it repeats the calculation and then compares that result with what is recorded in chapter32.sum. If the results are different, then there must have been a transmission error.

Whether you create a checksum file is optional.

See [R] **checksum**.

❏

32.3   Official updates

We distribute updates every other month, in January, March, May, July, September, and November. You do not have to update every other month—although we recommend you do. There are two ways to check whether your copy of Stata is up to date:

type

```
. update from http://www.stata.com
```

or

Pull down **Help** and select **Official Updates**
Click on *http://www.stata.com*

After that, you will either

type: or:
```
. update ado
```
click on *update ado-files*

or

> type: or:
> . update executable   click on *update executable*

or

> type: or:
> . update all   click on *update ado-files and executable*

and which, if any of those things need doing will be obvious.

After you have updated your Stata, to find out what has changed

> type: or:
> . help whatsnew   Pull down **Help** and select **What's New?**

32.3.1 Example

When you type `update from http://www.stata.com` or when you pull down **Help**, select
Official Updates, and click on *http://www.stata.com*, Stata presents a report:

```
. update from http://www.stata.com
(contacting http://www.stata.com)

Stata executable
    folder:               C:\STATA\
    name of file:         wstata.exe
    currently installed:  04 Jan 1999
    latest available:     04 Jan 1999

Ado-file updates
    folder:               C:\STATA\ado\updates\
    names of files:       (various)
    currently installed:  01 Mar 1999
    latest available:     01 Mar 1999

Recommendation
    Do nothing; all files up-to-date.
```

There are two components of official Stata: the binary Stata executable and the ado-files that we
shipped with it. Ado-files are just programs written in Stata. For instance, when you use `generate`,
you are using a command that was compiled into the Stata executable. When you use `stcox`, you
are using a command that was implemented as an ado-file.

Both components of our Stata are up to date.

32.3.2 Updating ado-files

When you obtain the above report, you might see

```
. update from http://www.stata.com
(contacting http://www.stata.com)

Stata executable
    folder:               C:\STATA\
    name of file:         wstata.exe
    currently installed:  04 Jan 1999
    latest available:     04 Jan 1999
```

```
Ado-file updates
    folder:                C:\STATA\ado\updates\
    names of files:        (various)
    currently installed:   01 Mar 1999
    latest available:      01 May 1999
Recommendation
    Type -update ado-
```

If you go with the point-and-click alternative, at the bottom of the screen you will see

```
Recommendation
    update ado-files
```

where *update ado-files* is in green and therefore clickable.

Anyway, what you are to do next is type '`update ado`' or click on *update ado-files*. Either way, you will see something like the following:

```
. update ado
(contacting http://www.stata.com)
Ado-file update log
    1.  verifying C:\STATA\ado\updates\ is writeable
    2.  obtaining list of files to be updated
    3.  downloading relevant files to temporary area
        downloading filename.ado
        downloading filename.hlp
        ...
        downloading filename.ado
    4.  examining files
    5.  installing files
    6.  setting last date updated
Updates successfully installed.
Recommendation
    Type -help whatsnew- to learn about the new features
```

That is all there is to it, but do type `help whatsnew` to learn about the new features. (If you go the point-and-click path, you click on *whatsnew*.)

Here is what happens if you type '`update ado`' and you are already up to date:

```
. update ado
(contacting http://www.stata.com)
ado-files already up to date
```

32.3.3 Frequently asked questions about updating the ado-files

1. Is there any danger that something could go wrong and my Stata become unusable?

   No. The updates are copied to a temporary place on your computer, Stata examines them to make sure they are complete and whole before copying them to the official place. Thus, the updates are either installed or they are not.

2. I do not believe you. Pretend something you did not anticipate goes wrong such as the power fails at the instant Stata is doing the local disk to local disk copy.

   If the improbable should happen, you can erase the update directory and then your Stata is back to being just as it was shipped. Updates go into a different directory than the originals and the originals are never erased.

   Stata tells you where it is installing your updates. You can also find out by typing `sysdir`. The directory you want is the one listed opposite UPDATES.

(By the way, power failure should not cause a problem; the marker that the update is applied is set last, so you could also just type 'update ado' again and Stata would refetch the partially installed update.)

3. How much is downloaded?

A typical update is 100k to 300k. Ado-files are small; the biggest file that is copied is probably the database for search.

4. I am using Unix or a networked version of Stata. When I try to update ado, I am told that the directory is not writable. Can I copy the updates into another directory and then copy them to the official directory myself?

Yes, assuming you are a system administrator. Type 'update ado, into(*dirname*)'. Stata will download the updates just as it would ordinarily but place them in the directory you specify. We recommend *dirname* be a new, empty directory, because later you will need to copy the entire contents of the directory to the official place. The official place is the directory listed next to UPDATES if you type 'sysdir'. When you copy the files, copy over any existing files. Previously existing files in the official update directory are just previous updates. Also remember to make the files globally readable if that is necessary. See [R] **update**.

32.3.4   Updating the executable

Ado-file updates are released every other month; updates for the executable are rarer than that and, in fact, they are rare indeed. If the executable needs updating, Stata will mention it when you type update:

```
. update from http://www.stata.com
(contacting http://www.stata.com)

Stata executable
      folder:              C:\STATA\
      name of file:        wstata.exe
      currently installed: 04 Jan 1999
      latest available:    12 Feb 1999

Ado-file updates
      folder:              C:\STATA\ado\updates\
      names of files:      (various)
      currently installed: 01 Mar 1999
      latest available:    01 May 1999

Recommendation
      Type -update executable-
```

Here is what happens when you type 'update executable':

```
. update executable
(contacting http://www.stata.com)

Executable update log
      1.   verifying C:\STATA\ is writeable
      2.   downloading new executable

New executable successfully downloaded

Instructions
      1.   Exit Stata
      2.   Change to C:\STATA\
      3.   Rename wstata.exe to wstata.old
      4.   Rename wstata.bin to wstata.exe
      5.   Try Stata
      6.   Later, erase wstata.old if satisfied
           or rename wstata.old back to wstata.exe
```

Just follow the instructions, which instructions will vary depending on the computer you are using.

32.3.5   Frequently asked questions about updating the executable

1. If I understand this, `update executable` does not really install the update; it just copies one file onto my computer, and that one file happens to be the new executable, right?

   Probably. There can be more than one file such as a DLL. All the files are copied to the same place.

2. How big is the downloaded file?

   Roughly 1 megabyte.

3. What happens if I type '`update executable`' and my executable is already up to date?

   Nothing. You are told "executable already up to date".

4. I am using Unix or a networked version of Stata. When I try to `update executable`, I am told that the directory is not writable. Can I download the updated executable to another directory and then copy it to the official directory myself?

   Yes, assuming you are a system administrator.
   Type '`update executable, into(`*dirname*`)`'. We recommend *dirname* be a new, empty directory, because there may be more than one file and later you will need to copy all of them to the official place. The official place is the directory listed next to STATA if you type '`sysdir`'. When you copy the files, copy over any existing files; we recommend you make a backup of the originals first. See [R] **update**.

32.3.6   Updating both ado-files and the executable

When you type `update`, you may be told you need to update both the ado-files and executable:

```
. update from http://www.stata.com
(contacting http://www.stata.com)

Stata executable
    folder:               C:\STATA\
    name of file:         wstata.exe
    currently installed:  04 Jan 1999
    latest available:     22 Apr 1999
Ado-file updates
    folder:               C:\STATA\ado\updates\
    names of files:       (various)
    currently installed:  01 Mar 1999
    latest available:     01 May 1999
Recommendation
    Type -update all-
```

Typing '`update all`' is the same as typing '`update ado`' followed by typing '`update executable`'. You could type the separate commands if you preferred. The order in which you do them does not matter.

Note, you could skip the `update from` step. You could just type '`update all`' and follow the instructions. If nothing needed updating, you would see

```
. update all
-------------------------------------------------------------------------------
> update ado
(contacting http://www.stata.com)
ado-files already up to date
-------------------------------------------------------------------------------
> update executable
(contacting http://www.stata.com)
executable already up to date

. _
```

32.4   Downloading and managing additions by users

Try the following:

type

> . net from http://www.stata.com

or

> Pull down **Help** and select **STB and User-written Programs**
> Click on *http://www.stata.com*

32.4.1   Downloading files

We are not the only ones developing additions to Stata. Stata is supported by a large and competent user community. An important part of this is the *Stata Technical Bulletin* (STB). The STB is published bimonthly—in January, March, May, July, September, and November—and consists of two parts—a written publication and corresponding software. You must subscribe to the journal if you want the written publication, but the software is freely available over the Internet. Stata can download and install that software.

Let us show you how easy this is. Stata has no command called `concord`. Convince yourself of that:

```
. concord
unrecognized command
r(199);
```

Now do one of the following. If you are using Stata for Windows or Stata for Macintosh,

1. Pull down **Help** and select **STB and User-written Programs**.

2. Click on *www.stata.com*.

3. Click on *stb*.

4. Click on *stb43*.

5. Go to the bottom of the list and click on *sg84*.

6. Click on *click here to install*.

You just installed an STB insert; you now have the `concord` command. If you use Stata for Unix or just prefer command mode, instead of pointing and clicking, do the following:

1. Type: `. net from http://www.stata.com`

2. Type: `. net cd stb`

3. Type: `. net cd stb43`

4. Type: `. net describe sg84`

5. Type: `. net install sg84`

32.4.2　Managing files

You now have the `concord` command because we just downloaded and installed it. Convince yourself of this by typing

```
. help concord
```

and you might try it out, too. Let's now list the additions you have installed—that is probably just `concord`—and then get rid of `concord`.

In command mode, you can type

```
. ado dir
[1] package sg84 from http://www.stata.com/stb/stb43
      STB-43 sg84.  Concordance correlation coefficient.
```

If you had more additions installed, they would be listed. Now knowing that you have *sg84* installed, you can obtain a more thorough description by typing

```
. ado describe sg84
----------------------------------------------------------------------------
[1] package sg84 from http://www.stata.com/stb/stb43
----------------------------------------------------------------------------
TITLE
      STB-43 sg84.  Concordance correlation coefficient.
DESCRIPTION/AUTHOR(S)
      STB insert by
      Thomas J. Steichen, RJRT,
      Nicholas J. Cox, University of Durham, UK.
      After installation, see help concord.
INSTALLATION FILES
      c/concord.ado
      c/concord.hlp
INSTALLED ON
      5 Oct 1999
----------------------------------------------------------------------------
```

You can erase *sg84* by typing

```
. ado uninstall sg84
package sg84 from http://www.stata.com/stb/stb43
      STB-43 sg84.  Concordance correlation coefficient.
(package uninstalled)
```

You can do all of this from the point-and-click interface, too. Pull down **Help** and select **STB and User-written Programs** and then click on *List*. From there, you can click on *sg84* to see the detailed description of the package and from there you can click on *click here to uninstall* if you want to erase it.

For more information on the `ado` command and the corresponding pulldown, see [R] **net**.

32.4.3   Finding files to download

There are two ways to find useful files to download. One is simply to thumb through sites. That is inefficient but entertaining. If you want to do that,

1. Pull down **Help** and choose **STB and User-written Programs**.

2. Click on *http://www.stata.com*.

3. Click on *links*.

What you are doing is starting at our download site and then working out from there. We maintain a list of a few other sites and those sites will have more links. You can do this from command mode, too:

```
. net from http://www.stata.com
. net cd links
```

The efficient way—at least if what you are looking for has been published in the STB—is to use Stata's `search` facility. For instance, you could type

```
. search concordance correlation
```

or you could pull down **Help** and select **Search**. Either way, you will learn about *sg84* and, if you choose the point-and-click method, you can even click to install.

32.5   Making your own download site

There are two reasons you may wish to create your own download site:

1. You have datasets and the like, you want to share them with colleagues, and you want to make it easier for colleagues to download the files.

2. You have written Stata programs, etc., that you wish to share with the Stata user community.

Making a download site is easy; the full instructions are found in [R] **net**.

At the beginning of this chapter we pretended you had a dataset you wanted to share with colleagues. We said you just had to copy the dataset onto your server and then let your colleagues know the dataset is there.

Let's now pretend that you had two datasets, `ds1.dta` and `ds2.dta`, and you wanted your colleagues to be able to learn about and fetch the datasets using the `net` command or by pulling down **Help** and selecting **STB and User-written Programs**.

First, you would copy the datasets to your homepage just as before. Then you would create three more files, one to describe your site named `stata.toc` and one more each to describe the "packages" you want to provide:

```
———————————————————————— top of stata.toc ————————
d My name and affiliation (or whatever other title I choose)
d Datasets for the PAR study
p ds1 The base dataset
p ds2 The detail dataset
———————————————————————— end of stata.toc ————————
```

```
———————————————————————— top of ds1.pkg ————————
d ds1.  The base dataset
d My name or whatever else I wanted to put
d This dataset contains the baseline values for ...
———————————————————————— end of ds1.pkg ————————
```

```
                                                   ── top of ds2.pkg ──────
 d ds1.  The detail dataset
 d My name or whatever else I wanted to put
 d This dataset contains the follow-up information ...
 p ds2.dta
                                                   ── end of ds1.pkg ──────
```

Here is what users would see when they went to your site:

```
. net from http://www.myuni.edu/hande/~aparker
-------------------------------------------------------------------------
http://www.myuni.edu/hande/~aparker
My name and whatever else I wanted to put
-------------------------------------------------------------------------
Datasets for the PAR study

PACKAGES you could -net describe-:
     ds1               The base dataset
     ds2               The detail dataset
-------------------------------------------------------------------------

. net describe ds1
-------------------------------------------------------------------------
package ds1 from http://www.myuni.edu/hande/~aparker
-------------------------------------------------------------------------
TITLE
     ds1.  The base dataset
DESCRIPTION/AUTHOR(S)
     My name and whatever else I wanted to put
     This dataset contains the baseline values for ...
ANCILLARY FILES                                (type net get ds1)
     ds1.dta
-------------------------------------------------------------------------

. net get ds1
checking ds1 consistency and verifying not already installed...

copying into current directory...
     copying  ds1.dta
ancillary files successfully copied.

. _
```

See [R] **net**.

Author Index

This is the combined author index for the *Stata Reference Manual* and the *Stata User's Guide*.

A

Aalen, O. O., [R] st sts
Abramowitz, M., [U] **16 Functions and expressions**, [R] **orthog**
Afifi, A. A., [R] **anova**, [R] **sw**
Agresti, A., [R] **tabulate**
Aitchison, J., [R] **ologit**, [R] **oprobit**
Aitken, A. C., [R] **reg3**
Akaike, H., [R] st **streg**
Aldrich, J. H., [R] **logit**, [R] **mlogit**, [R] **probit**
Allen, M. J., [R] **alpha**
Altman, D. G., [R] **anova**, [R] **fracpoly**, [R] **kappa**, [R] **kwallis**, [R] **meta**, [R] **nptrend**, [R] **oneway**, [R] st **stcox**
Ambler, G., [R] **regress**
Amemiya, T., [R] **glogit**, [R] **tobit**
Anagnoson, J. T., [U] **2 Resources for learning and using Stata**
Andersen, E. B., [R] **clogit**
Anderson, J. A., [R] **ologit**
Andrews, D. F., [R] **ml**, [R] **rreg**
Arbuthnott, J., [R] **signrank**
Armitage, P., [R] **ltable**, [R] **means**
Armstrong, R. D., [R] **qreg**
Arthur, B. S., [R] **symmetry**
Atkinson, A. C., [R] **boxcox**, [R] **nl**
Azen, S. P., [R] **anova**

B

Baker, R. J., [R] **glm**
Balanger, A., [R] **sktest**
Baltagi, B. H., [R] **hausman**, [R] **ivreg**, [R] **xtreg**
Bancroft, T. A., [R] **sw**
Bartlett, M. S., [R] **factor**, [R] **oneway**, [R] **wntestb**
Barnwell, B. G., [R] **svytab**
Basmann, R. L. [R] **ivreg**
Bassett, G., Jr., [R] **qreg**
Beale, E. M. L., [R] **sw**, [R] **test**
Beaton, A. E., [R] **rreg**
Beck, N., [R] **xtgls**
Becketti, S., [U] **1 Read this—it will help**, [R] **corrgram**, [R] **fracpoly**, [R] **pause**, [R] **runtest**, [R] **spearman**
Begg, C. B., [R] **meta**
Belle, G. van, [R] **dstdize**, [R] **epitab**
Belsley, D. A., [R] **regression diagnostics**
Bendel, R. B., [R] **sw**
Beniger, J. R., [R] **cumul**,
Berk, K. N., [R] **sw**
Berk, R. A., [R] **rreg**
Berkson, J., [R] **logit**, [R] **probit**

Bernstein, I. H., [R] **alpha**, [R] **loneway**
Berry, G., [R] **ltable**, [R] **means**
Beyer, W. H., [R] **qc**
Bickel, P. J., [R] **ml**, [R] **rreg**
Bickenböller, H., [R] **symmetry**
Bieler, G. S., [R] **svytab**
Binder, D. A., [U] **23 Estimation and post-estimation commands**, [R] **_robust**, [R] **svy estimators**
Birdsall, T. G., [R] **logistic**
Bland, M., [R] **signrank**
Bliss, C. I., [R] **probit**
Bloch, D. A., [R] **brier**
Bloomfield, P., [R] **qreg**
BMDP, [R] **symmetry**
Boice, J. D., [R] **bitest**, [R] **epitab**
Bollen, K. A., [R] **regression diagnostics**
Bollerslev, T., [R] **arch**
Bortkewitsch, L. von, [R] **poisson**
Bowker, A. H., [R] **symmetry**
Box, G. E. P., [R] **anova**, [R] **arima**, [R] **boxcox**, [R] **corrgram**, [R] **cumsp**, [R] **lnskew0**, [R] **pergram**, [R] **pperron**, [R] **wntestq**, [R] **xcorr**
Boyd, N. F., [R] **kappa**
Bradburn, M. J., [R] **meta**
Brady, A. R., [U] **1 Read this—it will help**, [R] **logistic**, [R] **spikeplt**
Brant, R., [R] **ologit**
Breslow, N. E., [R] **clogit**, [R] **dstdize**, [R] **epitab**, [R] st **stcox**, [R] st **sts test**, [R] **symmetry**
Breusch, T., [R] **mvreg**, [R] **sureg**, [R] **xtreg**
Brier, G. W., [R] **brier**
Brook, R., [R] **brier**
Brown, D. R., [R] **anova**, [R] **oneway**
Brown, S. E., [R] **symmetry**
Buchner, D. M., [R] **ladder**
Burnam, M. A., [R] **lincom**, [R] **mlogit**
Burr, I. W., [R] **qc**

C

Campbell, M. J., [R] **logistic**
Carlile, T., [R] **kappa**
Carlin, J., [U] **1 Read this—it will help**, [R] **means**
Carroll, R. J., [R] **rreg**
Chadwick, J., [R] **poisson**
Chamberlain, G., [R] **clogit**
Chambers, J. M., [U] **9 Stata's on-line tutorials and sample datasets**, [R] **diagplots**, [R] **grmeanby**, [R] **ksm**
Charlett, A., [R] **fracpoly**
Chatfield, C., [R] **corrgram**, [R] **pergram**
Chatterjee, S., [R] **eivreg**, [R] **poisson**, [R] **prais**, [R] **regress**, [R] **regression diagnostics**
Chiang, C. L., [R] **ltable**
Clark, V. A., [R] **ltable**
Clarke, M. R. B., [R] **factor**
Clarke, R. D., [R] **poisson**

Subject Index

This is the combined subject index for the *Stata Reference Manual* and the *Stata User's Guide*. Readers interested in graphics topics should see the index in the *Stata Graphics Manual.*

Semicolons set off the most important entries from the rest. Sometimes no entry will be set off with semicolons; this means all entries are equally important.

& (and), *see* logical operators

| (or), *see* logical operators

~ (not), *see* logical operators

! (not), *see* logical operators

!, *see* shell command

== (equality), *see* relational operators

!= (not equal), *see* relational operators

~= (not equal), *see* relational operators

< (less than), *see* relational operators

<= (less than or equal), *see* relational operators

> (greater than), *see* relational operators

>= (greater than or equal), *see* relational operators

A

Aalen–Nelson cumulative hazard, [R] **st sts**, [R] **st sts generate**, [R] **st sts graph**, [R] **st sts list**

abbreviations, [U] **14.2 Abbreviation rules**; [U] **14.1.1 varlist**, [U] **14.4 varlists**
    unabbreviating variable list, [R] **syntax**, [R] **unab**

aborting command execution, [U] **12 The Break key**, [U] **13 Keyboard use**

about command, [R] **about**

abs() function, [U] **16.3.1 Mathematical functions**, [R] **functions**

absolute value function, *see* abs() function

absorption in regression, [R] **areg**

ac command, [R] **corrgram**

accelerated failure-time model, [R] **st streg**

Access, Microsoft, reading data from, [U] **24.4 Transfer programs**

accum matrix subcommand, [R] **matrix accum**

acos() function, [U] **16.3.1 Mathematical functions**, [R] **functions**

acprplot command, [R] **regression diagnostics**

actuarial tables, [R] **ltable**

added-variable plots, [R] **regression diagnostics**

addition across observations, [U] **16.3.6 Special functions**, [R] **egen**

addition across variables, [R] **egen**

addition operator, *see* arithmetic operators

adjust command, [R] **adjust**

adjusted Kaplan–Meier survivor function, [R] **st sts**

adjusted partial residual plot, [R] **regression diagnostics**

.ado filename suffix, [U] **14.6 File-naming conventions**

ado-files, [U] **2.4 The Stata Technical Bulletin**, [U] **20 Ado-files**, [U] **21.11 Ado-files**; [R] **sysdir**, [R] **version**, *also see* programs
    editing, [R] **doedit**
    display version of, [R] **which**
    downloading, [U] **32 Using the Internet to keep up to date**
    installing, [U] **20.8 How do I obtain and install STB updates?**, [R] **net**
    location, [U] **20.5 Where does Stata look for ado-files?**
    long lines, [U] **21.11.2 Comments and long lines in ado-files**, [R] **#delimit**
    official, [U] **32 Using the Internet to keep up to date**, [R] **update**
    verifying installation, [R] **which**, *also see* verinst command

adopath command, [U] **20.5 Where does Stata look for ado-files?**, [R] **sysdir**

adosize parameter, [R] **sysdir**; [U] **21.11 Ado-files**, [R] **macro**

aggregate functions, [R] **egen**

aggregate statistics, dataset of, [R] **collapse**

agreement, interrater, [R] **kappa**

algebraic expressions, functions, and operators, [U] **16 Functions and expressions**, [U] **16.3 Functions**, [R] **matrix define**

_all, [U] **14.1.1 varlist**

alpha coefficient, Cronbach's, [R] **alpha**

alpha command, [R] **alpha**

alphabetizing
    observations, [R] **sort**; [R] **gsort**
    variable names, [R] **order**

alphanumeric variables, *see* string variables

analysis of covariance, *see* ANCOVA

analysis of variance, *see* ANOVA

analysis-of-variance test of normality, [R] **swilk**

analytic weights, [U] **14.1.6 weight**, [U] **23.13.2 Analytic weights**

and operator, [U] **16.2.4 Logical operators**

ANCOVA, [R] **anova**

ANOVA, [R] **anova**, [R] **loneway**, [R] **oneway**
    Kruskal–Wallis, [R] **kwallis**
    repeated measures, [R] **anova**

anova command, [R] **anova**; *also see* estimation commands
    with string variables, [R] **encode**

aorder command, [R] **order**

append command, [R] **append**; [U] **25 Commands for combining data**

_append variable, [R] **append**

appending data, [R] **append**; [U] **25 Commands for combining data**

appending files, [R] **copy**

appending rows and columns to matrix, [R] **matrix define**

AR, *see* autocorrelation

deleting
    casewise, [R] **egen**
    files, [R] **erase**
    variables or observations, [R] **drop**
#**delimit** command, [R] #**delimit**
delimiter for lines, [R] #**delimit**
delta beta influence statistic, [R] **logistic**, [R] **regression diagnostics**
delta chi-squared influence statistic, [R] **logistic**
delta deviance influence statistic, [R] **logistic**
density estimation, kernel, [R] **kdensity**
derivatives, numerical, [R] **range**, [R] **testnl**
describe command, [R] **describe**; [U] **15.6 Dataset, variable, and value labels**
descriptive statistics,
    creating dataset containing, [R] **collapse**
    creating variables containing, [R] **egen**
    displaying, [R] **summarize**, [R] **table**, [R] **tabsum**; [R] **codebook**, [R] **lv**, [R] **pctile**, [R] **xtsum**, [R] **xttab**
design effects, [R] **loneway**, [R] **svy estimators**, [R] **svymean**
determinant of matrix, [R] **matrix define**
deviance residual, [R] **cox**, [R] **glm**, [R] **logistic**, [R] **st stcox**, [R] **st streg**, [R] **weibull**
dfbeta command, [R] **regression diagnostics**
DFBETAs, [R] **regression diagnostics**
DFITS, [R] **regression diagnostics**
dfuller command, [R] **dfuller**
diagnostic plots, [R] **diagplots**, [R] **logistic**, [R] **regression diagnostics**
diagnostics, regression, [R] **regression diagnostics**
diagonals of matrices, [R] **matrix define**
dialog box, [R] **window**, [R] **window control**, [R] **window dialog**, [R] **window fopen**, [R] **window stopbox**
dichotomous outcome models, [R] **logistic**; [R] **brier**, [R] **clogit**, [R] **cloglog**, [R] **glm**, [R] **glogit**, [R] **hetprob**, [R] **logit**, [R] **probit**, [R] **scobit**, [R] **xtclog**, [R] **xtgee**, [R] **xtlogit**, [R] **xtprobit**
dictionaries, [R] **infile**, [R] **infile (fixed format)**, [R] **infix (fixed format)**, [R] **outfile**
diff() egen function, [R] **egen**
difference of estimated coefficients, *see* linear combinations of estimators
difference of subpopulation means, [R] **svylc**, [R] **svymean**
digamma() function, [R] **functions**
digits, controlling the number displayed, [U] **15.5 Formats: controlling how data is displayed**, [R] **format**
dir command, [R] **dir**
direct standardization, [R] **dstdize**
directories, [U] **14.6 File-naming conventions**, [U] **21.3.9 Constructing Windows filenames using macros**
    changing, [R] **cd**
    creating, [R] **mkdir**
    listing, [R] **dir**

directories, *continued*
    location of ado-files, [U] **20.5 Where does Stata look for ado-files?**
discard command, [U] **21.11.3 Debugging ado-files**, [R] **discard**
dispCns matrix subcommand, [R] **matrix constraint**
dispersion, measures of, [R] **summarize**, [R] **table**; [R] **centile**, [R] **lv**, [R] **pctile**, [R] **xtsum**
display command, [R] **display (for programmers)**; [R] **macro**
    as a calculator, [R] **display**
display formats, [U] **15.5 Formats: controlling how data is displayed**; [U] **27.2.3 Displaying dates**, [R] **describe**, [R] **format**, [R] **macro**, *also see Stata Graphics Manual*
display saved results, [R] **saved results**
display-width and length, [R] **log**
displaying
    contents, [R] **describe**
    data, [R] **edit**, [R] **list**
    macros, [R] **macro**
    matrix, [R] **matrix utility**
    previously typed lines, [R] #**review**
    saved results, [R] **saved results**
    scalar expressions, [R] **display (for programmers)**, [R] **scalar**
    also see printing
distributions,
    diagnostic plots, [R] **diagplots**
    examining, [R] **centile**, [R] **kdensity**, [R] **lv**, [R] **pctile**, [R] **stem**, [R] **summarize**, *also see Stata Graphics Manual*
    testing equality, [R] **ksmirnov**, [R] **kwallis**, [R] **signrank**
    testing for normality, [R] **sktest**, [R] **swilk**
division operator, *see* arithmetic operators
do command, [U] **19 Do-files**, [R] **do**
.**do** filename suffix, [U] **14.6 File-naming conventions**
do-files, [U] **19 Do-files** [U] **21.2 Relationship between a program and a do-file**, [R] **do**, [R] **version**, *also see* programs
    editing, [R] **doedit**
    long lines, [U] **21.11.2 Comments and long lines in ado-files**, [R] #**delimit**
documentation, keyword search on, [U] **8 Stata's on-line help and search facilities**, [R] **search**
documenting data, [R] **codebook**, [R] **notes**
doedit command, [R] **doedit**
dofb(), **dofd()**, **dofm()**, **dofq()**, **dofw()**, and **dofy()** functions, [U] **16.3.4 Time-series functions**, [U] **27.3.8 Translating between time-series units**, [R] **functions**
domain sampling, [R] **alpha**
dose–response models, [R] **glm**, [R] **logistic**
dotplot command, [R] **dotplot**
double, [U] **15.2.2 Numeric storage types**, [R] **data types**
double-precision floating point number, [U] **15.2.2 Numeric storage types**

Continued on next page

M

Continued on next page

Q

R

U

who command, *see Getting Started with Stata for Unix*

Wilcoxon rank-sum test, [R] **signrank**

Wilcoxon signed-ranks test, [R] **signrank**

Wilcoxon test (Wilcoxon–Breslow, Wilcoxon–Gehan, Wilcoxon–Mann–Whitney, [R] **st sts test**

window control command, [R] **window control**; [R] **window**

window dialog command, [R] **window dialog**; [R] **window**

window fopen command, [R] **window fopen**; [R] **window**

window manage command, [R] **window manage**; [R] **window**

window menu command, [R] **window menu**; [R] **window**

window push command, [R] **window push**; [R] **window**

window stopbox command, [R] **window stopbox**; [R] **window**

Windows, *also see Getting Started with Stata for Windows* manual

    dialog box, [R] **window control**, [R] **window dialog**

    filenames [U] **21.3.9 Constructing Windows filenames using macros**

    help, [R] **help**

    identifying in programs, [U] **21.3.10 Built-in global system macros**

    keyboard use, [U] **13 Keyboard use**

    pause, [R] **sleep**

    shell, [R] **shell**

    specifying filenames, [U] **14.6 File-naming conventions**

    STB installation, [U] **20.8 How do I obtain and install STB updates?**, [R] **net**, [R] **stb**

winexec command, [R] **shell**

within-cell means and variances, [R] **xtsum**

within estimators, [R] **xtreg**

wmf (Windows Metafile), *see Getting Started with Stata for Windows* manual

wntestb command, [R] **wntestb**

wntestq command, [R] **wntestq**

wofd() function, [U] **16.3.4 Time-series functions**, [U] **27.3.8 Translating between time-series units**, [R] **functions**

Woolf confidence intervals, [R] **epitab**

writing data, [R] **outfile**, [R] **outsheet**, [R] **save**

www.stata.com web site, [U] **2.2 The http://www.stata.com web site**

X

X Windows, *see Stata Graphics Manual, also see Getting Started with Stata for Unix* manual

xchart command, [R] **qc**

xcorr command, [R] **xcorr**

xi command, [R] **xi**

xpose command, [R] **xpose**

xt (panel or cross-sectional data), [R] **clogit**, [R] **quadchk**, [R] **xt**, [R] **xtclog**, [R] **xtgee**,

    [R] **xtgls**, [R] **xtintreg**, [R] **xtlogit**, [R] **xtnbreg**, [R] **xtpois**, [R] **xtprobit**, [R] **xtrchh**, [R] **xtreg**, [R] **xttobit**

xtclog command, [R] **xtclog**; *also see estimation commands*

xtcorr command, [R] **xtgee**

xtdata command, [R] **xtdata**; [R] **xt**

xtdes command, [R] **xtdes**; [R] **xt**

xtgee command, [R] **xtgee**; *also see estimation commands*

xtgls command, [R] **xtgls**; *also see estimation commands*

xthaus command, [R] **xtreg**; [R] **xt**

xtile command, [R] **pctile**

xtintreg command, [R] **xtintreg**; *also see estimation commands*

xtlogit command, [R] **xtlogit**; *also see estimation commands*

xtnbreg command, [R] **xtnbreg**; *also see estimation commands*

xtpois command, [R] **xtpois**; *also see estimation commands*

xtprobit command, [R] **xtprobit**; *also see estimation commands*

xtrchh command, [R] **xtrchh**; *also see estimation commands*

xtreg command, [R] **xtreg**; [R] **xt**; *also see estimation commands*

xtsum command, [R] **xtsum**; [R] **xt**

xttab command, [R] **xttab**; [R] **xt**

xttest0 command, [R] **xtreg**; [R] **xt**

xttobit command, [R] **xttobit**; *also see estimation commands*

xttrans command, [R] **xttab**; [R] **xt**

Y

y() function, [U] **16.3.4 Time-series functions**, [U] **27.3.5 Specifying particular times (time literals)**, [R] **functions**

year() function, [U] **16.3.3 Date functions**, [U] **16.3.4 Time-series functions**, [U] **27.2.4 Other date functions**, [U] **27.3.7 Extracting components of time**, [R] **functions**

yearly() function, [U] **16.3.4 Time-series functions**, [U] **27.3.4 Creating time variables**, [R] **functions**

yofd() function, [U] **16.3.4 Time-series functions**, [U] **27.3.8 Translating between time-series units**, [R] **functions**

Z

Zellner's seemingly unrelated regression, [R] **sureg**; [R] **reg3**

zero-altered Poisson regression, [R] **zip**

zero-inflated Poisson regression, [R] **zip**

zero matrix, [R] **matrix define**

zip command, [R] **zip**; *also see estimation commands*